T0413542

Molecular Electronics

Molecular Electronics

An Experimental and Theoretical Approach

edited by
Ioan Bâldea

Pan Stanford Publishing

Published by

Pan Stanford Publishing Pte. Ltd.
Penthouse Level, Suntec Tower 3
8 Temasek Boulevard
Singapore 038988

Email: editorial@panstanford.com
Web: www.panstanford.com

British Library Cataloguing-in-Publication Data
A catalogue record for this book is available from the British Library.

Molecular Electronics: An Experimental and Theoretical Approach

ISBN 978-981-4613-90-3 (Hardcover)
ISBN 978-981-4613-91-0 (eBook)

Printed in the USA

Contents

Preface

The field of molecular electronics, to which the present book is devoted, has rapidly evolved into a very active interdisciplinary field of research at the interface between nanotechnology, chemistry, and physics. It aims at fabricating devices with sizes of nanometers under atomic control, by developing novel bottom-up approaches, as opposed to the classical top-bottom approaches.

It is of course impossible, within a reasonable volume, to discuss or even list all of the topics in a field of science whose size is quickly approaching the "thermodynamic limit." One of the most difficult decisions one is faced with in editing a book on molecular electronics is the selection of the material. I proceeded by sampling a few definite topics, which are discussed in rather great detail in separate chapters by experts in their fields. Adopting a pedagogical style, with own introduction and written in a self-contained manner, the chapters mainly aim to provide guidelines for young scientists (physicists, chemists, engineers) planning to actively contribute, as experimentalists or theorists, to molecular electronics. Still, a series of results as well as the manner of presentation are new and can be inspiring and of interest to specialists in the field.

There are, of course, important problems that are practically not touched upon. This inherently reflects the state-of-the-art of a vivid field, which is very far from being in a "steady state."

I thank Pan Stanford Publishing for having invited me to edit such a book and all the authors for having accepted to contribute to it. I hope that the readers will find its content both useful and enjoyable.

Ioan Bâldea
Heidelberg, September 2015

Chapter 1

Single-Molecule Devices

Kai Sotthewes and Harold J. W. Zandvliet

Physics of Interfaces and Nanomaterials, MESA + Institute for Nanotechnology, University of Twente, 7500AE Enschede, The Netherlands

1.1 Introduction to Molecular Electronics

Molecular electronics is the research field that deals with the design and implementation of electronic devices that rely on a single or a few molecules. The idea to use single molecules as elementary electronic building blocks has been put forward in 1974 by Aviram and Ratner in a seminal paper [1]. It is evident that a proper understanding of the properties of an individual molecule is of utmost importance for molecular electronics. The transport properties of a single molecule are evidently the most relevant properties for basically all molecular electronics applications [2–9]. Measuring the resistance or conductance of a single molecule seems trivial: One connects both ends of the molecule to macroscopic electrical contacts and records a current-voltage (I–V) trace. However, there are several challenging hurdles that have to be overcome before a successful measurement can be executed. One cannot simply take two alligator clips and connect them to both ends of the molecule. The size of a molecule is of the order

Molecular Electronics: An Experimental and Theoretical Approach
Edited by Ioan Bâldea

ISBN 978-981-4613-90-3 (Hardcover), 978-981-4613-91-0 (eBook)
www.panstanford.com

of a nanometer, and therefore one has to apply clever tricks to capture a molecule between two macroscopic electrical contacts. In the next section, we will briefly discuss several methods that have been applied to "catch" a single molecule between two macroscopic electrical contacts. Once the molecule is properly contacted, the transport experiment is rather straightforward; however, the interpretation of the current–voltage traces is far from trivial. The molecular orbitals of the molecule can hybridize with the electronic states of the contacts leading in general to shifts and broadening of the molecular orbitals. Rather than measuring the conductance of a single molecule, one measures the conductance of the complete contact-molecule-contact junction. As pointed out by K. W. Hipps in 2001 in a *Science* article entitled "It's all about the contacts" molecular electronics is mainly a "contact" problem [10]. Regarding the properties of single molecules, we will restrict ourselves in this contribution to transport properties only. It should be pointed out here that recently many studies in the field of molecular electronics have been performed that go beyond the electronic transport characterization of single molecules. These studies involve thermoelectric, optoelectronic, mechanical, and spintronic phenomena. We would like to emphasize that these studies fall outside the scope of this chapter.

In the next section, we will provide the reader with a brief update of the various methods that are applied to capture a single molecule between two macroscopic electrodes. Subsequently, in Section 1.3 we will, in a rather scholarly manner, address the physical ingredients that are useful to understand electronic transport through a single molecule. In Section 1.4, we will discuss two elementary molecular electronic devices: a single-molecule switch and a single-molecule transistor. This chapter concludes with a short outlook.

1.2 The Art of Catching and Probing a Single Molecule

The available methods to capture an individual molecule between two electrical contacts can roughly be divided into two approaches.

In the first approach, one uses a quantum mechanical break junction to capture a single molecule [11–18]. In a quantum mechanical break junction experiment, a thin metallic wire with an incision is carefully stretched by using piezo actuators. During stretching, the wire gets thinner and thinner until it eventually has a cross section of only one atom and breaks. After breaking one usually attempts to capture a single molecule from either the gas or the liquid phase. Unfortunately, it is rather difficult to figure out whether only a single molecule or a bundle of molecules is captured between the electrical contacts. Since the electrical contacts are mounted on piezo actuators one can easily "repair" the break junction by moving both electrodes towards each other. By repeating this process many times, a conductance histogram can be obtained that provides valuable information on the preference for certain conductance values.

The second method makes use of a scanning tunneling microscope or an atomic force microscope and a substrate [6, 9, 19–31]. A small number of molecules are deposited on a well-defined single crystal surface and subsequently imaged with the scanning probe microscope. One can attempt to pick up a single and pre-selected molecule by parking the scanning probe microscope tip on top of the molecule, open the feedback loop, and move the tip towards the surface. In close vicinity of the molecule one usually applies a short voltage pulse to the tip in the hope that the molecule is picked up by the apex of the tip. After this picking up attempt, one has to scan the same area again in order to check if the molecule has really been picked up by the tip of the scanning probe microscope. If this is not the case, one has to repeat the same procedure again. Molecules that have well-chosen end groups that can anchor to the substrate and tip respectively are the molecules that can be captured most easily. The end groups are often chemisorbed to the electrical contacts usually resulting into stable contacts with low contact resistances. There are, however, also a few examples where one of the end groups is only physisorbed to either the substrate or the tip [20, 24, 25, 32]. Figure 1.1 shows an artist's impression of the capturing process of a single molecule between the tip of a scanning tunneling microscope and a substrate.

Figure 1.1 Artist's impression of a single molecule that is captured between the tip of a scanning tunneling microscope and a substrate. Image courtesy René Heimbuch.

Another method that has been applied by various research groups involves the preparation of inserted monolayers. A surface is covered with a self-assembled monolayer of insulating organic molecules (for instance alkanemonothiols) and subsequently another organic molecule (for instance a conjugated mono- or dithiol) is "inserted" into the insulating self-assembled monolayer. The insertion technique relies on an exchange process, where a molecule of the self-assembled monolayer is replaced by another molecule from the second solution that contains the other molecules. These exchange processes mainly occur at positions where the molecules are not so firmly bound to the substrate, i.e., defects, impurities, vacancies, and anti-phase boundaries [33–41].

1.3 Transport Properties of a Single Molecule

1.3.1 Quantization of Conductance

Ohm's law states that the resistance, R, of an object is given by V/I, where V is the voltage applied across the object and I the current that flows through the object. The resistance of a macroscopic metallic wire of length L, cross section A and specific resistance ρ is given by $\rho L/A$. The resistance of such a macroscopic wire scales linearly with the length of the wire. One would probably naively assume that this relation holds down to the nanoscale. However, as we will show below, this is not necessarily true. In case that the mean free path between successive collisions of the charge carriers, λ, is larger than the length of the wire, the resistance turns out to be quantized in units of $h/2e^2$ and independent of the actual length of the wire, at least, as long as the condition $\lambda > L$ is satisfied.

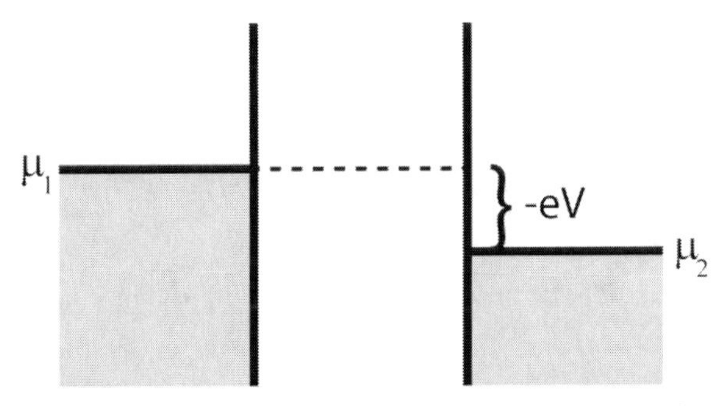

Figure 1.2 Energy diagram of two reservoirs with chemical potentials μ_1 and μ_2, respectively. The two reservoirs are adiabatically connected via a one-dimensional wire.

It is not so difficult to derive this result. Consider a one-dimensional wire that connects adiabatically two reservoirs with chemical potentials μ_1 and μ_2, respectively (see Fig. 1.2). The connections are assumed to be non-reflecting. The current, I, that flows through the wire is given by

$$I = -nev_F = -ev_F(\mu_1 - \mu_2)D(E_F), \tag{1.1}$$

where v_F is the Fermi velocity of the electrons, n the electron density, e the charge of the electrons and $D(E_F)$ the density of states at the Fermi level. The difference in chemical potential, i.e., $(\mu_1 - \mu_2)$, is $-eV$. Equation (1.1) represents the current per mode. The voltage difference V between the two reservoirs could involve more modes. The current is equally distributed among the N modes. This equipartition is due to the fact that electrons at the Fermi level in each mode have different group velocities v. However, this difference in group velocities is canceled by the density of states, which is inversely proportional to the group velocity. The density of states, $D(E)$, is given by

$$D(E)=\frac{2\left(\frac{1}{2\pi}\right)}{\left(\frac{dE}{dk}\right)}=\frac{2}{hv}. \tag{1.2}$$

In expression (1.2), a factor of 2 for the spin degeneracy has been taken into account. The current I carried per mode is then

$$I=e^2 v_F V\frac{2}{hv_F}=\frac{2e^2}{h}V. \tag{1.3a}$$

The total current is found by summing over all modes,

$$I=N\frac{2e^2}{h}V \tag{1.3b}$$

The resistance quantum is $\frac{h}{2e^2}$, where the spin degree of freedom is included. The conductance quantum is $\frac{e^2}{h}$ per channel and per spin. Interestingly, the conductance quantum is independent of the material properties.

A conductor is referred as ballistic if the mean free path of the charge carriers is larger than the length of the conductor. The conductance of a ballistic wire in quantized in units of $\frac{e^2}{h}$. Since no scattering takes place in a ballistic conductor it is a natural question to ask what the actual cause of the resistance is. It turns out that the quantum resistance is in fact a contact resistance because the incoming electron waves have to "find" the entrance of the wire. Only a small integer number $N \approx \frac{2w}{\lambda_F}$ of transverse modes can propagate at the Fermi level (λ_F is the Fermi wavelength and w is the width of the wire). In experiments, small deviations from the exact quantization are found. These deviations, typically of the order of 1% or so, are caused by the series resistance of the contacts and backscattering at the entrance and exit of the wire. It should be pointed out here that the deviations in the quantization of the Hall conductance are much smaller, i.e., deviations as small as 0.00001% can be achieved.

The first experimental evidence of the quantization of conductance came from papers in 1988 by a paper from a Delft-

Philips collaboration [42] and a paper from a Cambridge team [43]. These authors realized a quantum point contact in a two-dimensional electron gas. By tuning the width of the construction using a split gate they could show that the conductance is indeed quantized in units of $\frac{2e^2}{h}$. In molecular transport, one typically deals with only one, or at most a few conduction channels.

1.3.2 Coherent and Incoherent Transport

The transport through a one-dimensional wire can be described within the framework of the Landauer theory. In 1957, Landauer proposed that electrical transport in a one-dimensional system could be considered as a transmission problem [44]. An incoming electron wave has a probability T to be transmitted through the one-dimensional channel. The probability that the electron wave is reflected is represented by R ($T + R = 1$). Since a molecule that is contacted by electrical contacts can be considered as quasi one-dimensional object, its conductance is given by

$$G = \frac{2e^2}{h}T, \tag{1.4}$$

where T represents the average probability that an electron injected at one end of the molecule will make it to the other end of the molecule (for the sake of simplicity we assume that we are dealing with a single conduction channel) and $2e^2/h$ = 77.5 μS. In the case of perfect transmission, i.e., T = 1, we are dealing with a ballistic conductor. Thus a ballistic conductor has a non-zero resistance even though there are no impurities! This key result is at variance with the classical picture where one expects infinite conductance in the absence of impurity scattering.

Here we proceed by considering a wire that has several scattering centers. The latter is a good starting point for an electrode-molecule-electrode junction. In practice both ends of the molecule need to be connected to macroscopic electrical contacts and therefore the total conductance of a metal-molecule-metal junction depends on at least three transmission probabilities, the transmission probabilities of the left (T_L) contact, right contact (T_R) and molecule (T_M). We first consider the case of incoherent transport, i.e., we assume that the phase information is lost during the transport through the molecule. For the sake of clarity, we assume we only have two incoherent scattering centers with transmission probabilities T_1 and T_2. One might naively assume that the total conductance is given by $\frac{2e^2}{h}T_1T_2$. However, this is wrong since the multiple scattering events where the incoming electron wave bounces back and forth between the two scattering centers also contribute to the total transmission (see Fig. 1.3). The total transmission is given by

$$G = \frac{2e^2}{h}[T_1T_2 + T_1T_2R_1R_2 + T_1T_2(R_1R_2)^2 + T_1T_2(R_1R_2)^3 + \cdots] = \frac{2e^2}{h}\left[\frac{T_1T_2}{1-R_1R_2}\right], \tag{1.5a}$$

where $R_1 = 1 - T_1$ and $R_2 = 1 - T_2$ are the reflection probabilities. Equation (1.5a) can be written as

$$G = \frac{2e^2}{h}\left[\frac{T_1T_2}{T_1 + T_2 - T_1T_2}\right]. \tag{1.5b}$$

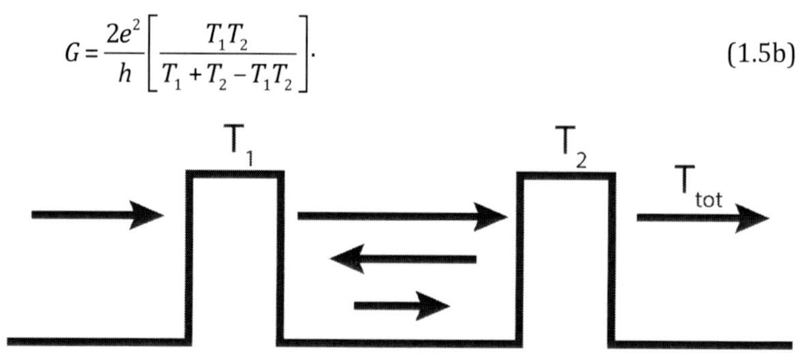

Figure 1.3 Two incoherent scattering centers in series. The total transmission is given by $T_{\text{total}} = T_1T_2 + T_1T_2R_1R_2 + T_1T_2(R_1R_2)^2 + T_1T_2(R_1R_2)^3 + \ldots$.

For three incoherent scattering centers in series one finds

$$G = \frac{2e^2}{h}\left[\frac{T_\text{L}T_\text{M}T_\text{R}}{T_\text{L}T_\text{M} + T_\text{L}T_\text{R} + T_\text{M}T_\text{R} - 2T_\text{L}T_\text{M}T_\text{R}}\right]. \tag{1.6}$$

In the case of coherent scattering, the total transmission also depends on the phase difference between both scattering centers. We introduce the complex transmission and reflection coefficients, $t_{1,2} = |t_{1,2}|e^{i\phi 1,2}$ and $r_{1,2} = |r_{1,2}|e^{i\theta 1,2}$. The transmission probability of the first and second scattering centers is $T_{1,2} = |t_{1,2}|^2$. The total transmission is then,

$$G = \frac{2e^2}{h}\left[\frac{|t_1|^2\,|t_2|^2}{1+|r_1|^2\,|r_2|^2\,-2|r_1||r_2|\cos(\theta)}\right], \tag{1.7}$$

where $\theta = 2kL + \theta_1 + \theta_2$, k the wave vector and L the separation between the scattering centers (see Fig. 1.4). Interestingly, the total transmission probability can become unity despite the fact that both scattering centers have a transmission probability smaller than 1. A total transmission of unity (i.e., resonant coherent transport) is achieved for a phase difference that is an integer multiple of 2π, i.e., $\theta = 2\pi n$, irrespective of the actual values T_1 and T_2. For three scattering centers in series, one can derive a similar, albeit a bit more difficult, relation.

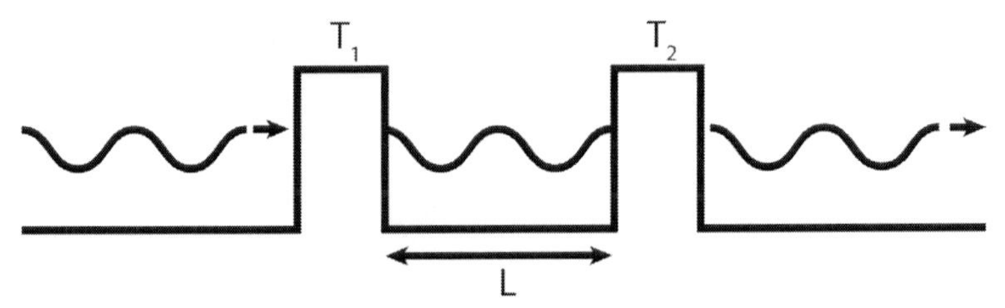

Figure 1.4 Two coherent scattering centers in series.

1.3.3 Coulomb Blockade

In this section, we will briefly touch upon another transport mechanism that is applicable to a subset of molecules. There are molecules, such as metal phthalocyanines, that have a metallic core surrounded by an organic shell [45–50]. Electrons that are transported through these molecules can reside on the metallic core leading to charging of the molecule. The charging energy of such a small entity is usually large since the capacitance with respect to its environment can be very small. These charging effects will only show up in the *I–V* curves if the charging energy exceeds the thermal energy. In case the latter condition is satisfied, the transport through the molecule will be fully blocked if the energy of the electrons is insufficient to overcome the charging energy.

In order to explain the essence of this transport mechanism we consider a simple system that consists of two tunnel junctions in series (see Fig. 1.5). The tunneling resistances of these junctions, R_1 and R_2, are assumed to be substantial, i.e., much larger than the quantum resistance. Furthermore, we assume that the total capacitance of the region in between the two tunnel junctions, $C = C_1 + C_2$, is small enough so that the charging energy, $\frac{e^2}{C}$, is larger than the thermal energy k_BT. We apply a voltage $V = V_1 + V_2$ across both junctions and assume that at $V = 0$ the region in between the two tunnel junction does not contain any charge. Upon increasing the voltage V, electrons will tunnel across junctions 1 and 2. The total number of electrons that have passed junction 1 is denoted by n_1, whereas the total number of electrons that have passed junction 2 is denoted by n_2. The total charge on the region between the two tunnel junctions (from now on referred as the quantum dot) $Q_{\text{dot}} = -n_1e + n_2e = -ne$, where $n = (n_1 - n_2)$ is the number of electrons on the quantum dot. The voltage drops across junction 1 and 2 are given by

$$V_1 = \frac{(C_2V + ne)}{(C_1 + C_2)} \tag{1.8a}$$

$$V_2 = \frac{(C_1V - ne)}{(C_1 + C_2)}. \tag{1.8b}$$

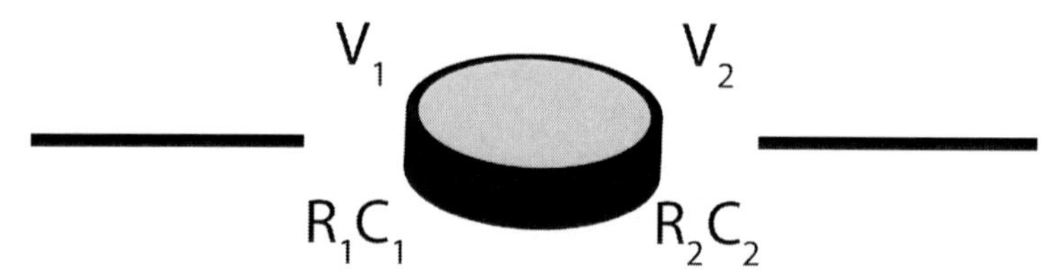

Figure 1.5 Two tunnel junctions in series with tunneling resistances and capacitances $R_{1,2}$ and $C_{1,2}$, respectively.

The total static energy stored in the quantum dot, E_s, can be written as

$$E_s = \frac{1}{2}C_1V_1^2 + \frac{1}{2}C_2V_2^2 = \frac{C_1C_2V^2 + n^2e^2}{2(C_1 + C_2)}. \tag{1.9}$$

In addition to this static energy, we also have to consider the energy transferred by the voltage source. The voltage source provides not only the energy to transfer an integer number of electrons across the tunnel junctions, but it also provides the energy that is required to compensate for the polarization charge when an electron tunnels across one of the junctions. In order to derive the total energy of the system we consider an electron on its journey from electrode 1 to electrode 2. When the electron tunnels across junction 1 the total charge on the quantum dot increases from $-ne = -(n_1 - n_2)e$ to $-(n+1)e = -((n_1+1)-n_2)e$. Due to the transfer of an electron across junction 1 also the voltage drop across junction 2 will change with an amount $\Delta V_2 = -\frac{e}{(C_1+C_2)}$. This potential change leads to polarization charge flow given by $\Delta Q_2 = -\frac{eC_2}{(C_1+C_2)}$ In order to transfer n_1 electrons across junction 1 and n_2 electrons across junction 2, an energy of E_V is required.

$$E_\text{v} = -\frac{eV(n_1C_2 + n_2C_1)}{(C_1+C_2)} \tag{1.10}$$

The total energy of the system $E_\text{tot}(n_1, n_2)$ is then,

$$E_\text{tot}(n_1,n_2) = E_\text{s} - E_\text{v} = \frac{C_1C_2V^2 + n^2e^2}{2(C_1+C_2)} + \frac{eV(n_1C_2+n_2C_1)}{(C_1+C_2)} \tag{1.11}$$

To transfer an electron across junction 1, the following requirement should be met:

$$\Delta E_1^{\pm} = E_\text{tot}(n_1 \pm 1, n_2) - E_\text{tot}(n_1, n_2) \le 0 \tag{1.12}$$

After some simple math this leads to

$$V = \mp\frac{e}{C_2}\left(\frac{1}{2} \pm n\right). \tag{1.13}$$

Similarly one finds for the second junction,

$$V = \mp\frac{e}{C_1}\left(\frac{1}{2} \mp n\right). \tag{1.14}$$

In case $C_2 > C_1$ the first electron will tunnel through junction 1, whereas the first electron will tunnel through junction 2 if $C_1 > C_2$. It is clear that the largest capacitance will determine the size of the Coulomb gap. Let us assume, for the sake of simplicity, that $C_1 = C_2 = C$ and $T = 0$ K. For $|V| < \frac{e}{2C}$, the transport through the quantum dot is fully blocked. This regime is referred as Coulomb blockade. For $|V| > \frac{e}{2C}$ there is a net current flowing through the quantum dot. Under certain conditions, a Coulomb staircase can be observed. In order to observe Coulomb blockade the charging energy should be larger than the thermal energy, i.e.,

$$\frac{e^2}{2C} > k_B T. \quad (1.15)$$

However, there is another important requirement that should be fulfilled; the tunneling resistances should be larger than a certain threshold value. This threshold value can be determined using Heisenberg's uncertainty relation.

$$\Delta E \Delta t \geq \frac{\hbar}{2} \quad (1.16)$$

The energy spacing between subsequent energy is $\Delta E = \frac{e^2}{2C}$ and the tunneling time $\Delta t = RC$. Inserting ΔE and Δt gives

$$R \geq \frac{\hbar}{e^2} \quad (1.17)$$

In order to resolve the energy levels, the tunneling resistances must be larger than $\frac{\hbar}{e^2}$. If requirements (1.15) and (1.17) are met, the Coulomb gap will be observed. The Coulomb staircase can only be observed for asymmetric tunnel junctions, i.e., $R_1C_1 \neq R_2C_2$ (see Fig. 1.6). So, one of the tunnel junctions should be the rate-limiting step in the transfer of electrons.

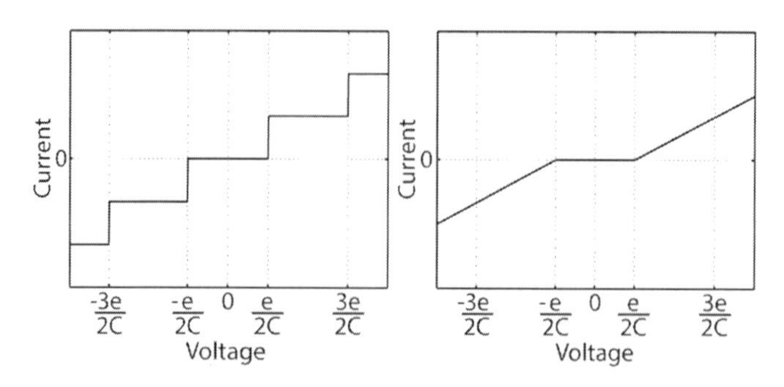

Figure 1.6 Schematic I–V curves of a double tunnel junction at $T = 0$ K. Left: asymmetric junction, $R_1C_1 \neq R_2C_2$ (Coulomb staircase). Right: symmetric junction, $R_1C_1 = R_2C_2$ (only a Coulomb gap).

The last aspect of the Coulomb blockade that we want to address deals with the fractional charge. Due to polarization effects the net charge on the quantum dot is not necessarily an integer value of e. The presence of a fractional charge, δe, will shift the Coulomb gap and Coulomb staircase.

We now consider the case of a quantum dot that has a fractional charge, δe, with $-1 < \delta < 1$. The energy of the quantum dot that contains n electrons and a fractional charge is given by

$$E_n = \frac{(n+\delta)^2 e^2}{2C}. \tag{1.18}$$

When an additional electron is added to the quantum dot, the energy increases with an amount,

$$\Delta E = E_{n+1} - E_n = \frac{2(n+\delta)e^2 + e^2}{2C} \tag{1.19}$$

An additional electron will be added to the quantum dot for $\Delta E = eV$. The threshold voltages for transferring an additional electron to the quantum dot are

$$V = \frac{e}{2C}(2n + 2\delta + 1). \tag{1.20}$$

A fractional charge will shift the Coulomb gap from $\left[-\frac{e}{2C}, \frac{e}{2C}\right]$ to $\left[-\frac{(2\delta-1)e}{2C}, \frac{(2\delta+1)e}{2C}\right]$, but the size of the Coulomb gap remains unaltered (see Fig. 1.7).

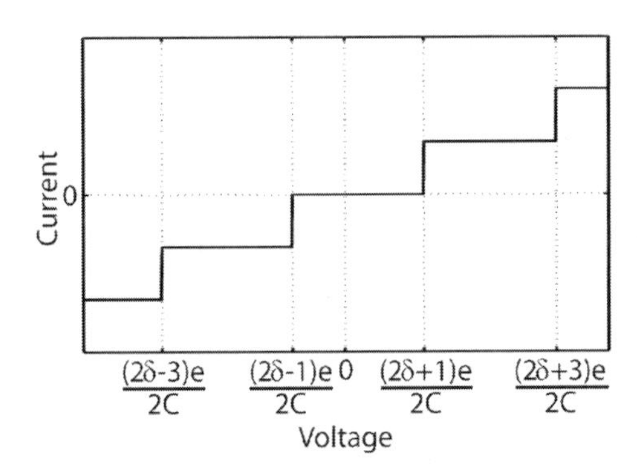

Figure 1.7 Coulomb staircase of a quantum dot that has a fractional charge at $V = 0$ ($T = 0$ K).

In Fig. 1.8a, Coulomb staircase recorded on a small Pd cluster deposited on a decanethiol self-assembled monolayer is shown [50]. The experimental results are compared with the orthodox theory of single electron tunneling [51].

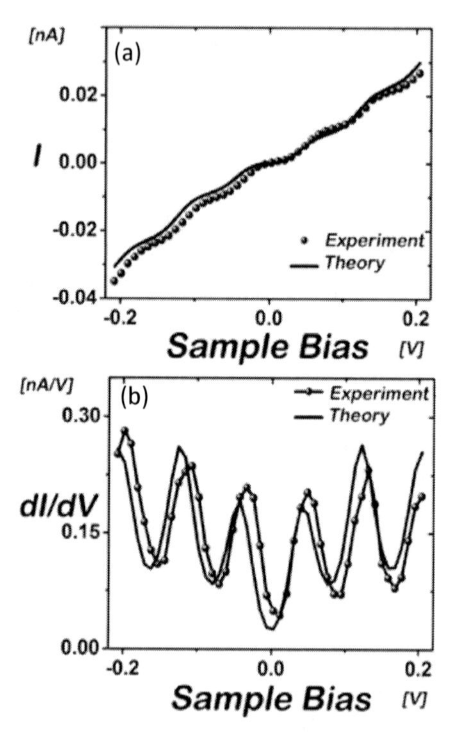

Figure 1.8 (a) Experimental (dotted line) and theoretical (solid line) *I–V* curve recorded at 77 K on a small Pd cluster deposited on a decanethiol self-assembled monolayer. (b) *dI/dV* versus *V*. Each oscillation corresponds to the addition or subtraction of one electron. Copyright AIP reprinted with permission from Oncel, N., *J. Chem. Phys.*, **123**, 044703 (2005).

1.3.4 Transport Mechanisms in Molecules

There are several conduction mechanisms for molecules, such as resonant and non-resonant coherent tunneling, incoherent diffusive tunneling, thermally induced hopping, Fowler–Nordheim tunneling, and thermionic emission [3, 5, 11–13, 52–56]. The properties of the electrical contacts can affect the transport through the molecule significantly. Electrical contacts can be chemically attached to the molecule or physically bound. Chemical binding leads to hybridization of the molecular orbitals with the electronic states of the contacts. This affects the position of the molecular orbitals, which, in turn, can influence the transport through the molecule. Physisorbed contacts lead to less interaction and only to minor changes in the positions of the molecular orbitals.

Below we briefly summarize the possible transport mechanisms in molecules.

1.3.4.1 Coherent tunneling

Coherent or classical tunneling dictated by quantum mechanics relies on the probability of an electron to tunnel through a barrier. The rate of coherent tunneling decays exponentially with the width of the barrier. The current, I, is given by

$$I = \frac{CV\sqrt{\phi - \frac{eV}{2}}}{d} e^{-2\frac{\sqrt{2m}}{\hbar}d\sqrt{\phi - \frac{eV}{2}}} \tag{1.21}$$

where V is the voltage, d the width of the barrier, ϕ the effective barrier height, $\beta = 2\frac{\sqrt{2m}}{\hbar}\sqrt{\phi - \frac{eV}{2}}$ the inverse decay length, and C a constant.

Quantum mechanical tunneling is temperature independent and the phase of the electron is preserved during the tunneling process [57]. The inverse decay length is ~2 $Å^{-1}$ for tunneling between two metal electrodes in vacuum. The inverse decay length of molecules is usually, however, much smaller, i.e., for alkanethiols the inverse decay length is ~0.5–1 $Å^{-1}$ resulting in an effective barrier height that is substantially smaller than the work function [4, 20, 28]. This value is much smaller than the typical highest occupied molecular orbital (HOMO)–lowest unoccupied molecular orbital (LUMO) gap of alkanethiols, which amounts ~8 eV (see Fig. 1.9). A commonly accepted explanation for this, much smaller than expected, inverse decay length is "superexchange." Interaction of the electron with the orbitals and electronic structure of the molecule enhance the tunneling rate, making "through bond" tunneling more efficient than "through space" tunneling. Another important effect that can lower the tunnel barrier and reduces its width is the presence of an image charge (see Fig. 1.10). About half a century ago, Simmons showed that for two planar electrodes the barrier height reduces to [58, 59],

$$\phi(x,d) = \phi_0 - \frac{eV}{2}\frac{\Delta x}{d} - \frac{1.15e^2\ln 2}{16\pi\varepsilon}\frac{d}{x(d-x)} \tag{1.22}$$

$$\Delta x = d\sqrt{\left(1 - \frac{1.15e^2\ln(2)}{16\pi\varepsilon\phi_0 d}\right)}, \tag{1.23}$$

resulting in a mean barrier height,

$$\phi_{\text{eff}} = \varphi_0 - \frac{eV}{2}\frac{\Delta x}{d} - \frac{1.15e^2\ln 2}{8\pi\varepsilon\Delta x}\ln\left(\frac{d+\Delta x}{d-\Delta x}\right), \tag{1.24}$$

where ε is the dielectric constant of the material in the junction, $x_{1,2}$ are the zero's of the potential $\phi(x_1) = \phi(x_2) = 0$ and $\Delta x = x_2 - x_1$ is the effective barrier width. It should be pointed out here that the formulas that we used for image charge effect assumes that we are dealing with two planar electrodes, which is of course for a scanning tunneling microscopy junction not correct. The results obtained here should therefore be considered as an upper bound for the image charge potential.

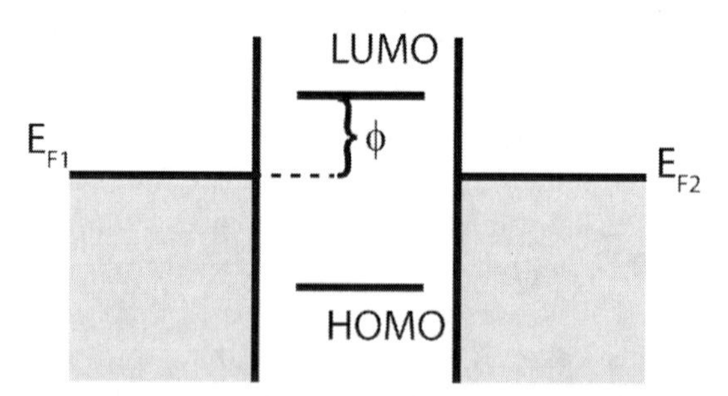

Figure 1.9 Schematic energy diagram of a molecular junction. The metallic electrodes constitute a continuum of electronic states filled with electrons up to the Fermi level. The highest occupied Molecular orbital (HOMO) and lowest unoccupied molecular orbital (LUMO) bands of the molecule are shown.

Another aspect of molecular transport that we want to highlight here is *quantum interference*. Quantum interference can occur if the length scale of the molecule becomes comparable to the electronic phase coherence length. In case that the electron wave arrives at a joint, where it can propagate via two (or more) different routes that eventually cross each other again, constructive or destructive interference effects can severely affect the conductance [30, 60, 61]. The length of these paths should of course be smaller than the electronic phase coherence length. For example, quantum interference can substantially lower the conductance of a cross-conjugated molecule when compared to its linearly conjugated configuration. The reduction of the conductance of the cross-conjugated molecule is due to an anti-resonance in the transmission function [30].

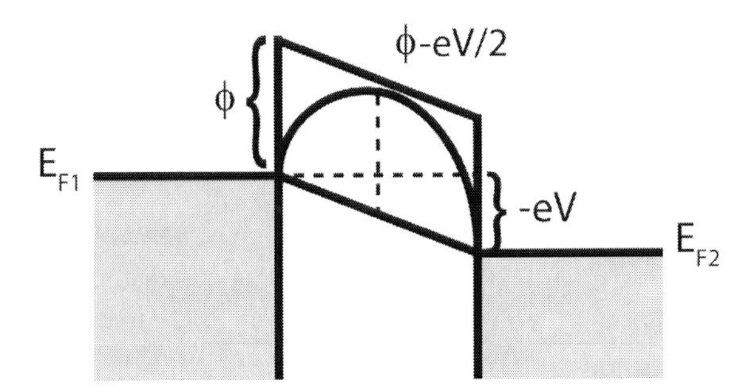

Figure 1.10 The effective barrier decreases with increasing bias voltage, $\phi = \phi_0 - \frac{eV}{2}$. The reduction of the effective barrier height and the reduction of the effective width of the tunneling barrier due to the image charge are not shown.

1.3.4.2 Incoherent tunneling

In the case of incoherent tunneling, the electron tunnels via a series of sites, which are characterized by potentials wells. The residence time in these wells is often long enough to disturb the phase of the electron. Also this tunneling process is in principle temperature independent. In case the "tunneling resistances" from site to site are larger than the quantum resistance, Coulomb charging and blockade effects can occur. In addition, during its journey the electron can also excite one of the vibronic modes of the molecule. The latter will only occur if the energy of the electron exceeds the threshold for excitation, i.e., $V > \hbar\omega/e$.

1.3.4.3 Hopping

Hopping is an Arrhenius activated process and thus strongly temperature dependent [62]. Hopping involves electron motion over the barrier, while tunneling involves electron transport through the barrier. Since hopping involves a series of transfers between relatively stable sites, it does not exhibit an exponential distance dependence characteristic for coherent tunneling, but instead varies as ~1/distance, i.e.,

$$I = \frac{GV}{d} e^{-E/kT}, \tag{1.25}$$

where E is the diffusion barrier for hopping, T the temperature, k Boltzmann's constant, and G a constant [52]. Because of the temperature dependence, hopping conduction is likely to happen at elevated temperatures.

1.3.4.4 Thermionic emission

A barrier (usually referred as Schottky barrier) can arise due to partial charge transfer from one phase to another phase at the interface, resulting in a depletion layer and an electrostatic barrier (as in semiconductor/metal contacts) [63]. This electrostatic barrier is affected by the local (applied) field resulting in a non-linear I–V characteristic. The Schottky–Richardson relation has been invoked to explain the I–V characteristic of a few molecular junctions [64, 65]. Thermoionic emission plays an important role for high temperatures and low barrier heights. The Schottky–Richardson relation is given by

$$I = AT^2 e^{-\phi/kT} e^{(B\sqrt{V}/kT\sqrt{d})}, \tag{1.26}$$

where A and B are constants.

1.3.4.5 Fowler–Nordheim tunneling or field emission

Fowler–Nordheim tunneling occurs when the applied voltage exceeds the barrier height. In field emission, electrons tunnel through a potential barrier, rather than escaping over the barrier as in thermionic emission. The effect is purely quantum-mechanical, with no classical analog. Due to the applied voltage the barrier, which is rectangular for $V = 0$, has a triangular shape facilitating the tunneling of the electrons. The field emission process is temperature independent and decreases exponentially with distance [52],

$$I = DV^2 e^{-(F\varphi^{3/2}/V)d}, \tag{1.27}$$

where D and F are constants. An elegant approach to obtain more detailed information on the transport process in a molecular junction is transition voltage spectroscopy (TVS) [54, 55, 66–71]. To extract meaningful information from the high-bias regime, it is useful to linearize Eq. (1.27):

$$\ln\left(\frac{I}{V^2}\right) \propto D - F\phi^{3/2} d \frac{1}{V}, \tag{1.28}$$

Plotting $\ln(I/V^2)$ versus $1/V$, a so-called Fowler–Nordheim plot, will show a linear decay in the high-bias regime. Equations (1.21) gives the relation for the low-bias regime and when it is rewritten and simplified in terms of $\ln(I/V^2)$ and $1/V$ one finds

$$\ln\left(\frac{I}{V^2}\right) \propto \ln\left(\frac{1}{V}\right) - \beta d, \tag{1.29}$$

where β is the inversed decay length. Equation (1.29) exhibits a logarithmic dependence in the low-bias regime and therefore a transition is observed which corresponds to the voltage where the barrier transforms from a trapezoidal (low-bias regime) to a triangular shape (high-bias regime). The transport mechanism changes from quantum mechanical tunneling (low-bias regime) to field emission (high-bias regime). The transition point is referred as the transition voltage (V_t) and gives an experimental estimate of the energy spacing between the Fermi level and the LUMO (or HOMO for hole tunneling) orbital, i.e., the barrier height ϕ. The specific value for the transition voltage remains a crude estimate because the original tunneling equation does not explicitly accounts for voltage drops over the contacts, image potential, potential profile across the junction and symmetry/asymmetry in the molecular junction [72–74]. Although the exact interpretation is still under debate, it is clear that TVS is an interesting spectroscopic tool in the field of molecular electronics. In vacuum tunnel junctions field-emission typically occurs at voltages that exceed the work function, i.e., at voltage larger than 4–5 V. In distance–voltage (z–V) or I–V traces well defined oscillations or resonances can be observed, which are interpreted as electronic standing waves patterns that can occur in triangular shaped potential wells. These field emission resonances are sometimes referred as Gundlach oscillations, after Gundlach who first discussed these resonances in 1966 [75, 76].

1.4 Molecular Devices

In the final section of this chapter, we will present a few examples of single-molecule devices. These devices all rely on a single octanethiol molecule. We will show that by simultaneously varying the separation and voltage difference between the macroscopic electrodes an octanethiol molecule can be captured controllably between a substrate and the apex of a scanning tunneling microscope tip. The method is so robust that it allows to open and close the molecular junction with a high accuracy over a temperature range from cryogenic temperatures all the way up to room temperature. This robustness not only allows one to measure the temperature dependence of the electronic transport through a single octanethiol molecule, but it also provides a simple an elegant route towards a single-molecule switch. In addition, by varying the contact's interspace once the octanethiol molecule is captured the electronic transport through the octanethiol molecule can be manipulated. This approach allows one to realize a single-molecule transistor that requires only two, rather than the conventional three terminals. The role of the gate terminal is replaced by a mechanical gate that can be tuned by varying the contact's interspace.

1.4.1 Contacting of a Single Octanethiol Molecule

Figure 1.11 shows a scanning tunneling microscope image of a germanium (001) surface covered with metallic platinum (Pt) nanowires. The Pt nanowires have a cross section of only one atom and are kink- and defect-free. This substrate has been exposed to 60 Langmuir of octanethiol. The large white protrusions, which are almost exclusively adsorbed on the Pt nanowires, are octanehiol molecules. The head of the octanthiol molecule, i.e., the sulfur

(S) atom, binds to Pt whereas the carbon tail of the octanethiol molecule is lying flat down on the Pt nanowire. Upon the adsorption of the SH group the hydrogen atom is released and the octanethiol becomes an octanethiolate. In the remainder of this Chapter we refer to an adsorbed octanethiol, whereas we formally dealing with an octanethiolate.

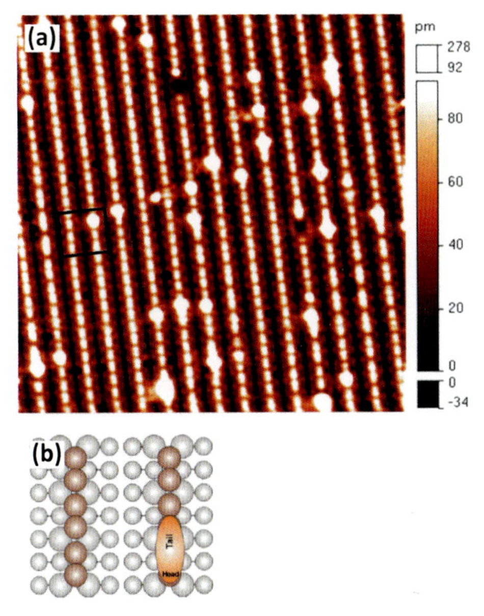

Figure 1.11 Scanning tunneling microscopy image (25 nm × 25 nm; sample bias −0.90 V and tunneling current 0.5 nA) of a platinum-modified germanium (001) surface after exposure to 60 Langmuir of octanethiol, recorded at 77 K (a). The octanethiol molecules (circular white spots) almost exclusively adsorb on the platinum atomic chains. In panel (b), we show a model of the region enclosed by the square in panel A. Grey dumbbells are substrate dimers, dark dumbbells are platinum dimers, and the adsorbed molecule is shown in orange. Copyright ACS reprinted with permission from Kockmann, D., et al., *Nano Lett.*, **9**, 1147 (2009).

In Fig. 1.12 a current–time trace with the feedback loop disabled recorded on top of a pre-selected octanethiol molecule is shown. The set point current is 1 nA, but after about 6 s the current jumps up to a value of around 12 nA and 15 s later the current jumps back to its original value of 1 nA. The only viable explanation for these abrupt and huge changes in the tunnel current is that the carbon tail of the octanethiol molecule flips up and attaches to the apex of the scanning tunneling microscope tip. The vacuum junction is thus replaced by a molecular junction and the electrons flow through the molecule rather than that they tunnel through the vacuum barrier. The length of the octanethiol molecule is ~1 nm and therefore nicely fits in the vacuum junction. One could argue that this system behaves as a molecular switch; however, the jump in and out of contact occurs randomly and the lack of control makes that this system does not resemble a molecular switch.

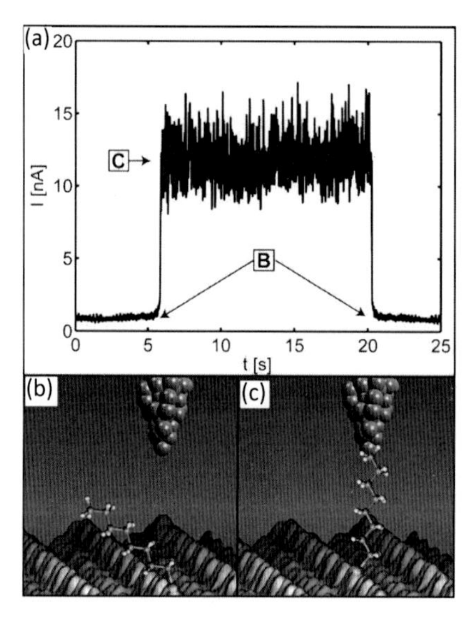

Figure 1.12 (a) Current-time trace recorded on top of an octanethiolate molecule at 77 K. The sample bias was 1.5 V and the setpoint current 1 nA. The cartoons in (b) and (c) show the octanethiolate molecule absorbed at a Pt atom chain and an octanethiolate molecule captured between a Pt atom chain and the apex of an STM tip. Copyright AIP reprinted with permission from Sotthewes, K., *Appl. Phys. Lett. Mater.*, **2**, 010701 (2014).

1.4.2 Single Molecule Switch

Unfortunately, we do not have any control over the Ge(001)/Pt-octanethiol-STM tip junction. This situation drastically alters if a single octanethiol molecule is attached to the apex of the scanning tunneling microscope tip. The sulfur head of the octanethiol can be attached to the apex of the scanning tunneling microscope tip by parking the tip onto a pre-selected octanethiol molecule that is adsorbed on a Pt nanowire. Subsequently, the feedback loop is disabled and the tip is moved a few Ångstroms towards the surface where a short voltage pulse in applied to the tip. When the sulfur atom of the octanethiol makes contact with the tungsten STM tip, it can form a strong bond and therefore the tail of the octanethiol is usually fully released from the surface upon retraction of the tip. In order to check if the octanethiol molecule is picked up from the substrate we performed two tests. First, an regular scanning tunneling microscopy image is recorded in order to verify if the octanethiol molecule has indeed disappeared. Second, an *I–V* curve is recorded. *I–V* curves of the tunnel junctions recorded using a tip decorated with an octanethiol molecule are significantly different from *I–V* curves recorded with a clean, i.e., molecule-free, tip. In Fig. 1.13 *I–V* curves recorded with a clean tip and a tip decorated with an octanethiol molecule are shown. As a set points we have taken a bias of 1.5 V and a tunnel current of 0.5 nA and therefore the asymmetry of the *I–V* curves only shows up a negative sample biases.

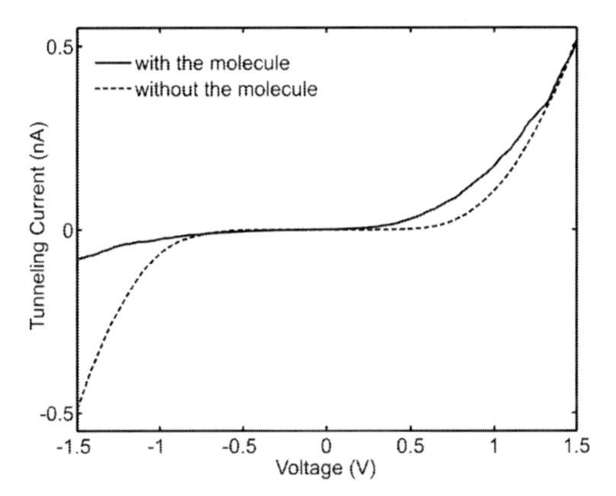

Figure 1.13 Tunneling current–voltage (*I–V*) curves of the tunnel junctions recorded at 77 K before and after the STM tip has picked up an octanethiol molecule. For both traces we have used a set point value of 0.5 nA at 1.5 V. Copyright IOP reprinted with permission from Kumar, A., et al., *J. Phys. Cond. Matter*, **24**, 082201 (2012).

Once the octanethiol molecule is attached to the apex of the scanning tunneling microscope tip a series of current–distance measurements, on various locations at the sample surface, has been recorded. The sample bias was set to +1.5 V and the tunneling current to 0.2 nA, respectively (see Fig. 1.14). After bringing the scanning tunneling microscope tip closer to the substrate by a distance ΔZ = 0.15–0.18 nm (ΔZ refers to the Z-displacement of the tip towards the surface with respect to the set point height), the tail of the octanethiol molecule flips into contact with the substrate and the current jumps to a much higher value of 35 ± 5 nA. The slight variation in the conductance can be attributed to the various contact geometries that the molecule can have with the scanning tunneling microscope tip and the substrate. For a sample bias of −1.5 V the tail of the octanethiol molecule never flips into contact and the tunneling current shows an exponential dependence on distance. The position of the transition from "off" to "on" (inset in Fig. 1.14) depends on the actual value of the applied bias voltage. The current–voltage (I–V) and current–distance (I–Z) spectroscopy data provide strong evidence for a successful attachment of a single octanethiole molecule to the apex of the tip.

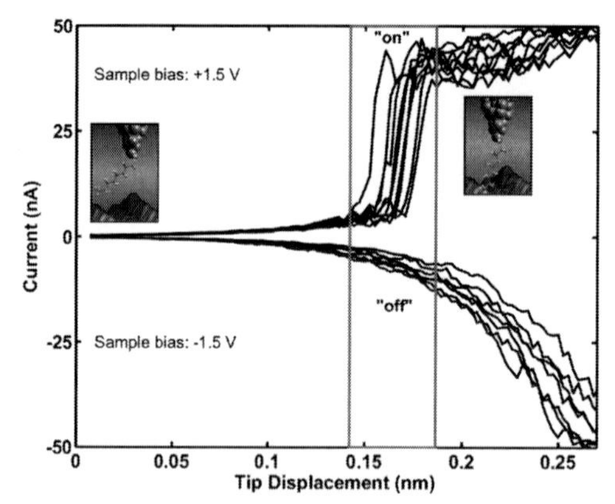

Figure 1.14 Current–distance traces recorded with an octanethiol molecule attached to the apex of the STM tip. Top: The sample bias is 1.5 V and the tunneling current is set to 0.2 nA. After the STM tip has approached, the substrate to ~0.15 nm the molecule makes contact and the current jumps to 35–40 nA. Bottom: The sample bias is −1.5 V and the tunneling current is set to 0.2 nA. The octanethiol molecule does not jump into contact. Copyright IOP reprinted with permission from Kumar, A., et al., *J. Phys. Cond. Matter*, **24**, 082201 (2012).

Figure 1.15 shows a series I–V and I–Z measurements for three different octanethiol molecules (red, blue, and green curves). A series of ten I–V curves recorded at different sample-tip distances has been recorded. The sample-tip distance is changed in increments of 0.5 Ångstroms. In the first two traces (labeled 1 and 2) the distance between the tip and substrate is too large for the tail of the octanethiol molecule to bridge the gap between tip and substrate. In trace 3 the sample-tip distance is 1 Ångstrom smaller as compared to trace 1. In two of the three cases, the tail of the octanethiol molecule flips into contact with the substrate at the starting voltage of 1.5 V. However, at a sample bias of about

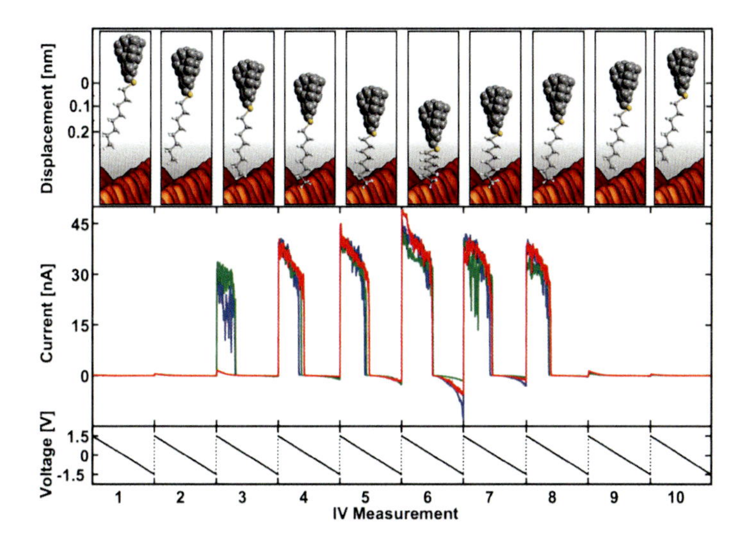

Figure 1.15 A set of three I–V curves (middle section, red, blue, and green curves) recorded in series, with varying tip-substrate distance and the feedback loop disabled. The top section shows a cartoon of the molecule attached to the apex of the tip and its relative position with respect to the substrate. The bottom section shows a series of voltage ramps from +1.5 V to −1.5 V as the tip has moved, in steps of 0.05 nm. Traces 1 to 6 correspond to the tip's relative position from the set point height (i.e., 0.2 nA and 1.5 V) to 0.25 nm, while traces 7 to 10 correspond to the relative position from 0.20 nm to 0.05 nm. After approaching the substrate by 0.1–0.15 nm (traces 3–4), the octanethiol molecule jumps into contact and for *IV* curve no. 8 the molecule jumps out of contact. Copyright IOP reprinted with permission from Kumar, A. et al., *J. Phys. Cond. Matter*, **24**, 082201 (2012).

0.5 V the octanethiol molecule flips out of contact again. A similar behavior is found for traces 4–8. It should be pointed out here that the reduction of sample-tip distance leads to (1) a higher current, i.e., a larger conductance and (2) lower threshold voltage at which the tail of the octanethiol molecule flips out of contact. The higher conductance at smaller sample-tip distances is due to the fact that the tail of the octanethiol molecule is bended during compression of the molecule [7] rather than that the tail molecule is sliding along the contact. If the octanethiol molecule would slide along the contact the conductance would increase in exponentially with decreasing sample-tip distance. From data presented in Fig. 1.15, it is clear that the molecular junction can be controllably closed and opened by varying the tip-sample distance and the bias voltage. Therefore, the substrate–octanethiol–molecule junction behaves as a molecular switch that can be operated with high precision.

1.4.3 Transport through a Single Octanethiol Molecule Junction

Since we can open and close the octanethiol junction controllably, it is very straightforward to measure the conductance of a single octanethiol molecule as a function of the temperature. The octanethiol switch works from cryogenic temperatures all the way up to room temperature. Figure 1.16 shows a plot of the conductance of a single octanethiol molecule measured in units of $\frac{2e^2}{h}$, i.e., G_0, versus temperature. The conductance of the molecule has been measured at a sample bias of +1.5 V, i.e., well below the reported tunneling barrier of an octanethiol molecule [4]. The conductance of the sample-octanethiol-tip junction remains throughout the experiments at a constant value of ~30 nS. Fowler–Nordheim tunneling and quantum-mechanical, or direct, tunneling are the only two transport mechanisms which are temperature independent. Fowler–Nordheim tunneling only occurs at voltages, V, that exceed the work function, i.e., $V > \phi/e$. Since the voltages applied in our experiments are substantially smaller than the 4 eV barrier (assuming that the Fermi edge lies somewhere in the middle of the 8–9 eV gap between highest occupied and lowest unoccupied molecular orbitals of the octanethiol molecule), Fowler–Nordheim tunneling has to be excluded.

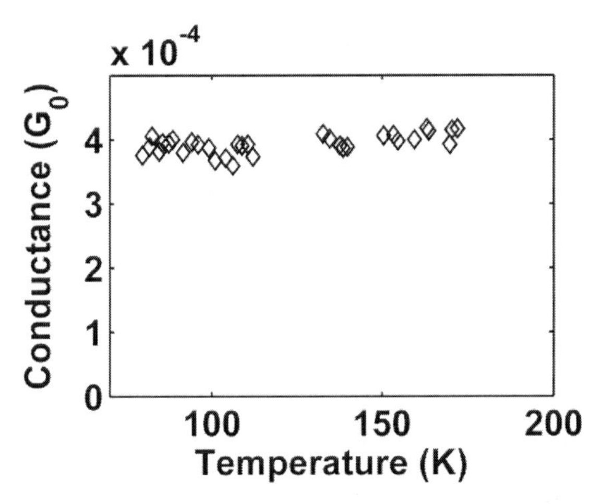

Figure 1.16 Conductance of an electrode single octanethiol molecule electrode junction versus temperature. $G_0 = \frac{2e^2}{h}$ is the conductance quantum. Copyright APS reprinted with permission from Heimbuch, R., et al., *Phys. Rev. B*, **86**, 075456 (2012).

These experiments reveal a single-molecule conductance, that is about a factor of 3 larger than obtained by Kockmann et al. [23] using a method where the sulfur atom of the octanethiol binds to the substrate rather than to the tip. We believe that this difference can be ascribed to the fact that in our experiment one carbon atom more is involved in the contact as compared to Kockmann's experiment. This interpretation is in agreement with the fact that the conductance of an alkanethiol molecule decreases with a factor of 3 per carbon atom.

1.4.4 Single-Molecule Transistor

In order to realize a single-molecule transistor one needs, in principle, three electrical contacts, a source, a drain and a gate. Capturing a single molecule between two contacts is already quite a challenging enterprise, but adding a third electrode is virtually impossible. A way out of this problem would be try to replace the third electrode, i.e., the gate, by another stimulus that could alter the transport through the molecular junction. In the previous section, we already saw that the conductance of the sample-octanethiol-tip junction can be manipulated by compressing or stretching the octanethiol molecule. This mechanical gating approach has already been applied in several experimental studies [7, 8, 13, 32, 77–83].

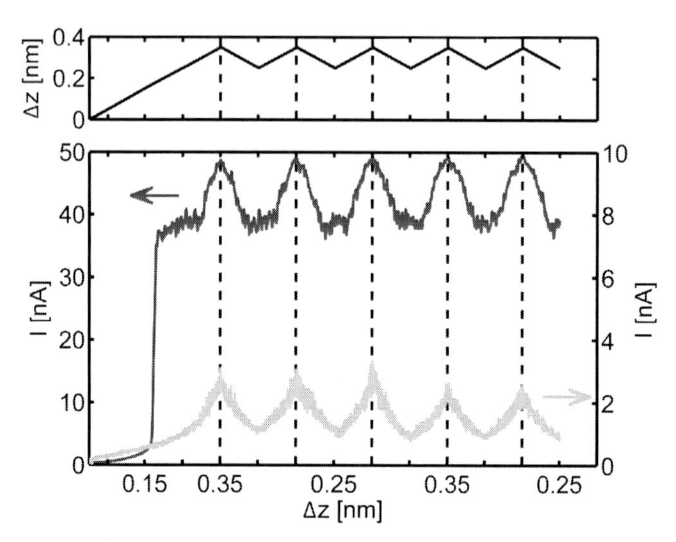

Figure 1.17 The current response of an electrode–octanethiolate–electrode junction (bottom graph, electrode–octanehiolate–electrode junction, black line, left axis) and a vacuum junction (bottom graph, vacuum junction, gray line, right axis) to a varying tip-substrate distance (top graph) at 77 K. The sample bias is +1.5 V and the starting current is 0.25 nA. The vacuum junction exhibits an exponential behavior, which is a hallmark for tunneling. The junction with a molecule initially shows the same exponential behavior, until the molecule jumps into contact (here ~0.17 nm). Upon further reducing the contact's interspace the conductance first marginally increases followed by a faster, but still non-exponential, increase. Copyright AIP reprinted with permission from Sotthewes, K., et al., *J. Chem. Phys.*, **139**, 214709 (2013).

In Fig. 1.17, the effect of mechanical gating of a sample-octanethiol-tip junction is shown. In the top graph the *z*-piezo displacement as a function of time is depicted. In the bottom graph, two curves are shown. The black curve refers to an experiment where an octanethiol molecule is attached the apex of the scanning tunneling microscope tip, whereas the grey curve refers to a clean, i.e., molecule-free, scanning tunneling microscope tip. The vacuum junction (gray curve) displays the expected exponential dependence of the current with sample-tip distance, whilst the molecular junction (black curve) exhibits a much weaker dependence on the sample-tip distance. After trapping the octanethiol molecule between tip and substrate the z-piezo is modulated with an amplitude of only 100 pm (from 0.25 nm to 0.35 nm). Each $I(z)$ trace consists of five compression/stretching cycles. In total we have measured 1200 of these $I(z)$ traces. The reproducibility of the experiments is extremely high, i.e., the variation from $I(z)$ trace to $I(z)$ trace is very small.

There is a small variation in the exact moment at which the tail of the octanethiol molecule flips into contact with the substrate. For more than 90% of the experiments the molecule jumps into contact between 0.16 nm and 0.19 nm. The variation of the conductance upon compressing or stretching of the octanethiol molecule is about 20%.

1.5 Outlook

Molecular electronics has a longstanding history that dates back to its invention in 1974 by Aviram and Ratner, but the progress in the first few decades after its birth has been rather modest due to the lack of experimental techniques that allowed a detailed study at the scale of an individual molecule. The advent of scanning probe microscopy in the 1980s has, however, spurred the field dramatically. Scanning probe microscopy has revolutionized our ability to explore and manipulate atoms and molecules on the size scale of atoms. Besides its unparalleled spatial power, scanning probe microscopy is also capable of a detailed spectroscopic study of the properties of single atoms and molecules. In this chapter, we have provided the reader only with a brief introduction to molecular electronics and a few very elementary examples of single-molecule devices. We believe the field is still its infancy and are convinced that the best has yet to come.

References

1. Aviram, A., and Ratner, M. A. (1974). Molecular rectifiers, *Chem. Phys. Lett.*, **29**, 277–283.
2. Petty, M. C. (2007). *Molecular Electronics: From Principles to Practice* (Wiley, Chichester, UK).

3. McCreery, R. L. (2004). Molecular electronic junctions, *Chem. Mater.*, **16**, 4477–4496.

4. Akkerman, H. B., and de Boer, B. (2008). Electrical conduction through single molecules and self-assembled monolayers, *J. Phys. Cond. Matter*, **20**, 013001.

5. Karthäuser, S. (2011). Control of molecule-based transport for future molecular devices, *J. Phys. Cond. Matter*, **23**, 013001.

6. Guo, S., Zhou, G., and Tao, N. J. (2013). Single molecule conductance, thermopower, and transition voltage, *Nano Lett.*, **13**, 4326–4332.

7. Sotthewes, K., Geskin, V., Heimbuch, R., Kumar, A., and Zandvliet, H. J. W. (2014). Research Update: Molecular electronics: The single-molecule switch and transistor, *APL Mater.*, **2**, 010701.

8. Aradhya, S. V., and Venkataraman, L. (2013). Single-molecule junctions beyond electronic transport, *Nat. Nanotechnol.*, **8**, 399–410.

9. Haiss, W., Wang, C. S., Grace, I., Batsanov, A. S., Schiffrin, D. J., Higgins, S. J., Bryce, M. R., Lambert, C. J., and Nichols, R. J. (2006). Precision control of single-molecule electrical junctions, *Nat. Mater.*, **5**, 995–1002.

10. Hipps, K. W. (2001). Molecular electronics: It's all about contacts, *Science*, **294**, 536–537.

11. Smit, R. H. M., Noat, Y., Untiedt, C., Lang, N. D., van Hemert, M. C., and van Ruitenbeek, J. M. (2002). Measurement of the conductance of a hydrogen molecule, *Nature*, **419**, 906–909.

12. Cai, L. T., Cabassi, M. A., Yoon, H., Cabarcos, O. M., McGuiness, C. L., Flatt, A. K., Allara, D. L., Tour, J. M., and Mayer, T. S. (2005). Reversible bistable switching in nanoscale thiol-substituted oligoaniline molecular junctions, *Nano. Lett.*, **5**, 2365–2372.

13. Bruot, C., Hihath, J., and Tao, N. J. (2012). Mechanically controlled molecular orbital alignment in single molecule junctions, *Nat. Nanotechnol.*, **7**, 35–40.

14. Perrin, M. L., Verzijl, C. J. O., Martin, C. A., Shaikh, A. J., Eelkema, R., van Esch, J. H., van Ruitenbeek, J. M., Thijssen, J. M., van der Zant, H. S. J., and Dulic, D. (2013). Large tunable image-charge effects in single-molecule junctions, *Nat. Nanotechnol.*, **8**, 282–287.

15. Xiang, D., Jeong, H., Kim, D., Lee, T., Cheng, Y. J., Wang, Q. L., and Mayer, D. (2013). Three-terminal single-molecule junctions formed by mechanically controllable break junctions with side gating, *Nano. Lett.*, **13**, 2809–2813.

16. Xiang, D., Jeong, H., Lee, T., and Mayer, D. (2013). Mechanically controllable break junctions for molecular electronics, *Adv. Mater.*, **25**, 4845–4867.

17. Ballmann, S., and Weber, H. B. (2012). An electrostatic gate for mechanically controlled single-molecule junctions, *New J. Phys.*, **14**.

18. Martin, C. A., Smit, R. H. M., van der Zant, H. S. J., and van Ruitenbeek, J. M. (2009). A nanoelectromechanical single-atom switch, *Nano Lett.*, **9**, 2940–2945.

19. Reichert, J., Ochs, R., Beckmann, D., Weber, H. B., Mayor, M., and von Lohneysen, H. (2002). Driving current through single organic molecules, *Phys. Rev. Lett.*, **88**, 176804.

20. Venkataraman, L., Klare, J. E., Tam, I. W., Nuckolls, C., Hybertsen, M. S., and Steigerwald, M. L. (2006). Single-molecule circuits with well-defined molecular conductance, *Nano Lett.*, **6**, 458–462.

21. Meszaros, G., Kronholz, S., Karthauser, S., Mayer, D., and Wandlowski, T. (2007). Electrochemical fabrication and characterization of nanocontacts and nm-sized gaps, *Appl. Phys. A*, **87**, 569–575.

22. Temirov, R., Lassise, A., Anders, F. B., and Tautz, F. S. (2008). Kondo effect by controlled cleavage of a single-molecule contact, *Nanotechnology*, **19**, 065401.

23. Kockmann, D., Poelsema, B., and Zandvliet, H. J. W. (2009). Transport through a Single Octanethiol Molecule, *Nano. Lett.*, **9**, 1147–1151.

24. Lafferentz, L., Ample, F., Yu, H., Hecht, S., Joachim, C., and Grill, L. (2009). Conductance of a single conjugated polymer as a continuous function of its length, *Science*, **323**, 1193–1197.

25. Leary, E., Gonzalez, M. T., van der Pol, C., Bryce, M. R., Filippone, S., Martin, N., Rubio-Bollinger, G., and Agrait, N. (2011). Unambiguous one-molecule conductance measurements under ambient conditions, *Nano. Lett.*, **11**, 2236–2241.

26. Toher, C., Temirov, R., Greuling, A., Pump, F., Kaczmarski, M., Cuniberti, G., Rohlfing, M., and Tautz, F. S. (2011). Electrical transport through a mechanically gated molecular wire, *Phys. Rev. B*, **83**, 155402.

27. Heimbuch, R., Wu, H. R., Kumar, A., Poelsema, B., Schon, P., Vancso, J., and Zandvliet, H. J. W. (2012). Variable-temperature study of the transport through a single octanethiol molecule, *Phys. Rev. B*, **86**, 075456.

28. Kumar, A., Heimbuch, R., Poelsema, B., and Zandvliet, H. J. W. (2012). Controlled transport through a single molecule, *J. Phys. Cond. Matter*, **24**, 082201.

29. Huang, T., Zhao, J., Peng, M., Popov, A. A., Yang, S. F., Dunsch, L., and Petek, H. (2011). A molecular switch based on current-driven rotation of an encapsulated cluster within a fullerene cage, *Nano Lett.*, **11**, 5327–5332.

30. Guedon, C. M., Valkenier, H., Markussen, T., Thygesen, K. S., Hummelen, J. C., and van der Molen, S. J. (2012). Observation of quantum interference in molecular charge transport, *Nat. Nanotechnol.*, **7**, 304–308.

31. Batra, A., Darancet, P., Chen, Q. S., Meisner, J. S., Widawsky, J. R., Neaton, J. B., Nuckolls, C., and Venkataraman, L. (2013). Tuning rectification in single-molecular diodes, *Nano Lett.*, **13**, 6233–6237.

32. Sotthewes, K., Heimbuch, R., and Zandvliet, H. J. W. (2013). Manipulating transport through a single-molecule junction, *J. Chem. Phys.*, **139**, 214709.

33. Donhauser, Z. J., Mantooth, B. A., Kelly, K. F., Bumm, L. A., Monnell, J. D., Stapleton, J. J., Price, D. W., Rawlett, A. M., Allara, D. L., Tour, J. M., and Weiss, P. S. (2001). Conductance switching in single molecules through conformational changes, *Science*, **292**, 2303–2307.

34. Salomon, A., Cahen, D., Lindsay, S., Tomfohr, J., Engelkes, V. B., and Frisbie, C. D. (2003). Comparison of electronic transport measurements on organic molecules, *Adv. Mater.*, **15**, 1881–1890.

35. Lewis, P. A., Inman, C. E., Yao, Y. X., Tour, J. M., Hutchison, J. E., and Weiss, P. S. (2004). Mediating stochastic switching of single molecules using chemical functionality, *J. Am. Chem. Soc.*, **126**, 12214–12215.

36. Cygan, M. T., Dunbar, T. D., Arnold, J. J., Bumm, L. A., Shedlock, N. F., Burgin, T. P., Jones, L., Allara, D. L., Tour, J. M., and Weiss, P. S. (1998). Insertion, conductivity, and structures of conjugated organic oligomers in self-assembled alkanethiol monolayers on Au{111}, *J. Am. Chem. Soc.*, **120**, 2721–2732.

37. Lussem, B., Muller-Meskamp, L., Karthauser, S., Waser, R., Homberger, M., and Simon, U. (2006). STM study of mixed alkanethiol/biphenylthiol self-assembled monolayers on Au(111), *Langmuir*, **22**, 3021–3027.

38. Muller-Meskamp, L., Lussem, B., Karthauser, S., Homberger, M., Simon, U., and Waser, R. (2007). Self assembly of mixed monolayers of mercaptoundecylferrocene and undecanethiol studied by STM, *J. Phys. Conf. Ser.*, **61**, 852–855.

39. Muller-Meskamp, L., Lussem, B., Karthauser, S., Prikhodovski, S., Homberger, M., Simon, U., and Waser, R. (2006). Molecular structure of ferrocenethiol islands embedded into alkanethiol self-assembled monolayers by UHV-STM, *Phys. Status Solidi A*, **203**, 1448–1452.

40. Sotthewes, K., Wu, H. R., Kumar, A., Vancso, G. J., Schon, P. M., and Zandvliet, H. J. W. (2013). Molecular dynamics and energy landscape of decanethiolates in self-assembled monolayers on Au(111) studied by scanning tunneling microscopy, *Langmuir*, **29**, 3662–3667.

41. Wu, H. R., Sotthewes, K., Kumar, A., Vancso, G. J., Schon, P. M., and Zandvliet, H. J. W. (2013). Dynamics of decanethiol self-assembled monolayers on Au(111) studied by time-resolved scanning tunneling microscopy, *Langmuir*, **29**, 2250–2257.

42. van Wees, B. J., Van Houten, H., Beenakker, C. W. J., Williamson, J. G., Kouwenhoven, L. P., Vandermarel, D., and Foxon, C. T. (1988). Quantized conductance of point contacts in a two-dimensional electron-gas, *Phys. Rev. Lett.*, **60**, 848–850.

43. Wharam, D. A., Thornton, T. J., Newbury, R., Pepper, M., Ahmed, H., Frost, J. E. F., Hasko, D. G., Peacock, D. C., Ritchie, D. A., and Jones, G. A. C. (1988). One-dimensional transport and the quantization of the ballistic resistance, *J. Phys. C*, **21**, 209–214.

44. Landauer, R. (1957). Spatial variation of currents and fields due to localized scatterers in metallic conduction, *Ibm J. Res. Dev.*, **1**, 223–231.

45. Mugarza, A., Robles, R., Krull, C., Korytar, R., Lorente, N., and Gambardella, P. (2012). Electronic and magnetic properties of molecule-metal interfaces: Transition-metal phthalocyanines adsorbed on Ag(100), *Phys. Rev. B*, **85**, 155437.

46. Schaffert, J., Cottin, M. C., Sonntag, A., Karacuban, H., Bobisch, C. A., Lorente, N., Gauyacq, J. P., and Moeller, R. (2013). Imaging the dynamics of individually adsorbed molecules, *Nat. Mater.*, **12**, 223–227.

47. Berkelaar, R. P., Sode, H., Mocking, T. F., Kumar, A., Poelsema, B., and Zandvliet, H. J. W. (2011). Molecular Bridges, *J. Phys. Chem. C*, **115**, 2268–2272.

48. Park, J., Pasupathy, A. N., Goldsmith, J. I., Chang, C., Yaish, Y., Petta, J. R., Rinkoski, M., Sethna, J. P., Abruna, H. D., McEuen, P. L., and Ralph, D. C. (2002). Coulomb blockade and the Kondo effect in single-atom transistors, *Nature*, **417**, 722–725.

49. Andres, R. P., Bein, T., Dorogi, M., Feng, S., Henderson, J. I., Kubiak, C. P., Mahoney, W., Osifchin, R. G., and Reifenberger, R. (1996). "Coulomb staircase" at room temperature in a self-assembled molecular nanostructure, *Science*, **272**, 1323–1325.

50. Oncel, N., Hallback, A. S., Zandvliet, H. J. W., Speets, E. A., Ravoo, B. J., Reinhoudt, D. N., and Poelsema, B. (2005). Coulomb blockade of small Pd clusters, *J. Chem. Phys.*, **123**, 044703.

51. Averin, D. V., and Likharev, K. K. (1991). Single electronics: A correlated transfer of single electrons and Cooper pairs in systems of small tunnel junctions. In: *Mesoscopic Phenomena in Solids* (eds. B. L. Altshuler, P. A. Lee, and R. A. Webb), Chapter 6, Elsevier, Amsterdam.
52. Wang, W. Y., Lee, T., and Reed, M. A. (2003). Mechanism of electron conduction in self-assembled alkanethiol monolayer devices, *Phys. Rev. B*, **68**, 035416.
53. Xiang, D., Zhang, Y., Pyatkov, F., Offenhausser, A., and Mayer, D. (2011). Gap size dependent transition from direct tunneling to field emission in single molecule junctions, *Chem. Commun.*, **47**, 4760–4762.
54. Wang, G., Kim, T. W., Jo, G., and Lee, T. (2009). Enhancement of field emission transport by molecular tilt configuration in metal-molecule-metal junctions, *J. Am. Chem. Soc.*, **131**, 5980–5985.
55. Song, H., Kim, Y., Jang, Y. H., Jeong, H., Reed, M. A., and Lee, T. (2009). Observation of molecular orbital gating, *Nature*, **462**, 1039–1043.
56. Hines, T., Diez-Perez, I., Hihath, J., Liu, H. M., Wang, Z. S., Zhao, J. W., Zhou, G., Muellen, K., and Tao, N. J. (2010). Transition from tunneling to hopping in single molecular junctions by measuring length and temperature dependence, *J. Am. Chem. Soc.*, **132**, 11658–11664.
57. Song, H., Kim, Y., Jeong, H., Reed, M. A., and Lee, T. (2010). Coherent tunneling transport in molecular junctions, *J. Phys. Chem. C*, **114**, 20431–20435.
58. Simmons, J. G. (1963). Generalized formula for electric tunnel effect between similar electrodes separated by a thin insulating film, *J. Appl. Phys.*, **34**, 1793.
59. Simmons, J. G. (1963). Electric tunnel effect between dissimilar electrodes separated by a thin insulating film, *J. Appl. Phys.*, **34**, 2581.
60. Aradhya, S. V., Meisner, J. S., Krikorian, M., Ahn, S., Parameswaran, R., Steigerwald, M. L., Nuckolls, C., and Venkataraman, L. (2012). Dissecting contact mechanics from quantum interference in single-molecule junctions of stilbene derivatives, *Nano Lett.*, **12**, 1643–1647.
61. Ballmann, S., Hartle, R., Coto, P. B., Elbing, M., Mayor, M., Bryce, M. R., Thoss, M., and Weber, H. B. (2012). Experimental evidence for quantum interference and vibrationally induced decoherence in single-molecule junctions, *Phys. Rev. Lett.*, **109**.
62. Marcus, R. A. (1993). Electron-transfer reactions in chemistry: Theory and experiment (Nobel lecture), *Angew. Chem. Int. Ed.*, **32**, 1111–1121.
63. Sze, S. M., and Ng, K. K. (1969). *Physics of Semiconductor Devices* (Wiley, New York).

64. Ouerghemmi, H. B., Kouki, F., Lang, R., Ben Ouada, H., and Bouchriha, H. (2009). Self-assembled monolayer effect on the characteristics of organic diodes, *Synthetic Met.*, **159**, 551–555.

65. Khare, P. K., Keller, J. M., Gaur, M. S., Singh, R., and Datt, S. C. (1994). Electrical-conductivity in iodine-doped ethyl cellulose, *Polym. Int.*, **35**, 337–343.

66. Beebe, J. M., Kim, B., Gadzuk, J. W., Frisbie, C. D., and Kushmerick, J. G. (2006). Transition from direct tunneling to field emission in metal-molecule-metal junctions, *Phys. Rev. Lett.*, **97**, 026801.

67. Beebe, J. M., Kim, B., Frisbie, C. D., and Kushmerick, J. G. (2008). Measuring relative barrier heights in molecular electronic junctions with transition voltage spectroscopy, *ACS Nano*, **2**, 827–832.

68. Tan, A. R., Balachandran, J., Dunietz, B. D., Jang, S. Y., Gavini, V., and Reddy, P. (2012). Length dependence of frontier orbital alignment in aromatic molecular junctions, *Appl. Phys. Lett.*, **101**, 243107.

69. Bâldea, I. (2013). Transition voltage spectroscopy reveals significant solvent effects on molecular transport and settles an important issue in bipyridine-based junctions, *Nanoscale*, **5**, 9222–9230.

70. Lennartz, M. C., Atodiresei, N., Caciuc, V., and Karthauser, S. (2011). Identifying molecular orbital energies by distance-dependent transition voltage spectroscopy, *J. Phys. Chem. C*, **115**, 15025–15030.

71. Huisman, E. H., Guedon, C. M., van Wees, B. J., and van der Molen, S. J. (2009). Interpretation of transition voltage spectroscopy, *Nano Lett.*, **9**, 3909–3913.

72. Mirjani, F., Thijssen, J. M., and van der Molen, S. J. (2011). Advantages and limitations of transition voltage spectroscopy: A theoretical analysis, *Phys. Rev. B*, **84**, 115402.

73. Vilan, A., Cahen, D., and Kraisler, E. (2013). Rethinking transition voltage spectroscopy within a generic Taylor expansion view, *ACS Nano*, **7**, 695–706.

74. Markussen, T., Chen, J. Z., and Thygesen, K. S. (2011). Improving transition voltage spectroscopy of molecular junctions, *Phys. Rev. B*, **83**.

75. Gundlach, K. H. (1966). Zur Berechnung Des Tunnelstroms Durch Eine Trapezformige Potentialstufe, *Solid-State Electron.*, **9**, 949–957.

76. Lin, C. L., Lu, S. M., Su, W. B., Shih, H. T., Wu, B. F., Yao, Y. D., Chang, C. S., and Tsong, T. T. (2007). Manifestation of work function difference in high order Gundlach oscillation, *Phys. Rev. Lett.*, **99**, 216103.

77. Zhou, J. F., Guo, C. L., and Xu, B. Q. (2012). Electron transport properties of single molecular junctions under mechanical modulations, *J. Phys. Cond. Matter*, **24**, 164209.

78. Zhou, J. F., Chen, G. J., and Xu, B. Q. (2010). Probing the molecule-electrode interface of single-molecule junctions by controllable mechanical modulations, *J. Phys. Chem. C*, **114**, 8587–8592.

79. Diez-Perez, I., Hihath, J., Hines, T., Wang, Z. S., Zhou, G., Mullen, K., and Tao, N. J. (2011). Controlling single-molecule conductance through lateral coupling of pi orbitals, *Nat. Nanotechnol.*, **6**, 226–231.

80. Fournier, N., Wagner, C., Weiss, C., Temirov, R., and Tautz, F. S. (2011). Force-controlled lifting of molecular wires, *Phys. Rev. B*, **84**, 016802.

81. Ternes, M., Gonzalez, C., Lutz, C. P., Hapala, P., Giessibl, F. J., Jelinek, P., and Heinrich, A. J. (2011). Interplay of conductance, force, and structural change in metallic point contacts, *Phys. Rev. Lett.*, **106**, 016802.

82. Xia, J. L., Diez-Perez, I., and Tao, N. J. (2008). Electron transport in single molecules measured by a distance-modulation assisted break junction method, *Nano Lett.*, **8**, 1960–1964.

83. Hong, W. J., Valkenier, H., Meszaros, G., Manrique, D. Z., Mishchenko, A., Putz, A., Garcia, P. M., Lambert, C. J., Hummelen, J. C., and Wandlowski, T. (2011). An MCBJ case study: The influence of pi-conjugation on the single-molecule conductance at a solid/liquid interface, *Beilstein J. Nanotech*, **2**, 699–713.

Chapter 2

Making Contact to Molecular Layers: Linking Large Ensembles of Molecules to the Outside World

Christina A. Hacker,[a] Sujitra Pookpanratana,[a] Mariona Coll,[a,b] and Curt A. Richter[a]

[a]*Semiconductor and Dimensional Metrology Division,*
Physical Measurement Laboratory, National Institute of Standards and Technology,
100 Bureau Drive, Mailstop 8120, Gaithersburg, MD 20899, USA
[b]*Present Address: Institut de Ciència de Materials de Barcelona (ICMAB-CSIC),*
Campus UAB 08193, Barcelona, Spain

Christina.hacker@nist.gov

In this chapter, we examine the various approaches to making electrical contact to large area molecular junctions. We will highlight the experimental concerns that one must consider and the various approaches to make reliable structures and characterize the junctions electrically, structurally, and chemically.

2.1 Introduction

Molecular electronics was proposed in the early 1970s with molecules performing electronic functions and the notion that

Molecular Electronics: An Experimental and Theoretical Approach
Edited by Ioan Bâldea

ISBN 978-981-4613-90-3 (Hardcover), 978-981-4613-91-0 (eBook)
www.panstanford.com

molecules could be incorporated into advanced electronic architectures as conventional silicon-based components continue to scale to smaller and smaller dimensions. However, understanding of the electronic properties of molecules and single molecular layers were slow until advances in nanotechnology enabled researchers to reliably fabricate and characterize properties on the nanoscale. Investigating the electrical properties of molecules remains an active area of research because molecules are among the smallest sized objects that can be mass-produced through synthetic chemistry with millions to choose from in high yield and high purity. The properties of molecules can also be easily changed through synthetic chemistry enabling researchers to probe the link between molecular structure and the electronic function. In addition, self-assembly can be used on many surfaces and surface reactions can be tailored for more reactive surfaces to make high-quality monolayers for electronic applications at lower cost and utilizing different materials than conventional electronics. Many molecules exhibit selectivity that can be utilized for sensors, catalysts, or other applications and can be incorporated with nanoparticles, microfluidics, flexible electrodes, or other specialized surfaces, for nanoscale engineering of optimal properties.

Research in molecular electronics has largely been conducted through two focus areas: investigation of single molecules and investigation of thousands of molecules. In the case of single molecules, these investigations typically involve a break junction or scanned probe technique where a small number of molecules (ranging from one to several hundreds) are trapped between two electrodes and the resulting properties are measured. These studies have the advantage of investigating the properties of molecules in a predefined environment, which provides a fundamental understanding of the electronic properties of molecules with proper data collection and analysis and have been the subject of several recent reviews [1, 2]. However, the transient nature of this measurement approach limits the ultimate utility of these structures, and it is not well understood how the properties of isolated molecules compare with the properties of many molecules in a thin film where many-body effects are present. The second experimental approach involves investigating thin molecular films where the active device area could range from 10^3 to 10^{10} molecules. Often these molecular electronic junctions are fabricated in a more permanent substrate-molecular layer-substrate approach that have the advantage that they are stable and can be measured multiple times and in multiple locations under differing conditions of temperature, pressure, and magnetic field. As such, they more closely align with conventional device structures and fabrication techniques with promise to be more easily integrated into existing technologies. However, making reliable structures on the nanoscale is non-trivial as the larger area structures are more likely to contain defects or other artifacts that alter the electronic properties of the structure. We will focus further on approaches to fabricating reliable electrical contacts on large area ensembles of molecules and characterization approaches to interrogate these structures.

2.2 Challenges Facing Ensemble Molecular Junctions

A molecule is typically 0.5 nm wide and 0.5 nm to several nm long depending on the specific molecule. In order to make a molecular junction, the bottom substrate, molecular layer, and top substrate all need to be controlled within these nm length scales over the entire range of the junction, which could extend laterally several micrometers to millimeters. Let us first consider the bottom electrode where evaporated metals can have a root-mean square roughness of several nanometers, making the grains of the substrate nearly equivalent to the molecular length. Thus, alternative approaches are needed to prepare substrates that are atomically smooth on large area length scales. For metals, this is often accomplished by using a template stripping approach where metal is initially evaporated onto a very smooth surface and the incident kinetic and thermal energy enable the impinging metal to form an intimate, smooth interface between the deposited metal and the smooth substrate. Often, another layer or polymer is attached to the top of this metal surface for ease of handling. Next, the metal is peeled off in a template stripping fashion to expose the smooth underside of the evaporated metal. Alternatively, there are carbon-based and polymer-based substrates that are typically quite smooth on these relevant length scales. A third approach is to use a crystal substrate, such as silicon, which can be processed to remove the oxide and form an atomically smooth surface both in solution and under ultrahigh vacuum (UHV) conditions.

The next challenge is the formation of a high-quality molecular layer. First, one must start with a clean surface. This means that any native oxide and adventitious hydrocarbons are removed prior to monolayer formation without roughening the surface. Often self-assembly on gold is investigated because gold surfaces do not form native oxides and residual hydrocarbons are easily cleaned by exposure to ultraviolet generated ozone. Self-assembly takes advantage of a special bond between the molecule and the substrate surface where the bond energy between the molecular functional group and substrate is strong enough that the molecule is tethered in place, but also weak enough that the molecules have some lateral degrees of freedom to "organize" and optimize molecular packing based on intermolecular interactions. The most well-studied case of this is the gold–thiol bond, particularly in the case of an aliphatic molecule such as octadecanethiol where the methylene chains can orient into an all-trans configuration and the resulting packing on the Au surface is quite high. Other substrates, such as semiconductors and oxides, typically have much larger bonding energies involved in the molecule–substrate interaction, which limit the lateral degrees of freedom and make achieving a densely packed monolayer a challenge. Dense monolayers can be formed on these substrates with careful control of reaction variables as the monolayer quality and reproducibility on surfaces with limited lateral diffusion is even more sensitive to the reaction conditions [3]. Experimental parameters, such as solvent, concentration, illumination, temperature, water content, and soaking time, are often adjusted to obtain the most-dense, highly ordered molecular layer on the substrate surface.

The largest challenge has remained reliable formation of the top contact. First of all, once the top contact is put in place, characterizing the molecular layer is exceedingly challenging since the top electrode is often not transparent to most optical and charged particle probes. More important, the means by which the top contact is formed often induces several changes into the carefully formed molecular layer and substrate as depicted in Fig. 2.1. This invalidates the assumption that the final electrode–molecule–electrode structure is the same as the, often carefully fabricated and characterized, starting electrode–molecule structure. A filamentary short can form in an area where there are step edges at the substrate (Fig. 2.1a) or domain boundaries within the monolayer (Fig. 2.1b). Alternatively, the roughness of the top electrode could be quite large or could partially penetrate the molecular layer to make the effective tunneling distance much less than the molecular layer and make the resulting electrical properties not representative of the molecular layer (Fig. 2.1c). Unfortunately, molecules can be damaged by the impinging top electrode where either the energy is sufficient to decompose the molecules and make intimate contact with the substrate (Fig. 2.1d), decompose the molecules to create intermediate species (Fig. 2.1f) or alter the chemical nature of substituents (Fig. 2.1g). All of these possible structures have been experimentally observed and are sufficiently different from the initial monolayer.

Figure 2.1 Applying a top contact to a molecular layer can result in many non-ideal structures, including a filamentary short at step edges (a) or monolayer grain boundaries (b); partial penetration of the metal through the monolayer to cause an effectively "thinner" molecular layer (c); destruction of the molecular layer resulting in intimate contact between the two electrodes (d), significant damage to the molecular layer that does not result in an obvious short (f), or minor altering of electrically active components of the molecule due to interaction with the top electrode (g). The ideal molecular junction (h) consists of a well-ordered monolayer between two atomically smooth electrodes. Adapted by permission of the Electrochemical Society from Hacker et al., *ECS Trans.*, 2010, **28**(2), 549–562.

2.3 Investigating Metal–Molecule Interactions

After observing many non-ideal structures, to construct reliable molecular junctions and reduce the amount of metal penetration, researchers have examined many innovative approaches, including the use of a thin reactive metal [4–10], such as Ti, Cr, and Ni, to form a diffusion barrier where traditional metals such as Au, Cu, and Ag could be deposited on top of this "skin" layer to form bulk electrical contacts without penetrating through the molecular layer. Typically a monolayer with a top reactive group [11–14] is used, such as a carboxylic acid, and the reactive metal, such as Ti, forms either a Ti–O or a Ti–C bond at the interface. Alternative metal deposition conditions have also been shown to create molecular junctions with little monolayer degradation [15–18]. Another approach is to limit the device size by creating a "nanopore," which often helps in preparing a sufficient yield of reliable devices but characterizing the chemical and physical structure is limited [19]. Successful fabrication approaches, however, hinge on the ability to characterize the structure of the molecular junction. All of these approaches rely on surface sensitive characterization, such as vibrational spectroscopy (infrared and Raman), atomic spectroscopy (X-ray photoelectron spectroscopy, near-edge X-ray fluorescence spectroscopy, etc.) or destructive secondary ion mass spectrometry or depth profiling X-ray photoelectron spectroscopy techniques to understand the interaction of the metal with the molecular layer. These techniques usually require very thin metal overlayers to characterize the underlying organic monolayer as metal layers thicker than 5 to 10 nm are opaque to most techniques. One exception to this thin metal requirement is inelastic tunneling spectroscopy (IETS) where the vibrational signature of the molecular layer can be acquired within a buried molecular junction [20, 21]; however, different spectra have been obtained for nearly identical molecular junctions prompting further experimental and theoretical investigations to fully understand the data [22, 23]. Often, researchers either extrapolate the thin-metal results to bulk conditions or use novel structures to investigate the monolayer properties under bulk electrode conditions. We will examine a prime example where the characterization of unique semiconductor–molecule–metal molecular junctions elucidates the interplay between device structure and fabrication approaches.

Vibrational spectroscopy of semiconductor-based monolayers under metal electrodes has been particularly useful in understanding the metal–molecule interactions as most semiconductors, e.g., silicon, are transparent in the infrared range of the electromagnetic spectrum. Infrared spectroscopy (Fig. 2.2) has been used to show the impact of metallization conditions by comparing the spectra obtained from gold-nitrobenzene-silicon samples prepared by evaporating directly on the monolayer with samples prepared by using a soft landing technique where the samples are facing away from the evaporation source and the bell-jar is backfilled with argon to reduce the kinetic energy of the metal impinging on the surface [24]. The spectra obtained following direct gold metallization of nitrobenzene differ significantly from the starting monolayer and indicate significant molecular decomposition. In contrast, the spectra obtained after indirect metallization closely resemble the starting molecular surface indicating the means of metallization impacts the resulting metal–molecule–silicon structure.

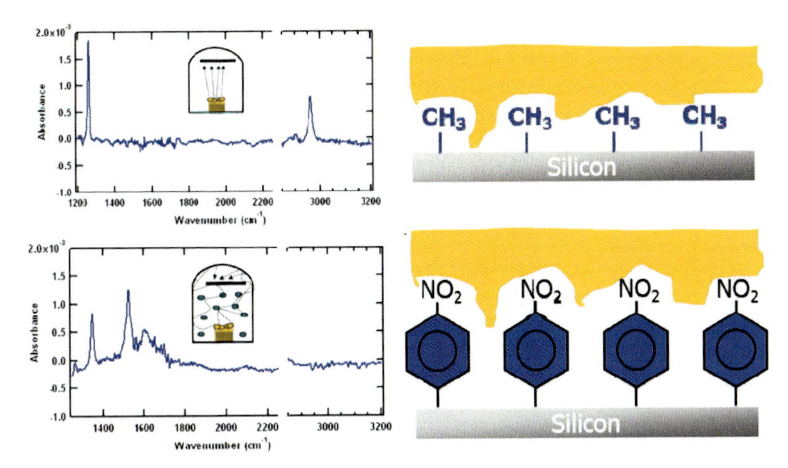

Figure 2.2 The method used to metalize organic monolayers has a large impact on the properties of the monolayer. Infrared spectra obtained from nitrobenzene monolayers on silicon after conventional metallization show the molecules have been destroyed with residual methyl stretches evident (top). Infrared spectra obtained after metallization using a soft landing technique contain many of the vibrational peaks that were observed prior to metallization. Adapted with permission from Scott et al., *JPCC*, 2008, **112**, 14021–14026. Copyright 2008 American Chemical Society.

Not only does the method of metallization impact the resulting molecular junction, but the entire system, including the monolayer and substrate, must be considered when fabricating nanoscale structures. Using the same metallization conditions on seemingly identical monolayers on silicon and silicon oxide substrates resulted in very different molecular junctions. Infrared studies [25, 26] comparing the metallization of aliphatic monolayers directly bonded to silicon and silicon oxide showed very different results for these two substrates as shown in Fig. 2.3. After metallization with Al, Au, and Ti, the vibrational signature of the aliphatic chains on Si are no longer evident, however, they are still observed for the aliphatic molecules on oxide surfaces. Because molecules on silicon bond through strong covalent linkages hindering the ability to self-assemble and form densely packed monolayers, the density of the molecules on these two surfaces differs slightly. Follow-up experiments performed on lower-density monolayers on the silicon oxide surface still exhibited the molecular vibrations following metallization indicating the feature-less silicon-based spectra could not be attributed solely to the less-dense monolayer enhancing metal penetration. These findings highlight the previously ignored substrate-monolayer interfacial chemistry as playing a critical role in the interaction of organic monolayers and top electrodes. Further work elucidated the mechanism for this observation of differing aliphatic molecular junctions is indeed linked to the substrate. Molecular junctions formed by evaporating silver onto molecules linked to silicon through Si–O and Si–C bonds were evident in the infrared spectra similar to the monolayers on silicon oxide as shown in Fig. 2.4 [27]. Depth profiling X-ray photoelectron spectroscopy data revealed that Au-based molecular junctions consisted of a gold silicide at the interface, while Ag-based molecular junctions consisted of metallic silver since it does not react with silicon to form a silicide. The monolayers attached to silicon oxide are observed after gold deposition because the silicide reaction is suppressed by the oxide surface. Due to the nanometer length scale and high energy of metallization, metal–substrate and metal–organic interactions are critical to consider when applying a top electrode on molecular electronic systems. Moreover, it is clear that the entire electrode–molecule–electrode structure should be characterized holistically since everything has an impact on the nanoscale interactions within the molecular junctions.

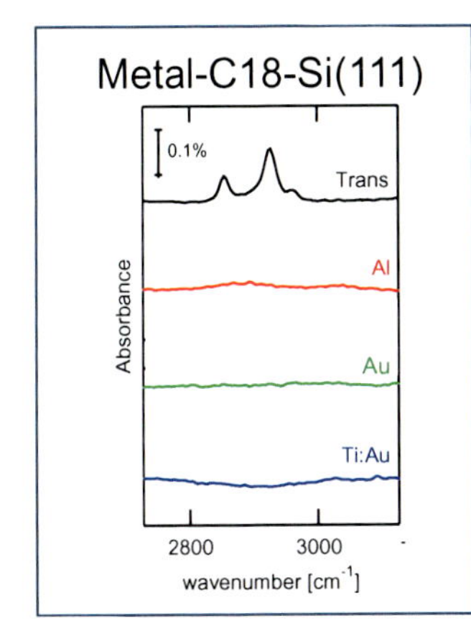

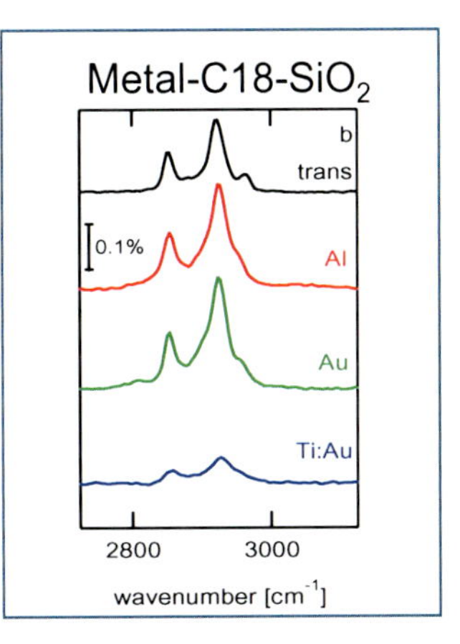

Figure 2.3 The effect of the substrate cannot be ignored when making molecular electronic junctions as metal (Al, Au, and Ti) evaporated on seemingly similar aliphatic chains on silicon (left) and silicon oxide (right) exhibit very different spectra. After metallization, the vibrational signature that was evident in the transmission spectrum (top) is no longer evident in the reflection spectra for the monolayers directly attached to silicon but is still observed for the monolayers on silicon oxide(right). Adapted with permission from Richter et al., *JPCB*, 2005, **109**, 21836–21841. Copyright 2005 American Chemical Society.

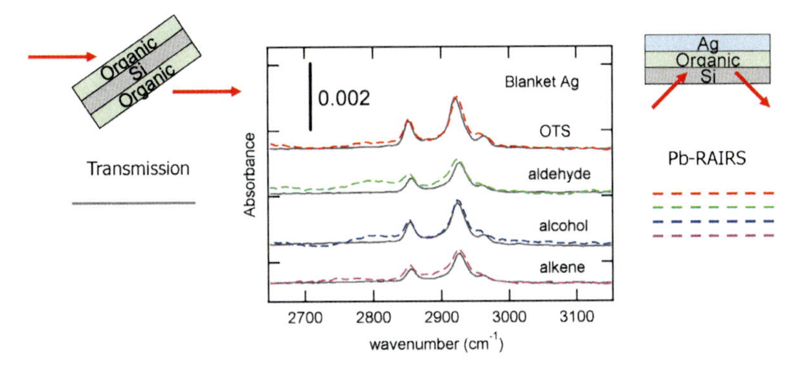

Figure 2.4 Infrared spectroscopy indicates the presence of molecular layers on silicon oxide (OTS, octadecyltrichlorosiloxane) and silicon (aldehyde, alcohol, alkene) after metallization with silver. Adapted with permission from Hacker et al., *JPCC*, 2007, **111**, 9384–9392. Copyright 2007 American Chemical Society.

2.4 Novel Fabrication Approaches

There are many intrinsic properties that lead to the challenges in forming reliable molecular electronic junctions and researchers have taken novel approaches to maximize favorable processes, such as self-assembly, and minimize unfavorable processes, such as metal evaporation. Often these alternative approaches are termed "soft" contact methods stemming from a bottom-up approach onto the molecular layer. These approaches to making reliable electrical contact to molecular monolayers include the use of a conducting polymer, liquid metal, crossed wire junctions, conducting-probe atomic force microscopy (CP-AFM), and placement of preformed metal. The next section will focus on these approaches in detail.

2.4.1 Conducting Polymer Top Contacts

A conducting polymer serves as an intermediate layer between the monolayer and metal contact. This approach allows for the organic components of a device to be produced by solution processing, which is appealing from an industrial manufacturing perspective as it is scalable and relatively cheap. These structures start with the junction size defined by pores (or wells) patterned within a dielectric layer (either photoresist or an inorganic material) by photolithographic methods. Often, the pore will have gold at the bottom for self-assembly. After the molecular layer self-assembles onto the gold surface, a conductive polymer is spun on top to cover the molecular layer. Finally, a metal is deposited by conventional methods, such as evaporation, over the polymer layer to enable measurement of the electrical properties as depicted in Fig. 2.5. The polymer forms a soft, intimate contact to the molecular layer and provides a protective barrier to prevent electrical shorts and direct contact between the top and bottom metal electrodes. The use of aqueous poly(3,4-ethylenedioxythiophene): poly(styrenesulfonate) (PEDOT:PSS) as a conducting polymer contact for molecular layers was first pioneered by the teams of Bert de Boer and Dago De Leeuw [28, 29], which demonstrated high yield of molecular junctions that successfully electrically connect to a SAM layer, which allowed for reliable, reproducible and potentially scalable molecular electronic components. Since then, a couple of other groups have successfully utilized PEDOT:PSS as a conducting interlayer between SAMs and the inorganic metal contacts [30, 31]. Flexible molecular electronic test beds can even be fabricated using this strategy and are able to withstand numerous flexing and bending cycles [30, 32]. Due to the wetting properties of PEDOT:PSS, it can only penetrate into pores of a diameter of 1 μm or larger, which limits the accessible size of test structures and hinders its integration with conventional electronics. Graphene has been investigated as a soft top contact electrode showing better thermal stability and resistivity than PEDOT:PSS molecular junctions [33]. Recently, another commercial conducting polymer, Aedotron P, has been used to contact SAMs in pores as small as 300 nm in diameter, which is promising for the scalability of the molecular junctions [34].

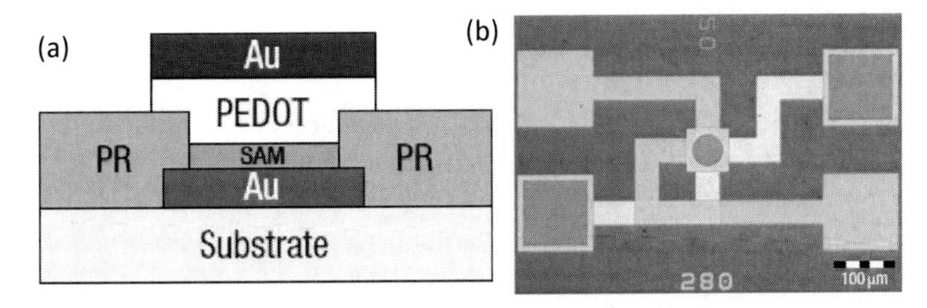

Figure 2.5 Polymer contacts are utilized in the following (a) device scheme, and (b) an optical image of the fabricated molecular electronic junction. Photoresist is abbreviated as PR. Images are adapted with permission from Van Hal et al., *Nat. Nano*, **3**(12), 749–754 (2008). Copyright 2008 Nature Publishing Group.

There are some limitations to the use of conductive polymer contacts within molecular junctions. First, at low temperatures, the transport mechanism of the polymer dominates [35, 34] in the direct current (DC) current–voltage (I–V) measurements, which is a challenge for interrogating the intrinsic electrical properties of the molecule of interest. Another shortcoming with polymer contacts is the interaction between the polymer layer and the molecular monolayer. With the use of aqueous-based PEDOT:PSS, the polymer likely does not penetrate through the monolayer when the tail group of the SAM presents a hydrophobic surface. The SAM properties are often varied, and the interaction at the polymer-SAM interface may not be negligible in many cases. Conductive polymer contacts offer a good route for manufacturing potentially scalable and reliable electrical contacts to molecular layers, but they complicate experiments for investigating intrinsic electrical properties of molecular layers due to the properties of the polymer.

2.4.2 Liquid Metal Top Contacts

The use of liquid metal to measure electrical properties of monolayers is among the first methods investigated to form a soft, non-destructive contact. In 1939, researchers used a mercury-drop approach to measure the dielectric properties of multilayer Langmuir–Blodgett films [36]. More recently, junctions are often made to SAMs by pushing the liquid metal from a syringe or another dispensing medium then the liquid metal is physically brought into contact with the molecular monolayer as depicted in the top portion of Fig. 2.6. Liquid metal contacts are appealing in the sense that they can be formed quickly (i.e., on demand) without any additional processing or manufacturing steps, which is an appealing route to investigate the intrinsic properties of molecules. Mercury has been the primary liquid metal to contact molecular layers [37–41]. Aside from its toxicity, another drawback of using mercury is that it forms an amalgam when it comes into physical contact with coinage metals such as gold and silver. Thus, the success of forming junctions is low if there are defects within the molecular layer that allow the mercury droplet to penetrate through and contact the bottom metal substrate (i.e., electrical shorts). One modification to reduce the likeliness of contact between mercury and the bottom metal, is to self-assemble another monolayer onto the mercury drop itself [38, 42]. However, this adds another processing step and layer in analyzing the electrical transport since an additional tunneling barrier needs to be considered, as the junction becomes metal–molecule–molecule–metal. However, the mercury drop contact method is suitable for measuring metal–molecule–semiconductor junctions [38, 40, 41, 43] since mercury does not react with conventional semiconducting (or oxide) materials. Mercury droplets tend to be somewhat large with a diameter of about 1 to 2 mm, and can limit the measurement bias range that can be applied to the molecular layer before the molecular (dielectric) breaks down.

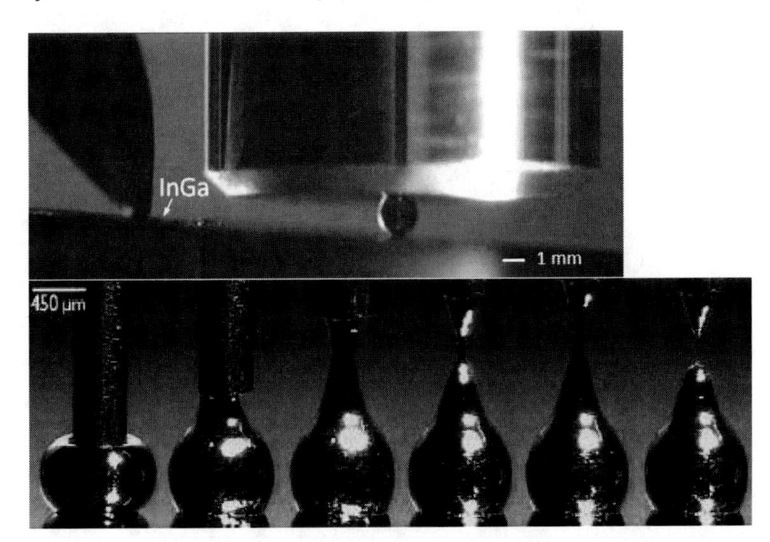

Figure 1.6 Optical images of liquid metal contacts consisting of a (top) mercury droplet, and (bottom) the sequence of forming an E-GaIn conical tip. Notice the difference in shapes that can be achieved with the different metals. The mercury droplet image is adapted with permission from reference [Vilan, *Langmuir*, **28**, 404] and the E-GaIn tip image is adapted with permission from reference [Chiechi, *Angew. Chem. Int. Ed.*, **47**, 142].

More recently, eutectic gallium indium (E-GaIn) metal has been demonstrated as a superior alternative to mercury for contacting SAMs [44–46]. E-GaIn offers many benefits that cannot be achieved by mercury. For instance, E-GaIn is non-toxic, chemically benign with other metals, and can be molded into a tip-shape such that smaller contact areas can be realized as depicted in Fig. 2.6. The non-Newtonian liquid behavior of E-GaIn is attributed to a native oxide skin layer [47, 48], which allows for the conical E-GaIn shape to be maintained in a range between 1–100 μm. After the E-GaIn tip is formed, it does not require any additional processing steps such as the SAM-coated Hg requirement. Thus, by using E-GaIn, the junction of metal–molecule–metal can be achieved to measure the impact of the molecules on the electrical properties. E-GaIn has been used extensively in forming back contacts for electrical- and magneto-based measurements of traditional solid-state devices. In combination with a very smooth silver substrate to provide a platform for the SAMs, reliable and statistically relevant measurements have been reported by using E-GaIn as a top contact [44]. Recently, the use of E-GaIn has enabled studies of the impact of odd-and even-numbered methylene units on the rectification properties of ferrocene-containing SAMs observed in the *I–V* measurements [46]. The impact of the native oxide on E-GaIn on electrical measurements is suggested to be negligible [49]. E-GaIn contacts are shaping up to be a superior alternative to mercury due to the ease of use and other desirable properties.

An obvious shortcoming with using liquid metal contacts for molecular electronic junctions is that this contact method cannot be adapted or implemented for manufacturing electronic components. This method currently has limitations for studying the fundamental electron transport through molecular layers by low temperature measurements, where the temperature range is limited by the freezing point of the metal. While there have been a couple of demonstrations of performing temperature dependent *I–V* measurements with a liquid metal as the top contact, [50, 51] the lack of widespread reports of this measurement suggest that the liquid nature of the metal make it difficult to perform. The contact area of the liquid metal can be difficult to estimate and control, which makes comparing the current density, *J*, of molecular electronic junctions between different molecules and laboratories difficult. Nevertheless, liquid metal contacts offer a route to interrogate the direct impact of SAMs in molecular electron junctions at room temperature in a relatively straightforward manner.

2.4.3 Crossed Wire Top Contacts

A variation on the metal–molecule–metal junction involves two thin metallic wires, one bare and one coated with a self-assembled monolayer. The crossed wire junction is formed by creating a crossed geometry with one wire perpendicular to an applied magnetic field (*B*) and the wire spacing is controlled through a Lorentz Force, as shown in Fig. 2.7. This approach eliminates the potential of metal atoms penetrating or reacting with the monolayer during junction fabrication and does not require advanced fabrication techniques to create nanometer-scale junctions. It is well suited to compare different molecules, end groups, and metal wires because the junction formation is relatively straightforward. However, the number of molecules within the junction is uncertain (approximately 10^3 molecules) and the orientation of the SAMs on a curved (and likely very rough) surface is unclear. Kushmerick and coworkers successfully studied the charge transport in both aliphatic and conjugated molecular layers observing a clear difference in conductance between them [52, 53]. Also, the influence of the metal–molecule contact has been studied with the formation of molecular junctions with symmetric (i.e., Au–molecule–Au) and asymmetric electrodes (Au–molecule–Pd). As predicted, the devices with Pd–molecule interface are more conductive than Au–molecule interfaces [54, 55]. The crossed wire approach was successfully used to create electronic switching elements by forming Ag filaments within the junction [56]. McCreery et al. also studied crossed-wire molecular junctions with a carbon/molecule/copper structure investigating the dependence of electron transport on both molecular structure and temperature [57].

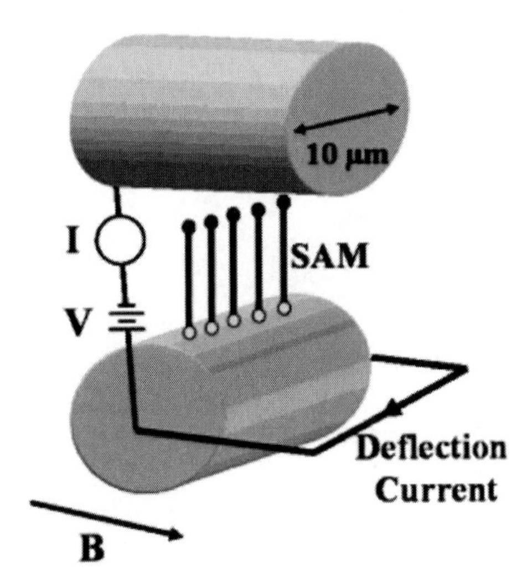

Figure 2.7 Simplified illustration of a crossed wire top contact molecular junction configuration. Image is adapted with permission from Kushmerick et al., *Phys. Rev. Lett.*, 2002, **89**.

2.4.4 Conducting Probe AFM Top Contacts

A molecular junction can be formed by approaching a metal-coated conductive atomic force microscope (C-AFM) tip onto a pre-formed SAM on a metal substrate. The electron transport through SAMs is measured by applying a DC bias between the AFM probe and the substrate as depicted in Fig. 2.8. Noteworthy, the force with which the tip contacts the SAMs can greatly affect the electrical properties [58]. Similar to the crossed-wire molecular junctions, C-AFM can be used to probe the conductance of various molecular layers, differing electrodes, and measurement conditions. C-AFM is useful to extract contact resistance information and to obtain information regarding the tunneling characteristics of the molecules. This top contact mode approach also offers the possibility to study the effect of chemisorbed and physisorbed contacts [59]. The conducting tip can be coated with different metallic layers (Ag, Au, Pd or Pt), which enable the study of the influence of metal work function on the electronic transport through the junction [60]. Using this architecture, the number of molecules under study is unknown, but likely range from 10^2 to 10^3 molecules. A modification of the C-AFM technique includes depositing a metal nanoparticle on the SAM or a SAM-coated "nanodot" and subsequently contacting with the C-AFM tip. These approaches better define the area and have been shown to be insensitive to

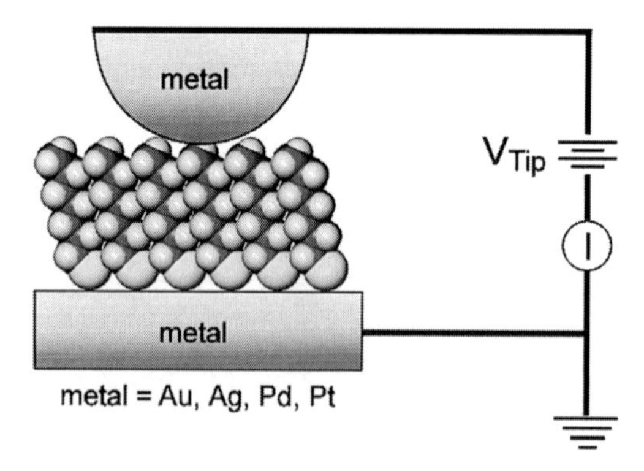

Figure 2.8 Simplified illustration of a conductive probe atomic force microscopy top contact molecular junction configuration. Image is adapted with permission from Beebe et al., *J. Am. Chem. Soc.*, 2002, **124**, 11268–11269. Copyright 2002 American Chemical Society.

the applied force making the number of molecules under study better known [61–63]. Finally, the combination of C-AFM and STM can be used to monitor the electrical properties of single-molecule junctions, with a tight control of the spacing and tilt angle of the molecule [64].

2.4.5 Deposition of Preformed Metal Contacts

One approach of using preformed electrodes is to use contact lithography to bring together preformed electrodes. A self-assembled monolayer is prepared onto a preformed template-striped electrode. Because the electrode is template-striped, the roughness is quite small thus minimizing grain boundaries and defects attributed to them [65]. If self-assembly is performed using a bifunctional molecule with a thiol at one end, the thiol will selectively self-assemble onto the Au surface and the second functional group can be chosen to react with the second electrode [66]. For example, molecular junctions created between Au and silicon electrodes were successfully fabricated with mercaptohexadecanoic acid molecules following a process depicted in Fig. 2.9 [67] as well as Au–molecule–GaAs junctions [68, 69].

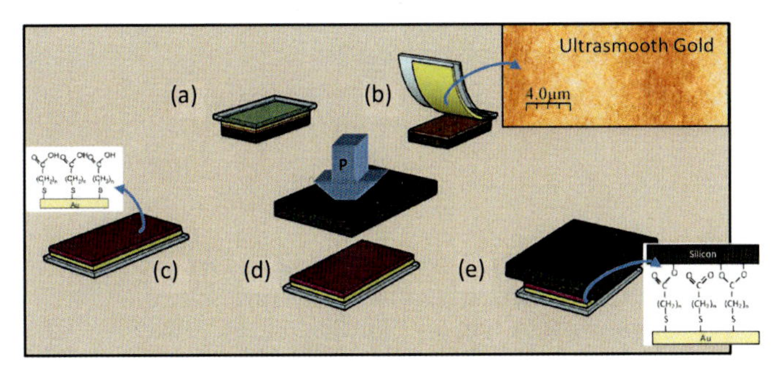

Figure 2.9 Flip chip lamination process. Gold is evaporated onto a substrate previously treated with a release layer and then lifted off to reveal the ultrasmooth underside (a and b). Thiol-containing molecules are self-assembled on the gold surface (c) and brought into contact with an H-terminated silicon surface under pressure and temperature (d) to form a chemical bond resulting in a semiconductor–molecule–metal junction (e). Image is adapted with permission from Coll et al., *JACS*, 2009, **131**, 12451–12457. Copyright 2009 American Chemical Society.

This method has the advantage of creating stable large area junctions chemically bonded to both electrodes. In addition, the junction is formed by applied temperature and pressure rather than in a solution environment minimizing the opportunity for trapped solvent within the junction. Because the junction is formed by the differential adhesion between two surfaces, the choice of molecules and electrodes must be carefully considered and may be somewhat of a limitation. That being said, however, successful junctions have been formed from a variety of molecules and functional groups [70], bilayer junctions incorporating metal ions [71], and unique materials such as graphene [72], organic crystals [73] and other carbon materials [74].

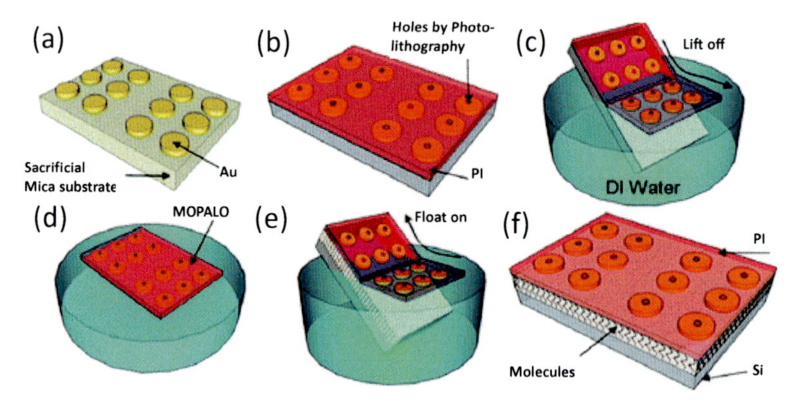

Figure 2.10 A modified polymer-assisted lift-off (MoPALO) process used to created molecular junctions from preformed metal contacts. First the metal is deposited on a sacrificial substrate (a) and coated with a polymer (b). Holes through the polymer layer are made to enable electrical contact and the metal-polymer substrate is immersed into solution to separate it from the substrate (c). The floating metal–polymer (d) is then carefully transferred onto a monolayer-terminated surface (e) to prepare the molecular junctions (f). Image is adapted with permission from Stein et al., *J. Phys. Chem. C*, 2010, **114**(29), 12769–12776. Copyright 2010 American Chemical Society.

An alternate approach of using preformed electrodes involves first forming the electrodes on a sacrificial substrate, such as glass or mica, where the adhesion is weak enough that when placed in solution, the metal separates from the sacrificial substrate and floats on top of the solution. The junction is then created by "floating" these electrodes onto a monolayer covered bottom electrode [75]. The metal can also be patterned into arrays or wires with the addition of a polymer top layer to maintain the structural integrity [76, 77]. A sample process flow is depicted in Fig. 2.10 of a modified version to enable electrical contact through the polymer top layer [78]. The active area of the junction can be smaller than those obtained with evaporated contacts because lithography can be used to pattern the contacts. Because the contacts are placed with the assistance of a liquid medium, this adds the challenge of electrode wrinkling and repulsive hydrophobic and hydrophilic interactions.

2.5 Junction Characterization Approaches

An important aspect to characterize the active molecular layer is to probe the buried junction by electrical and spectroscopic measurements. Electrical measurements are necessary to validate functionality and to connect molecules to the outside world.

2.5.1 Electrical Characterization

The majority of electrical measurements on ensemble-scale molecular electronic junctions are based on direct current (DC) current–voltage (I–V) measurements in a limited voltage bias window, typically in the range of ±1 V to limit the breakdown of the molecular layer. Molecular layers consisting of alkanes follow a transport mechanism of nonresonant, through-bond tunneling that can be described by the following equation:

$$J = A \times \exp(-\beta n),$$

where J is the current density, A the Richardson constant, β a decay parameter, and n the number of carbons within the molecular layer. Since the current density is expected to exponentially decay with increasing molecular length, new test structures and test beds are often validated by electrical measurements of SAMs with varying length. The beta parameter is experimentally determined and often used as a qualitative factor to determine the quality of the molecular layer (i.e., defects or surface roughness) and effectiveness of the electrical test platform. Beta values of ~1 per angstrom (for aliphatic systems) are considered ideal and have been reported for a variety of electrical measurement platforms in the solid state [29, 79–81]. There are also numerous reports of beta values less than 0.8 per angstrom measured in alkane-based SAMs [34, 71, 82, 83] and the cause for a relatively low β value in alkane-based SAMs is still under debate. Temperature-dependent DC I–V measurements are also used to confirm the transport mechanism within the molecular junction since J does not vary with temperature in the case of nonresonant tunneling [19, 84].

Recently, transition voltage spectroscopy (TVS) has been a promising technique to extract more qualitative information from DC I–V measurements. TVS is used to estimate the energy barrier at the electrode–molecular layer interface, which is the energy difference from the Fermi level of the electrode and a molecular orbital within the SAM. This is directly estimated by transforming the I–V data into a Fowler–Nordheim plot, which plots $\ln\left(\frac{1}{V^2}\right)$ versus $\frac{1}{V}$, where the minima corresponds to a transition from a direct tunneling regime to field emission regime. Beebe and co-workers first reported a correlation between the transition voltage and the highest occupied molecular orbital (HOMO) (with respect to Fermi energy) in metal–molecule–metal junctions and metal–molecule interfaces [85]. TVS has been applied to different types of molecular electronic junctions such as liquid metal junctions, [38, 86] conducting probe AFM, [85] crossed-wire junctions, [85] electromigration nanogap junctions, [81] and junctions with pre-formed contacts [67]. Disparity in the transition voltage obtained from these different test structures has highlighted the controversy in the origin of minimum in the Fowler–Nordheim plot and an accurate understanding of TVS remains an active area of research [1].

A drawback of DC I–V measurements is that the electrical current is inherently prone to find the least resistive electrical pathway between the electrodes sandwiching the molecular layer. This makes electrical techniques extremely sensitive to defects within the molecular layers (such as pinholes) and/or formed during the contact formation process (e.g., filaments). For this reason, researchers often try to limit the electrical contact area to limit the number of defects that would contribute to spurious electrical data. A less common technique is to use alternating current (AC) techniques such as capacitance–voltage (*C–V*). This method is less sensitive to defects within the molecular layer. This measurement technique has been applied to alkane-based SAMs in a variety of electrical test structures [87–91]. The junction follows a metal–SAM–semiconductor structure to mimic metal-insulator-semiconductor (MIS) structures. This technique is promising as another characterization tool to extract the impact of SAMs on changing the work function of an electrode of a completed junction [90] and density of interface traps in a junction [88].

The interpretation of these electrical measurements often relies on a simple tunneling theory applied to molecular junctions and this can be inaccurate in some cases. For example, SAMs do not form ideal tunneling barriers that are typically achieved by inorganic compounds. The application of metal–semiconductor theory on the electrostatic interaction between a metal and organic material is also not ideal from a band structure framework and the theory of charge transport within molecular systems continues to remain an active area of research [92].

2.5.2 Spectroscopic Characterization

Characterization of the free-standing molecular layer has been performed by a myriad of surface sensitive tools ranging from spectroscopies to imaging. A thorough review of these characterization techniques is beyond the scope of this chapter and has been the subject of countless research reports and many books. Characterization of the molecular junction is more specialized and warrants some attention here since the top electrode typically makes the junction unrecognizable by most surface sensitive techniques. However, it is clear that characterizing the holistic electrode–molecule–electrode junction is essential to accurately understand the structure and ultimately, the origin of electronic properties.

Early work by Czandera et al. [7, 93, 94] examined metal penetration through self-assembled monolayers with X-ray photoelectron spectroscopy (XPS) and ion scattering spectroscopy (ISS) as model systems for bonding and adhesion between metallic and organic surfaces. These techniques provide information regarding the identity and quantity of the elements present, and their oxidation state and are limited to semitransparent, thin metal layers. Later investigations by others branched out to include vibrational spectroscopy (infrared and Raman) and secondary ion mass-spectrometry investigations of the interaction of thin metal layers with organic monolayers [9, 95–98]. Again thin metal is key for most optical and charge-based systems since thick metal is reflective. Vibrational spectroscopy provides information on the chemical composition of the species present as well as the quantity and quality of the molecular orientation. Secondary-ion mass spectrometry provides the chemical and physical structure as a function of depth by destructively sputtering the sample and typically the sensitivity is sufficient to observe chemical bonding at the interfaces and penetration of metal through molecular layers. Researchers have been able to obtain the vibrational spectra from molecular junctions under bulk electrodes with the use of specialized junction design incorporating transparent electrodes, such as silicon, which is transparent in the infrared, and carbon, which is transparent in the visible region.

Internal photoemission (IPE) is a powerful technique used to extract the interfacial transport barrier [99] and has been applied to molecular junctions on silicon [100] and carbon [101]. This technique involves measuring the photocurrent generated as a function of applied bias and modulated optical illumination. Typically, the optical yield is plotted in a Fowler plot with the intercept related to the energy needed for a carrier (electron or hole) to overcome the interfacial electrode–molecule barrier and be transported through the junction. Application of positive and negative bias can be used to extract the barrier for holes and electrons, which is an important distinction and source of controversy for molecular electronic devices. Careful experimental design and interpretation are needed, however, to separate the effects of the two different electrodes and the molecules, which may interact with the light.

2.6 Conclusion and Outlook

Molecular electronics remains an active area of research after many of the foundational discoveries, and researchers are in a position to take advantage of this nanotechnology. Underpinning the utility of molecular electronics, however, is the ability of researchers to fabricate reliable molecular junctions. This means not only making the molecular junction, but also characterizing the junction to ensure that the chemical and physical structure is indeed what was intended and then to ultimately engineer the desired electrical properties.

References

1. D. Xiang, H. Jeong, T. Lee, D. Mayer, *Adv. Mater.,* **25** (2013) 4845.
2. S. V. Aradhya, L. Venkataraman, *Nat. Nanotechnol.,* **8** (2013) 399.
3. A. Ulman, *Chem. Rev.,* **96** (1996) 1533.
4. E. DeIonno, H. R. Tseng, D. D. Harvey, J. F. Stoddart, J. R. Heath, *J. Phys. Chem. B*, **110** (2006) 7609.
5. S. C. Chang, Z. Y. Li, C. N. Lau, B. Larade, R. S. Williams, *Appl. Phys. Lett.,* **83** (2003) 3198.
6. T. B. Tighe, T. A. Daniel, Z. H. Zhu, S. Uppili, N. Winograd, D. L. Allara, *J. Phys. Chem. B,* **109** (2005) 21006.
7. G. C. Herdt, A. W. Czanderna, *J. Vacuum Sci. Technol. A*, **17** (1999) 3415.
8. D. R. Jung, A. W. Czanderna, *Crit. Rev. Solid State Mater. Sci.,* **19** (1994) 1.
9. A. V. Walker, T. B. Tighe, B. C. Haynie, S. Uppili, N. Winograd, D. L. Allara, *J. Phys. Chem. B*, **109** (2005) 11263.
10. K. Konstadinidis, P. Zhang, R. L. Opila, D. L. Allara, *Surf. Sci.,* **338** (1995) 300.
11. A. Hooper, G. L. Fisher, K. Konstadinidis, D. Jung, H. Nguyen, R. Opila, R. W. Collins, N. Winograd, D. L. Allara, *J. Am. Chem. Soc.,* **121** (1999) 8052.
12. B. de Boer, M. M. Frank, Y. J. Chabal, W. R. Jiang, E. Garfunkel, Z. Bao, *Langmuir*, **20** (2004) 1539.
13. Y. Tai, A. Shaporenko, H. Noda, M. Grunze, M. Zharnikov, *Adv. Mater.*, **17** (2005) 1745.
14. D. R. Jung, A. W. Czanderna, G. C. Herdt, *J. Vacuum Sci. Technol. A*, **14** (1996) 1779.
15. H. Haick, M. Ambrico, T. Ligonzo, R. T. Tung, D. Cahen, *J. Am. Chem. Soc.,* **128** (2006) 6854.
16. R. M. Metzger, T. Xu, I. R. Peterson, *J. Phys. Chem. B,* **105** (2001) 7280.

17. S. Lodha, D. B. Janes, *J. Appl. Phys.*, **100** (2006) 024503.
18. L. T. Cai, H. Skulason, J. G. Kushmerick, S. K. Pollack, J. Naciri, R. Shashidhar, D. L. Allara, T. E. Mallouk, T. S. Mayer, *J. Phys. Chem. B*, **108** (2004) 2827.
19. W. Wang, T. Lee, M. A. Reed, *Phys. Rev. B*, **68** (2003) 035416.
20. W. Y. Wang, T. Lee, I. Kretzschmar, M. A. Reed, *Nano Lett.*, **4** (2004) 643.
21. J. G. Kushmerick, J. Lazorcik, C. H. Patterson, R. Shashidhar, D. S. Seferos, G. C. Bazan, *Nano Lett.*, **4** (2004) 639.
22. J. Hihath, N. J. Tao, *Prog. Surf. Sci.*, **87** (2012) 189.
23. N. Okabayashi, M. Paulsson, T. Komeda, *Prog. Surf. Sci.*, **88** (2013) 1.
24. A. Scott, C. A. Hacker, D. B. Janes, *J. Phys. Chem. C*, **112** (2008) 14021.
25. C. A. Richter, C. A. Hacker, L. J. Richter, *J. Phys. Chem. B*, **109** (2005) 21836.
26. C. A. Richter, C. A. Hacker, L. J. Richter, O. A. Kirillov, J. S. Suehle, E. M. Vogel, *Solid-State Electron.*, **50** (2006) 1088.
27. C. A. Hacker, C. A. Richter, N. Gergel-Hackett, L. J. Richter, *J. Phys. Chem. C*, **111** (2007) 9384.
28. H. B. Akkerman, P. W. M. Blom, D. M. de Leeuw, B. de Boer, *Nature*, **441** (2006) 69.
29. P. A. Van Hal, E. C. P. Smits, T. C. T. Geuns, H. B. Akkerman, B. C. De Brito, S. Perissinotto, G. Lanzani, A. J. Kronemeijer, V. Geskin, J. Cornil, P. W. M. Blom, B. De Boer, D. M. De Leeuw, *Nat. Nano*, **3** (2008) 749.
30. S. Park, G. Wang, B. Cho, Y. Kim, S. Song, Y. Ji, M.-H. Yoon, T. Lee, *Nat. Nano*, **7** (2012) 438.
31. Z. Ng, K. P. Loh, L. Li, P. Ho, P. Bai, J. H. K. Yip, *ACS Nano*, **3** (2009) 2103.
32. H. Jeong, D. Kim, G. Wang, S. Park, H. Lee, K. Cho, W.-T. Hwang, M.-H. Yoon, Y. H. Jang, H. Song, D. Xiang, T. Lee, *Adv. Funct. Mater.*, **24** (2014) 2472.
33. G. Wang, Y. Kim, M. Choe, T.-W. Kim, T. Lee, *Adv. Mater.*, **23** (2011) 755.
34. A. B. Neuhausen, A. Hosseini, J. A. Sulpizio, C. E. D. Chidsey, D. Goldhaber-Gordon, *ACS Nano*, **6** (2012) 9920.
35. A. J. Kronemeijer, E. H. Huisman, I. Katsouras, P. A. van Hal, T. C. T. Geuns, P. W. M. Blom, S. J. van der Molen, D. M. de Leeuw, *Phys. Rev. Lett.*, **105** (2010) 156604.
36. H. H. Race, S. I. Reynolds, *J. Am. Chem. Soc.*, **61** (1939) 1425.

37. K. Slowinski, H. K. Y. Fong, M. Majda, *J. Am. Chem. Soc.*, **121** (1999) 7257.
38. D. Guerin, S. Lenfant, S. Godey, D. Vuillaume, *J. Mater. Chem.*, **20** (2010) 2680.
39. J. D. Le, Y. He, T. R. Hoye, C. C. Mead, R. A. Kiehl, *Appl. Phys. Lett.*, **83** (2003) 5518.
40. Y.-J. Liu, D. M. Waugh, H.-Z. Yu, *Appl. Phys. Lett.*, **81** (2002) 4967.
41. Y. Selzer, A. Salomon, D. Cahen, *J. Am. Chem. Soc.*, **124** (2002) 2886.
42. R. Haag, M. A. Rampi, R. E. Holmlin, G. M. Whitesides, *J. Am. Chem. Soc.*, **121** (1999) 7895.
43. G. Nesher, A. Vilan, H. Cohen, D. Cahen, F. Amy, C. Chan, J. Hwang, A. Kahn, *J. Phys. Chem. B*, **110** (2006) 14363.
44. R. C. Chiechi, E. A. Weiss, M. D. Dickey, G. M. Whitesides, *Angew. Chem. Int. Ed.*, **47** (2008) 142.
45. C. A. Nijhuis, W. F. Reus, G. M. Whitesides, *J. Am. Chem. Soc.*, **131** (2009) 17814.
46. N. Nerngchamnong, Y. Li, D.-C. Qi, J. Li, D. Thompson, C. A. Nijhuis, *Nat. Nano*, **8** (2013) 113.
47. M. D. Dickey, R. C. Chiechi, R. J. Larsen, E. A. Weiss, D. A. Weitz, G. M. Whitesides, *Adv. Funct. Mater.*, **18** (2008) 1097.
48. Q. Xu, E. Brown, H. M. Jaeger, *Phys. Rev. E*, **87** (2013) 043012.
49. J. R. Barber, H. J. Yoon, C. M. Bowers, M. M. Thuo, B. Breiten, D. M. Gooding, G. M. Whitesides, *Chem. Mater.*, **26** (2014) 3938.
50. C. A. Nijhuis, W. F. Reus, J. R. Barber, M. D. Dickey, G. M. Whitesides, *Nano Lett.*, **10** (2010) 3611.
51. H. Shpaisman, O. Seitz, O. Yaffe, K. Roodenko, L. Scheres, H. Zuilhof, Y. J. Chabal, T. Sueyoshi, S. Kera, N. Ueno, A. Vilan, D. Cahen, *Chem. Sci.*, **3** (2012) 851.
52. J. G. Kushmerick, D. B. Holt, J. C. Yang, J. Naciri, M. H. Moore, R. Shashidhar, *Phys. Rev. Lett.*, **89** (2002) 086802.
53. J. G. Kushmerick, J. Naciri, J. C. Yang, R. Shashidhar, *Nano Lett.*, **3** (2003) 897.
54. A. S. Blum, J. G. Kushmerick, S. K. Pollack, J. C. Yang, M. Moore, J. Naciri, R. Shashidhar, B. R. Ratna, *J. Phys. Chem. B*, **108** (2004) 18124.
55. J. M. Seminario, C. E. De la Cruz, P. A. Derosa, *J. Am. Chem. Soc.*, **123** (2001) 5616.

56. J. M. Beebe, J. G. Kushmerick, *Appl. Phys. Lett.*, **90** (2007).
57. F. Anariba, J. K. Steach, R. L. McCreery, *J. Phys. Chem. B,* **109** (2005) 11163.
58. X. D. Cui, X. Zarate, J. Tomfohr, O. F. Sankey, A. Primak, A. L. Moore, T. A. Moore, D. Gust, G. Harris, S. M. Lindsay, *Nanotechnology*, **13** (2002) 5.
59. C. C. Kaun, H. Guo, *Nano Lett.*, **3** (2003) 1521.
60. J. M. Beebe, V. B. Engelkes, L. L. Miller, C. D. Frisbie, *J. Am. Chem. Soc.*, **124** (2002) 11268.
61. T. Morita, S. Lindsay, *J. Am. Chem. Soc.*, **129** (2007) 7262.
62. A. M. Rawlett, T. J. Hopson, L. A. Nagahara, R. K. Tsui, G. K. Ramachandran, S. M. Lindsay, *Appl. Phys. Lett.*, **81** (2002) 3043.
63. K. Smaali, N. Clément, G. Patriarche, D. Vuillaume, *ACS Nano*, **6** (2012) 4639.
64. W. Haiss, C. S. Wang, I. Grace, A. S. Batsanov, D. J. Schiffrin, S. J. Higgins, M. R. Bryce, C. J. Lambert, R. J. Nichols, *Nat. Mater.*, **5** (2006) 995.
65. E. A. Weiss, G. K. Kaufman, J. K. Kriebel, Z. Li, R. Schalek, G. M. Whitesides, *Langmuir*, **23** (2007) 9686.
66. Y.-L. Loo, R. L. Willett, K. W. Baldwin, J. A. Rogers, *J. Am. Chem. Soc.*, **124** (2002) 7654.
67. M. Coll, L. H. Miller, L. J. Richter, D. R. Hines, O. D. Jurchescu, N. Gergel-Hackett, C. A. Richter, C. A. Hacker, *J. Am. Chem. Soc.*, **131** (2009) 12451.
68. Y.-L. Loo, D. V. Lang, J. A. Rogers, J. W. P. Hsu, *Nano Lett.*, **3** (2003) 913.
69. R. Y. Wang, R. A. Segalman, A. Majumdar, *Appl. Phys. Lett.*, **89** (2006).
70. M. Coll, N. Gergel-Hackett, C. A. Richter, C. A. Hacker, *J. Phys. Chem. C*, **115** (2011) 24353.
71. S. Pookpanratana, J. W. F. Robertson, C. Jaye, D. A. Fischer, C. A. Richter, C. A. Hacker, *Langmuir*, **29** (2013) 2083.
72. E. H. Lock, M. Baraket, M. Laskoski, S. P. Mulvaney, W. K. Lee, P. E. Sheehan, D. R. Hines, J. T. Robinson, J. Tosado, M. S. Fuhrer, S. C. Hernandez, S. G. Waltont, *Nano Lett.*, **12** (2012) 102.
73. M. Coll, K. P. Goetz, B. R. Conrad, C. A. Hacker, D. J. Gundlach, C. A. Richter, O. D. Jurchescu, *Appl. Phys. Lett.*, **98** (2011) 3.
74. V. K. Sangwan, A. Southard, T. L. Moore, V. W. Ballarotto, D. R. Hines, M. S. Fuhrer, E. D. Williams, *Microelectron. Eng.*, **88** (2011) 3150.
75. A. Vilan, D. Cahen, *Adv. Funct. Mater.*, **12** (2002) 795.
76. K. T. Shimizu, J. D. Fabbri, J. J. Jelincic, N. A. Melosh, *Adv. Mater.*, **18** (2006) 1499.

77. S. O. Krabbenborg, J. G. E. Wilbers, J. Huskens, W. G. van der Wiel, *Adv. Funct. Mater.*, **23** (2013) 770.

78. N. Stein, R. Korobko, O. Yaffe, R. Har Lavan, H. Shpaisman, E. Tirosh, A. Vilan, D. Cahen, *J. Phys. Chem. C*, **114** (2010) 12769.

79. V. B. Engelkes, J. M. Beebe, C. D. Frisbie, *J. Am. Chem. Soc.*, **126** (2004) 14287.

80. C. C. Bof Bufon, J. D. Arias Espinoza, D. J. Thurmer, M. Bauer, C. Deneke, U. Zschieschang, H. Klauk, O. G. Schmidt, *Nano Lett.*, **11** (2011) 3727.

81. H. Song, Y. Kim, H. Jeong, M. A. Reed, T. Lee, *J. Phys. Chem. C*, **114** (2010) 20431.

82. E. A. Weiss, R. C. Chiechi, G. K. Kaufman, J. K. Kriebel, Z. Li, M. Duati, M. A. Rampi, G. M. Whitesides, *J. Am. Chem. Soc.*, **129** (2007) 4336.

83. R. Lovrinčić, O. Kraynis, R. Har-Lavan, A.-E. Haj-Yahya, W. Li, A. Vilan, D. Cahen, *J. Phys. Chem. Lett.*, **4** (2013) 426.

84. X. L. Li, J. He, J. Hihath, B. Q. Xu, S. M. Lindsay, N. J. Tao, *J. Am. Chem. Soc.*, **128** (2006) 2135.

85. J. M. Beebe, B. Kim, J. W. Gadzuk, C. Daniel Frisbie, J. G. Kushmerick, *Phys. Rev. Lett.*, **97** (2006) 026801.

86. G. Ricœur, S. Lenfant, D. Guérin, D. Vuillaume, *J. Phys. Chem. C*, **116** (2012) 20722.

87. C. A. Richter, C. A. Hacker, L. J. Richter, *J. Phys. Chem. B*, **109** (2005) 21836.

88. W. Peng, O. Seitz, R. A. Chapman, E. M. Vogel, Y. J. Chabal, *Appl. Phys. Lett.*, **101** (2012).

89. L. Kornblum, Y. Paska, H. Haick, M. Eizenberg, *J. Phys. Chem. C*, **117** (2012) 233.

90. L. Kornblum, Y. Paska, J. A. Rothschild, H. Haick, M. Eizenberg, *Appl. Phys. Lett.*, **99** (2011).

91. N. Clément, D. Guérin, S. Pleutin, S. Godey, D. Vuillaume, *J. Phys. Chem. C*, **116** (2012) 17753.

92. J. P. Bergfield, M. A. Ratner, *Phys. Status Solidi B*, **250** (2013) 2249.

93. G. C. Herdt, D. R. Jung, A. W. Czanderna, *J. Adhesion*, **60** (1997) 197.

94. G. C. Herdt, D. E. King, A. W. Czanderna, *Z. Phys. Chem.*, **202** (1997) 163.

95. A. V. Walker, T. B. Tighe, O. M. Cabarcos, M. D. Reinard, B. C. Haynie, S. Uppili, N. Winograd, D. L. Allara, *J. Am. Chem. Soc.*, **126** (2004) 3954.

96. C. L. McGuiness, A. Shaporenko, M. Zharnikov, A. V. Walker, D. L. Allara, *J. Phys. Chem. C*, **111** (2007) 4226.

97. R. L. McCreery, *Anal. Chem.*, **78** (2006) 3490.

98. P. D. Carpenter, S. Lodha, D. B. Janes, A. V. Walker, *Chem. Phys. Lett.*, **472** (2009) 220.

99. V. V. Afanas'ev, *Internal Photoemission Spectroscopy: Fundamentals and Recent Advances*, Elsevier, London, 2014.

100. D. Vuillaume, C. Boulas, C. Collet, G. Allan, C. Deleru, *Phys. Rev. B*, **58** (1998) 16491.

101. J. A. Fereiro, R. L. McCreery, A. J. Bergren, *J. Am. Chem. Soc.*, **135** (2013) 9584.

Chapter 3

Charge Transport in Dynamic Molecular Junctions

Stéphane Lenfant

Nanostructures nanoComponents & Molecules Group,
Institute of Electronics, Microelectronics and Nanotechnology,
CNRS & University of Lille,
B.P. 60069, 59652 Villeneuve d'Ascq, France

stephane.lenfant@iemn.univ-lille1.fr

3.1 Introduction

Recently the electrical characterization of molecular junctions studied in interaction with an external stimulus has gained considerable interest. Since 1995, date of the first electrical characterization of a single molecule by STM [1], many molecular junctions, formed by one or few molecules in a controlled configuration between two electrodes, have been studied electrically. Since recently, these junctions have been studied during a dynamic processed caused by an external stimulus which modifies the structure or conformation of the molecule. This modification induces a variation of the conductance measured in the molecular junction. This new approach in molecular electronics was made possible mostly with the help of a better control of the molecular

Molecular Electronics: An Experimental and Theoretical Approach
Edited by Ioan Bâldea

ISBN 978-981-4613-90-3 (Hardcover), 978-981-4613-91-0 (eBook)
www.panstanford.com

junction formed, and consequently a reduction of the electrical property dispersion. This low dispersion on the electronic properties make possible to differentiate between two (or sometime more) levels of conductance in the molecular junction. Usually two levels of conductance are observed: a high conductance (ON) and a low conductance (OFF) state, monitored by an external stimulus (see scheme in Fig. 3.1).

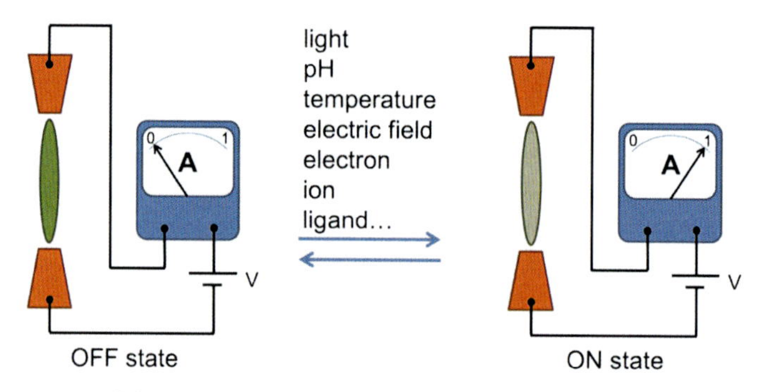

Figure 3.1 Schematic drawing illustrating the principle of the dynamic molecular junction: The molecule connected between two electrodes can change of conductance under application of an external stimulus.

From a chemist point of view, a large variety of molecules can be used for these dynamic molecular junctions. Called molecular switches, these molecules can shift reversibly between two or more stable states defined by their structural, electronic, and/or optical properties, in response to environmental stimuli. Generally demonstrated in liquid environment, these molecular switches are sensible to external stimuli such as light, pH, temperature, electric field, electron, ion or in presence of a ligand [2, 3]. Due to the richness of the chemistry, molecules should also enable a wide range of other promising functions. These functional molecules are of great fundamental interest and also recognized as a promising candidate for the integration in electronic devices and the development of future molecular electronics [4]. This approach by using functional molecule aims not to compete with complementary metal oxide semiconductor (CMOS) technology on the integration density due to the small size of molecule, but to complement existing electronic devices by bringing specific solutions at lower cost and simpler to produce. Sensor and mainly memory are recognized as one of the most promising applications for these molecular switches. This new memory generation is potentially less complex compare to CMOS technology where several transistors are required to store one bit of information, and consequently less expensive to produce.

Experimentally, incorporate molecular switches in a molecular junction in order to analyze the electronic properties remains a big challenge. The main difficulties are the usual and recurrent problem in molecular electronics: the molecular connection to macroscopic electrodes, and further here the interaction of external stimuli with the connected molecule in the junction. Until now, several examples of molecular switches integrated in a molecular junction have been demonstrated. Experimental demonstrations of dynamic molecular junction (detailed below) have been realized for various external stimuli: light, electric field, pH, biomolecule, ion, etc.

3.1.1 Light Stimulus

Photochromic molecules are the most studied molecular switches due to the well-known and numerous chemical functions sensible to light. The azobenzene function was frequently used as active unit in dynamic molecular junction. This compound has the interesting property to switch reversibly between the *cis* and *trans* isomers under, respectively, ultraviolet and blue light (Fig. 3.2). Due to the fact that the *trans* isomer is more stable, *cis* isomer can relax back to the *trans* with help of heat via *cis*-to-*trans* isomerization.

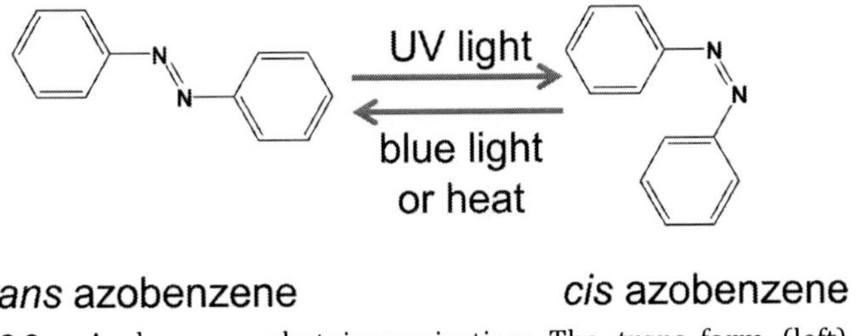

Figure 3.2 Azobenzene photoisomerization: The *trans* form (left) can switch to the *cis* form (right) using UV light (wavelength comprised between 300 and 400 nm) and reversibly switch to the *trans* form using visible blue light (wavelength superior to 400 nm) or heat.

Various experimental setups were used to demonstrate the correlation between the molecular structures of azobenzene derivatives and electronic transport properties, such as mechanically controlled break-junction (MCBJ) at 4.2 K [5], graphene molecular junction [6], scanning tunneling microscope (STM) [7, 8], Conductive Atomic Force Microscopy (C-AFM) [9, 10], sandwiched between two graphene electrodes [11] and mercury drop electrode [12, 13]. In these experiments some authors associate the "ON" state to the *trans* isomer [7, 9, 13, 14], while others conclude in favor of the *cis* form [8, 10, 11, 12, 15]. This difference can be explained by the different molecules studied in these works and also by a different geometrical orientation of the molecules on the surface [8]. About the electronic properties, the conductance ratio between the two isomers is around 10–20 at maximum at a voltage inferior to 1 V. Until now this difference of conductance was explained by a change in the length of the molecule during the photoisomerization [10, 12].

Another frequently studied photochromic systems is based on a diarylethene unit. This class of photochromic switches exists in two thermally stable forms: the so-called open, colorless form and the closed, colored form (Fig. 3.3). As the azobenzene molecules, this compound can reversibly switch between the close (conjugated) and open form (non-conjugated) under, respectively, ultraviolet light and blue light. The rearrangement of double and single bonds through photoisomerization is expected to be accompanied by a significant modification in molecular conductance. This point has been demonstrated on various dynamic molecular junctions: by STM [16–19], conductive atomic force microscopy (C-AFM) [20], micropore in a resist [21, 22], gap open in single-walled carbon nanotubes (SWNT) [23], mechanically controllable break-junction (MCBJ) [24], and 2D gold nanoparticles networks [25]. As expected, in these works authors observe systematically a higher conductance for the close form ("ON" state) than for the open form ("OFF" state). The "ON/OFF" conductance ratio at 1 V up to 2 or 3 orders of magnitude was reported in some cases [17, 24], for other cases cited before this ratio is weak (less than 1 order of magnitude).

More recently a promising new optical switch was demonstrated based on a dimethyldihydropyrene unit with "ON/OFF" conductance ratio at 1 V approaching 10^4, and measured with MCBJ [26].

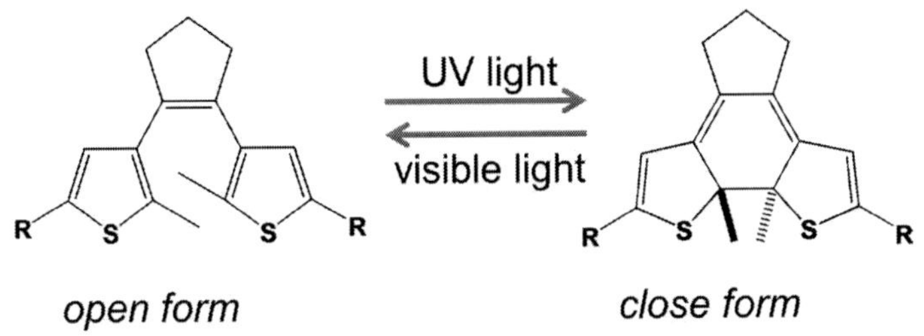

Figure 3.3 Switchable diarylethene molecules: from the open form to the close form by illumination with UV light (wavelength comprised between 300 and 400 nm), and reversibly from the close to the open form by illumination with visible light (wavelength comprised between 500 and 700 nm).

3.1.2 Electric Field Stimulus

Using an electric field to change the molecular conductance presents the important advantage to realize a fully integrated device; the stimuli can be applied with the electrodes connected to the molecules without interaction with the exterior. This approach is studied with the perspective of achieving memory devices.

For photochromic molecules, the electric field was also suggested to induce the isomerization of azobenzene unit, instead of the light exposure. Flip-flop motions were observed in *I–V* characteristic by STM on azobenzene derivative molecules grafted on gold surfaces. These oscillations between two states of conductance were obtained generally for positive bias superior to 0.5 V, and were associated by the authors to a change of the isomer conformation. The high and low-current states were attributed to the *trans* and *cis* conformations, respectively [14]. Also by STM, Alemani et al. induced reversibly the *trans-cis* isomerization of azobenzene derivative molecules grafted on gold by adjusting specifically the bias voltage for each isomerization *cis*-to-*trans* or *trans*-to-*cis* [27]. More recently, by incorporate the azobenzene derivative in devices with reduced graphene oxide contacts, the authors observed the *trans*-to-*cis* isomerization by applying a negative bias of –3 V and a *cis*-to-*trans* isomerization by applying a positive bias of +3 V [28]. This memory device showed clear ON/OFF states and nondestructive states over more than 20,000 readouts for 10,000 s.

Reversible switching with the electric field was also demonstrated on non-photochromic molecule. Bipyridyl-dinitro oligophenylene-ethynylene dithiol molecule connected with the mechanically controllable break-junction (MCBJ) technique exhibits clear change of conductance under threshold voltages [29]. After a positive threshold value for the voltage (around +0.8 V) the system switches from the initial "OFF" state to the "ON" state. This state was maintained for lower voltage applied. A negative voltage below the negative threshold value (around −1.0 V) resets the molecule again to the initial "OFF" state. Authors claimed the realization of a bit of memory with a bit separation ranged between 7 and 70 (I_{ON}/I_{OFF}) and stable within reading times of 30 s [29].

Rotaxane molecules that consist of a mechanically interlocked molecular architecture are frequently mentioned in the molecular electronics field as a promising candidate for the realization of molecular switches or memories. The principle is based on the oxidation of a moiety of the molecule at a threshold voltage, which leads to coulomb repulsion with another moiety. This repulsion induces a translation of the mobile part of the rotaxane molecule, which can change the molecular conductance. HP Labs have demonstrated different devices based on these molecules [30–33]. Unfortunately the switching effect observed was not clearly due to the rotaxane molecule [34, 35].

We can mention also that the oxidation of a molecule addressed in a junction can induce clearly and reversibly a modification of the conductance, as observed on thiol-terminated hepta aniline oligomer, electrically characterized by the scanning tunneling microscope break junction technique (STMBJ) [36].

3.1.3 Other Stimuli: pH, Biomolecule, Ion...

Variation of the pH can act as a stimulus on the molecular conductance of the junction. This was demonstrated for molecular wires integrated in a nanometric gap open in SWNT [6, 37]. In this system the molecular conductance changed by more than two orders of magnitude for several switching cycles between a low pH (pH 1) and a high pH (pH 12). These authors demonstrated also by using the gap open in SWNT technique the realization of single-molecule biosensors for directly detecting DNA hybridization, DNA–protein interaction, enzymatic activity, and DNA translocation in real time [38]. These devices are based on a specific reaction of the biomolecule connected in the SWNT gap with the target biomolecule. This reaction induces distinct conductance states in the molecular junction. A similar approach developed by the same group is based on a gap open in a CVD-grown graphene single layer. By this technique called "graphene point contacts" by the authors, reversible cycles of decomplexation and complexation of cobalt ions with a molecular bridge was shown [6, 39]. According to the authors, these systems made by covalent amide bond formation with graphene are quite robust and tolerate broad chemical treatments.

Molecules with tetrathiafulvalene redox units were inserted into two dimensional nanoparticles array to form functional networks of molecular junctions [40]. By chemical oxidation and reduction of the molecule, the conductance of the array oscillates between two conductance states separated by one order of magnitude. These oscillations of conductance were observed for 24 devices during four cycles of oxidation-reduction. For the authors, this system opens interesting perspectives for chemical sensing based on high surface area and chemical tunability of the interlinking elements for specific and selective reactions or interactions.

Other switching effect was demonstrated by STM break junction on 4,4′-bipyridine–gold single-molecule junctions, which can reversibly switch between two conductance states through repeated junction elongation and compression [41]. In this last example, the molecular junction is exposed to a mechanical stimulus.

In summary, these examples of dynamic molecular junction presented show the great potential of these devices for applications such as memory devices, switches, biosensors, and sensors. These examples demonstrate controllable molecular junction and diverse functionalizations with specific capabilities and detection of specific molecular scale activities.

In the following, we will present some examples of our approach for the realization of dynamic molecular junction. This approach is based on the use of self-assembled monolayers (SAMs) grafted on gold surfaces via a thiol function. The SAMs are composed of molecules sensible to an external stimulus: light and cobalt ion. Before detailing these two examples of dynamic molecular junction, we first present the problematic of the fixation of pi-conjugated molecules on gold surface via a thiol linkage.

3.2 Fixation of pi-Conjugated Molecules on Gold Surfaces via Thiol Bond

The structural control of molecules on surfaces is a key point in molecular electronics. The interpretations of the electronic properties of the molecule depend directly on the orientation of the molecule on the surface. In the majority of studies in molecular electronics, a quasi-vertical orientation of the molecule is adopted, perpendicular to the electrode surface. This approach furthers the study of electron transport along the molecules (see these recent reviews of molecular junctions [42–44]). The molecules are generally covalently bounded on the surface via a fixation group such as thiol (R-SH), sulfide (R-S-R), disulfide (R-S-S-R), thiocyanate (R-SCN), alcohols (R-OH), amine ($R\text{-}NH_2$), alkene ($R\text{-}CH{=}CH_2$) or silane ($SiCl_3$, $Si(OMe)_3$, $Si(OEt)_3$) [45]. Due to the promising properties, most of the molecules studied are pi-conjugate molecules. The conjugated system is frequently separated from the surface by a flexible linear alkyl chain. This alkyl spacer improves the cohesion and the density of molecule on the surface via van der Waals interactions between neighboring alkyl chains; and it favors also the electronic decoupling between the surface and the conjugated system. This interaction between molecule and surface corresponds in fact to the formation of SAMs on surfaces. This approach by SAMs has been very extensively study due to the great potential applications in various fields [45].

3.2.1 Systems Studied: With One or Two Anchor Groups

Before investigate the switching properties of molecules deposed on surfaces by SAM, the interaction of the molecule with the surface have to be well controlled and the orientation of the pi-conjugated system perfectly known. In order to illustrate this problematic we have studied the formation of two SAMs with quaterthiophene molecules referenced in the following **1** and **2** (Fig. 3.4), bearing, respectively, one and two alkanethiol chains attached at the internal β-position of the outermost thiophene ring by a sulfide linkage.

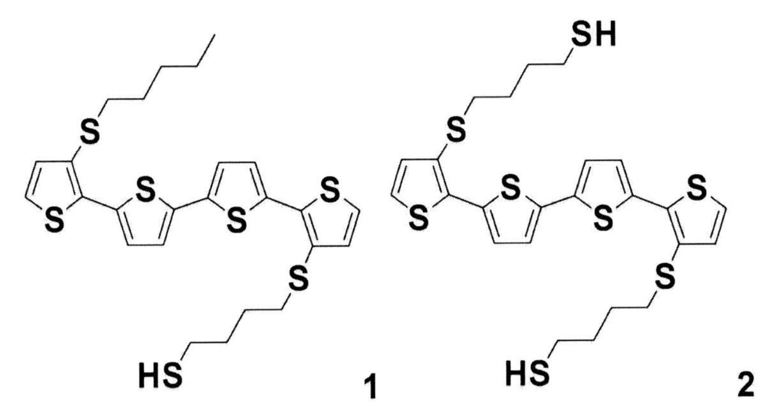

Figure 3.4 Scheme of the studied molecules for the SAM formation: quaterthiophene molecules bearing one (molecule **1**) and two (molecule **2**) alkanethiol chains attached at the internal β-position of the outermost thiophene ring by a sulfide linkage.

Synthesis of these molecules was done by our collaborators and described elsewhere [46]. The main interest of these molecules is to study the influence of the presence of one versus two thiol group(s) on the formation, structure, and properties of the SAM in order to control the orientation of the pi-conjugated system on the surface.

The SAMs were realized on flat gold surfaces by immersing a silicon wafer covered with a freshly evaporated gold layer in millimolar solutions of purified compounds, during 3 or 72 h for, respectively, molecule **2** and **1** in controlled atmosphere (Fig. 3.5). After the immersion, it is important to rinse profusely the sample in order to remove non-covalently bounded molecule; typically it was done by dipping the modified surface in dichloromethane and under ultrasound during at least 5 min. The structure of the two classes of SAMs formed is discussed on the basis of their characterization by cyclic voltammetry, ellipsometry, contact angle measurement, and X-ray photo-electron spectroscopy (XPS).

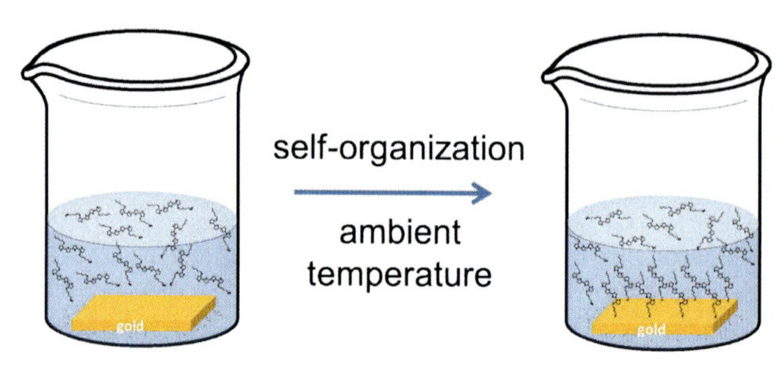

Figure 3.5 Scheme of SAM formation for the quaterthiophene monothiol (**1**) by self-organization process in dichloromethane solution and non-controlled atmosphere.

3.2.2 Structural Characterization: Orientation of the Molecule on the Surface

The surface coverage of both SAMs determined by cyclic voltammetry reveals typical values for the formation of SAMs on surface. The estimated surface coverage was measured for SAM of **1** and **2**, respectively, 42 Å^2 and 83 Å^2 [46].

After the SAM formation, the SAM of **1** was slightly more hydrophobic than the SAM of **2**, with a contact angle measured with deionized water at θ_{H2O} = 89 ± 1° and θ_{H2O} = 83 ± 1°, respectively. This difference suggests the presence more important of alkyl chains on the top of SAM for molecule **1** than for molecule **2**. These alkyl chains are well known to form highly hydrophobic surfaces (θ_{H2O} > 90°) [45].

Measurements of SAM thicknesses by spectroscopic ellipsometer indicate thicker SAM for molecule **2** than for molecule **1**, with, respectively, values between 11.8–14.0 and 9.3–11.7 Å. For the SAM **2** thickness values are compatible with the length of the molecules **2** estimated around 13 Å with MOPAC software (Chem3D software from CambridgeSoft) in the conformation for a double fixation on gold surface. For the molecule **1**, the relative low thickness value compared to the length of molecule calculated by MOPAC software in vertical conformation (~20 Å) suggests an important tilt angle of the molecule **1** on the surface (~45 Å).

The X-ray photo-electron spectroscopy (XPS) on both SAMs reveals clear spectra without presence of residual solvent trapped in the SAM. Three peaks associated to the different atoms of the SAM were observed by XPS: carbon (peak 1S at 284.5 eV) and sulfur (peaks S2s at 227.7 eV and S2p at 163.5 eV). The atomic ratios (number of sulfur atoms/number of carbon atoms) based on corrected peak area were very closed to those expected by observation of the molecular structure (see for details [46]). An interesting point in XPS is the deconvolution of the S2p peak. The curve-fitted high-resolution XPS spectra for the S2p region of both SAM show two doublets (Fig. 3.6 for molecule **1**, and Fig. 3.7 for molecule **2**). Each doublet results from spin–orbit splitting of the S2p level and consists of a high-intensity S2p 3/2 peak at lower energy and a low-intensity S2p 1/2 peak at higher energy, separated by 1.2 eV with an intensity ratio of 2:1 [47].

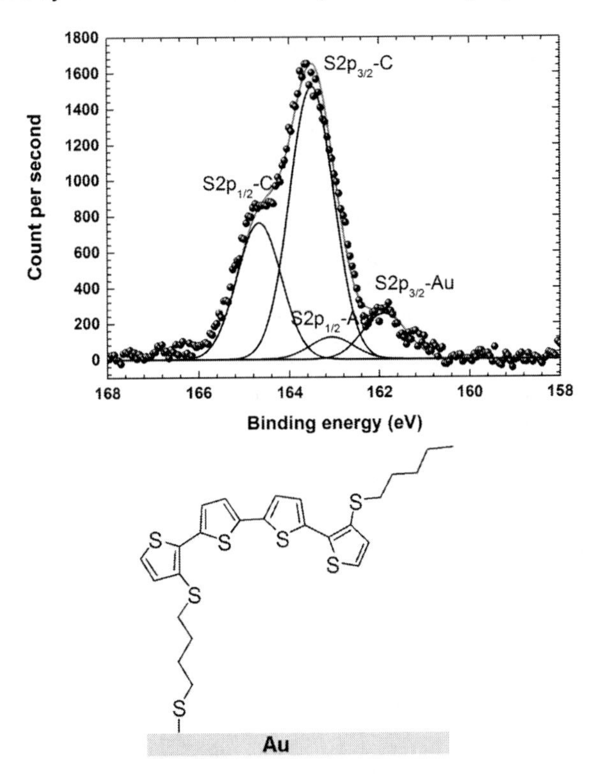

Figure 3.6 (Up) Curve-fitted high-resolution XPS spectra for the S2p region of SAM composed of molecules **1**. (Down) representation of the molecule **1** in interaction with the gold surface. Copyright © 2008 WILEY-VCH Verlag GmbH & Co. KGaA, Weinheim.

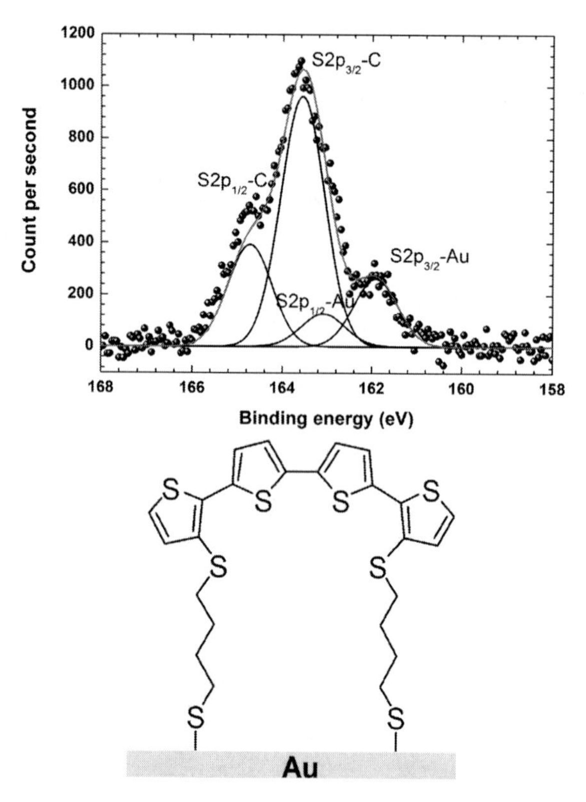

Figure 3.7 On the left, curve-fitted high-resolution XPS spectra for the S2p region of SAM composed of molecules **2**. On the right, representation of the molecule **2** in interaction with the gold surface. Copyright © 2008 WILEY-VCH Verlag GmbH & Co. KGaA, Weinheim.

The value of the S2p 3/2 peak at 161.9 eV for SAM of **1** and **2** is in excellent agreement with the binding energy for bounded alkanethiolate on gold (161.9 to 162.0 eV) [48, 49]. The lower energy doublet constituted of peaks at 161.9 and 163.1 eV is assigned to the thiol chemisorbed on the gold surface. The other doublet at higher energy with peaks positioned at 163.5 and 164.7 eV for SAM **1** and 163.6 and 164.7 eV for SAM **2** corresponds to the other sulfur atoms present in molecule and non-covalently attached with gold atom. Comparison of the areas of these two doublet signals allows estimating the S-C/S-Au ratio (number of sulfur atoms non-linked with gold atom/number of sulfur atoms linked with gold atom). These ratios were equal to 6.0 ± 0.5 and 3.7 ± 0.5 for, respectively, the SAM **1** and SAM **2**, in agreement with the expected value S-C/S-Au = 6/1 = 6 for a single fixation of molecule **1** and S-C/S-Au = 6/2 = 3 for a double fixation of molecule **2**. This analyze by XPS shows the formation of thiol bound with the surface for the molecule **1**. In the case of the molecule **2** the slightly superior value measured for the ratio suggests that around 83 ± 12% of the molecule in the SAM were doubly bounded with the gold surface (details in [46]).

Characterization by cyclic voltammetry, ellipsometry, contact angle measurement, and XPS of the SAM provides coherent results indicating that in the SAM derived from the doubly functionalized oligomer, most of the conjugated molecules are doubly attached on the surface with a horizontal orientation of the conjugated system.

This prior study illustrates the importance of the design of the molecule in order to control the structural orientation of this molecule on the surface, and presents the characterization tools available to analyze the structure of the SAM formed. In the following, we will present two examples of molecular switch: The first is based on a cation-binding switchable molecule doubly attached on gold surface and the second a photo-switch molecule single attached on gold surface.

3.3 A Cation-Binding Switchable Molecular Junction

The first example of molecular switch presented here is based on a cation-binding molecule doubly attached on a flat gold surface. This molecule is a dithiol quaterthiophene derivatized with a polyether loop (Fig. 3.8). The polyether loop attached at two fixed points of an oligothiophene acts as a cation-binding group [50–52]. This cation complexation by the polyether side chain induces a modification of the geometry and also of the electronic properties of the system [50]. The aim here is to detect a change of the conductance with the cation complexation, of the molecular junction formed with these molecules. Synthesis of these cation-binding molecules was done by our collaborators and described elsewhere [53].

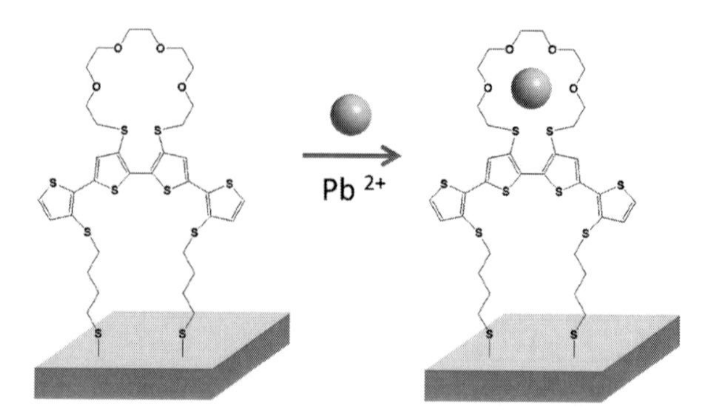

Figure 3.8 Scheme of the dithiol quaterthiophene derivatized with a polyether loop molecule in interaction with the gold surface before and after the complexation by a lead cation (Pb^{2+}).

3.3.1 Preparation and Characterization of the SAM

As described before (Fig. 3.5), the SAMs were realized on flat gold surfaces freshly evaporated by immersion in millimolar solution. After 48 h of immersion, samples were abundantly rinsed with dichloromethane and under ultrasound during at least 5 min. For the complexation with lead cation, the SAMs were dipped during 48 h in a $Pb(ClO_4)_2\ 3H_2O$ in acetonitrile and rinsed with acetonitrile. As done previously, the structure of the SAMs before and after exposition to lead cation is discussed on the basis of their characterization by cyclic voltammetry, ellipsometry, water contact angle measurement, UV-visible spectroscopy and XPS. Then, after the structure of the SAM known, the effect of the cation on the electronic properties of the molecular junction is investigated.

The surface coverage value (around 167Å^2 per molecule) determined by cyclic voltammetry is compatible with the formation of a SAM on gold [53]. The thickness measured by spectroscopic ellipsometer confirms the SAM formation. The thickness value of 14 ± 2 Å is close to the length of molecule doubly attached on gold (18 Å) calculated by MOPAC software. If we assume that the difference between the thickness and the length of molecule is due to the tilt of the molecule, we can estimate this tilt angle of the molecule with the surface normal around 39°. The relatively low values of water contact angle θ_{H2O} = 74 ± 2° shows the hydrophilic character of the surface compatible with the presence of conjugated moieties on top of the surface [54]. This value of contact angle and also the thickness measured remain unchanged after exposition to lead cation, respectively, θ_{H2O} = 71 ± 2° and 16 ± 2 Å. This stability of the measured values indicates that the integrity of the SAM is globally preserved upon Pb^{2+} complexation. This complexation of the SAM analyzed by UV-visible spectroscopy in solution shows an intensification of two maxima at 279 and 380 nm with the addition of Pb^{2+}, which suggests interconversion between two species [53]. Preliminary tests using UV-visible spectroscopy with various metal cations (Li^+, Na^+, Cs^+, Ba^{2+}, Sr^{2+}, Cd^{2+}, and Pb^{2+}) in contact with the SAM have shown that only Pb^{2+} is complexed by the SAM [53].

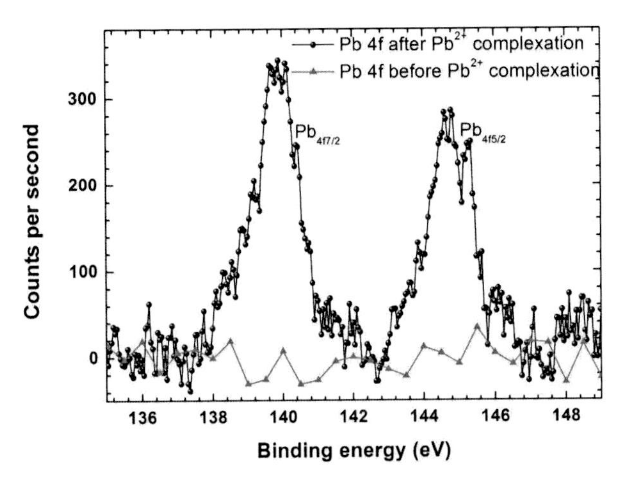

Figure 3.9 XPS spectra of the Pb 4f region before and after Pb^{2+} complexation. Copyright © 2013 WILEY-VCH Verlag GmbH & Co. KGaA, Weinheim.

The high-resolution XPS spectrum of the SAM before cation complexation shows peaks associated to the different atoms of the SAM: carbon (peak 1S), oxygen (peak 1s at 532.9 eV) and sulfur (peaks S2s at 228.4 eV and S2p). After deconvolution, the carbon peak presents two components at 285.5 eV and 287 eV associated, respectively, to C–C (carbon atom linked with another carbon atom) and C–O (carbon atom linked with oxygen atom). The area peak ratio C–C/C–O of 3.3 is close to the expected value of 26/8 = 3.25. As described previously (Section 3.2.2) the S2p peak is deconvoluted in four peaks, which leads to estimate the S-C/S-Au ratio of 3.8 ± 1 value close to the expected value for a double fixation (S–C/S–Au

= 8/2 = 4). This analysis shows the conservation of the integrity of the molecule grafted on gold, and that most of the molecules are doubly fixed on the gold surface. After Pb^{2+} complexation, the different peaks remain globally unchanged (shift in position inferior to 0.2 eV) showing the stability of the SAM during the complexation. The only difference is the presence of two peaks at 139.8 and 144.9 eV attributed, respectively, to the Pb $4f_{7/2,5/2}$ doublet (Fig. 3.9). The experimental ratio Pb/C (number of lead atom/number of carbon atom) leads to estimate that around 13% of the molecules in the SAM are complexed by a lead cation [53].

3.3.2 Electronic Transport Properties

The electronic properties of the SAMs after and before complexation were studied by the eutectic GaIn (eGaIn) technique. For this technique close to the one developed by Chiechi et al. [55], we formed a tip of eGaIn (99.99%, Ga:In; 75.5:24.5 wt%) at the extremity of a needle controlled by a micromanipulator. The extremity of the tip was gently brought into contact with the SAM, to form an electrical contact with a diameter estimated between 20 and 200 μm (Fig. 3.10). The voltage is applied on this eGaIn electrode and the current measured on the gold surface electrode.

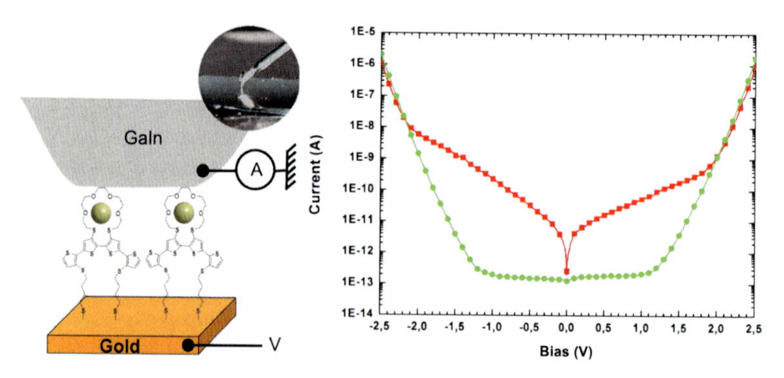

Figure 3.10 (Left) Scheme of the studied molecular junction after complexation with Pb^{2+} and electrically connected with the eGaIn technique. In inset a photography of the junction formed is presented. (Right) Typical *I–V* characteristic measured on the SAM before (●) and after (■) Pb^{2+} complexation. Copyright © 2013 WILEY-VCH Verlag GmbH & Co. KGaA, Weinheim.

The I–V characteristics measured on the junction (right of Fig. 3.10) show a clear and reproducible modification of the electronic properties with the Pb^{2+} complexation. At lower bias ($|V| < 2$ V), the current before the complexation is lower than the current after complexation with a ratio up to 280 at around 1.15 V and 1.6×10^3 at −1.2 V. We observe also a difference of symmetry; before complexation the I–V curve is rather symmetrical with the voltage (i.e., the ratio I_{-V}/I_{+V} is always around 1). However, after complexation we note a slight asymmetrical behavior with higher current at negative than at positive bias with a maximum ratio I_{-V}/I_{+V} of about 9 at 1.7 V. At higher voltages ($|V| > 2$ V) the two I–V curves overlap.

Applying the transition voltage spectroscopy (TVS) technique by plotting the same data of Fig. 3.10 in a plot $\ln(I/V^2)$ in function of $1/V$, we obtain a positive and negative transition voltage (Fig. 3.11). In a molecular junction, the voltage at which a minimum is observed (named voltage transition V_T) can give an estimation of the energy position of the molecular orbital (ϕ). Albeit, the fact that the exact relation between V_T and ϕ and the physical origin of V_T are still under debate [56–60], TVS has become a popular tool in molecular electronics. In our case, we observe a diminution of the V_T value with the complexation: from $V_{T+} = 1.09 \pm 0.02$ V to $V_{T+} = 0.086 \pm 0.04$ V for positive bias and from $V_{T-} = -1.11 \pm 0.02$ V to $V_{T+} = -0.52 \pm 0.04$ V for negative bias. These values with the help of the Bâldea's formulas [59] lead to give an estimation of the diagram energy of the molecular junction. In our case, we observe than before the complexation the molecular orbital involved in the electronic conduction is localized in the middle of the molecular junction (due to the fact that the values of $|V_T|$ are the same). And after complexation, the molecular orbital moves close an electrode accompanied with a variation of its energy (more details in [53]).

This first example of dynamical molecular junction sensible to lead cation via a complexation with molecule deposited in SAM, clearly shows a modification of the electronic properties of the junction with the complexation. The conductance ratio with the complexation can reach value up to 1.6×10^3 at −1.2 V. As shown by TVS, this complexation impacts on the position of the molecular orbitals of the molecule in the junction.

3.4 An Optical Switch Molecular Junction

The second example of molecular switch presented is based on an azobenzene moiety as active function in the molecule (see azobenzene presentation in Section 3.1.1). Synthesis of these molecules called AzBT (azobenzene-bithiophene derivative) was done by our collaborators and described elsewhere [61]. The molecule studied here is fixed on the gold substrate by only one fixation group (a thiol group) and principally oriented perpendicular to the substrate (Fig. 3.12). The electronic transport occurs mainly

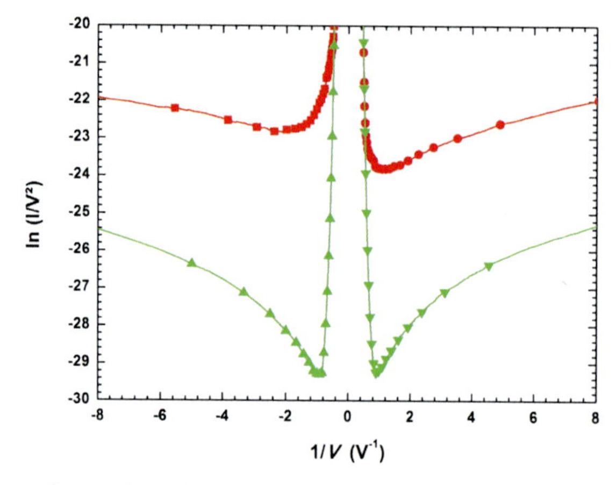

Figure 3.11 Same data than Fig. 3.10 plotted in a Transition Voltage Spectroscopy (TVS) plot before (▲) and after (●) Pb^{2+} complexation. Copyright © 2013 WILEY-VCH Verlag GmbH & Co. KGaA, Weinheim.

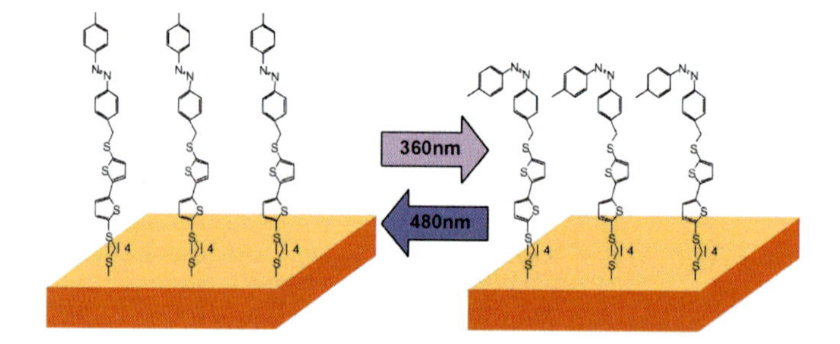

Figure 3.12 Schematic view of the AzBT molecules deposited in SAM with the reversible isomerization under light exposure. Copyright © 2010, American Chemical Society.

along the molecule. The electronic transport occurs mainly along the molecule. In order to improve the conductance ratio between the two states with this molecule, the photoisomerisable part is electronically decoupled from the metal surface by a bithiophene associated to a short alkyl chain (4 carbon atoms). As described in Section 3.1.1, the azobenzene group changes from a *cis* to a *trans* isomer under blue light exposure (480 nm) and reversibly from a *trans* to a *cis* isomer under UV light exposure (360 nm) (Fig. 3.12).

3.4.1 Preparation and Characterization of the SAM

The SAM formation was done with the same procedure described for the two precedent systems; the SAMs were formed on flat gold surfaces freshly evaporated by immersion in millimolar solution. After 3 days of immersion, samples were abundantly rinsed with dichloromethane and under ultrasound during at least 5 min.

Water contact angles and thickness measurements validate the formation of a dense SAM on the gold substrate. For a pristine SAM, i.e., after synthesis, the surface was hydrophobic with $\theta_{H20} = 94 \pm 2°$ compatible with the presence of methyl group on top of the surface. The thickness measured by spectroscopic ellipsometer 27 ± 1 Å is relatively close to length of the molecule determined by MOPAC (~26 Å for the *cis* isomer and ~31 Å for the *trans* isomer). This thickness measured shows the formation of a densely packed SAM on gold, as confirmed also by the surface coverage value determined by cyclic voltammetry around 42 Å^2 per molecule [61].

As expected, the high-resolution XPS spectrum of the SAM shows peaks associated to the different atoms of the SAM: carbon (peak 1s at 284.3 eV), nitrogen (peak 1s at 399.5 eV), and sulfur (peaks S2s at 227.4 eV and S2p). The experimental ratios C/N, C/S2s, and C/S2p (number of carbon atoms/number of, respectively, nitrogen, sulfur 2s and sulfur 2p atoms) were estimated, respectively, at 12.2 ± 1, 6.55 ± 1 and 6.56 ± 1. With regard to the molecule composition, these ratios are compatible with the expected values C/N = 26/2 = 13 and C/S = 26/5 = 5.2 (2 nitrogen atoms, 5 sulfur atoms, and 26 carbon atoms in the molecule). As described earlier (Section 3.2.1), the S2p peak is deconvoluted in four peaks (Fig. 3.13), which permit to estimate the S–C/S-Au ratio. Here this ratio was estimated at 6.2 ± 2, value relatively close to the expected value for a double fixation (S–C/S–Au = 4/1 = 4).

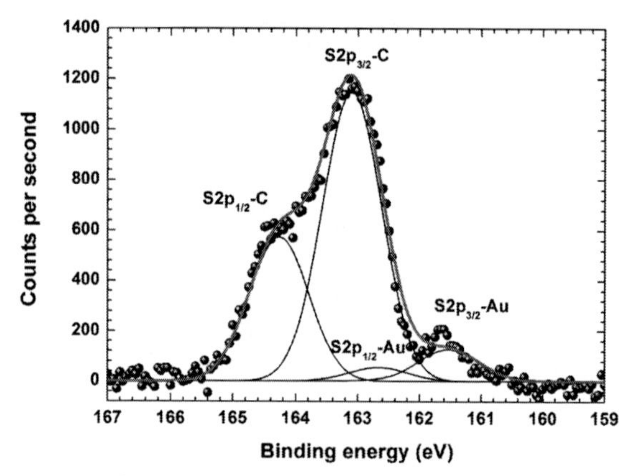

Figure 3.13 Curve-fitted high-resolution XPS spectra for the S2p region of AzBT grafted on gold surfaces. Copyright © 2010, American Chemical Society.

3.4.2 Characterization under Irradiation

The isomerization of the AzBT molecule in dichloromethane solution was studied by UV-visible spectroscopy under light irradiation [61]. The irradiation by UV light induces in the UV-visible absorption spectrum, a progressive decrease of the intensity of the 342 nm band with an increase of the absorbance in the 400–450 nm region. The presence of two isosbestic points at 297 and 380 nm indicates the coexistence of two chemical species in equilibrium, in full agreement with the isomerization of the azobenzene moiety. The irradiation at 480 nm produces the reversible isomerization with a complete return to the initial conditions of the UV-visible spectrum.

We followed the evolution of the water contact angle of the SAM as a function of the light irradiation (Fig. 3.14). A clear oscillation of the water contact angle value was observed with the irradiation between two values: ~92° after UV light (isomer *cis*) and ~98° after blue light (isomer *trans*). This variation of the contact angle with the isomers is mainly associated to the difference of the dipole moment between the two isomers [62]. An oscillation of the thickness value measured was also observed between two values: ~25 Å after UV light (isomer *cis*) and ~30 Å after blue light (isomer *trans*). These values are consistent with the estimated length of the two isomers determined by MOPAC: ~26 Å for the *cis* isomer and ~31 Å for the *trans* isomer. We can notice the intermediate values for the contact angle and thickness after the SAM formation, explained by the fact that the SAM after formation is composed of a mix of *cis* and *trans* isomers. These cycles of irradiation show the stability of the SAM under light exposure and also the reversibility of the isomerization for these molecules incorporated in dense SAMs on a surface. This last point was not obvious; some authors have suggested incorporating inert molecules (alkyl thiol) with the azobenzene derivatives in the SAM, in order to observe the photo-isomerization [14, 63]. Otherwise for densely packed SAMs with only the azobenzene derivatives, no switching effect has been observed [63].

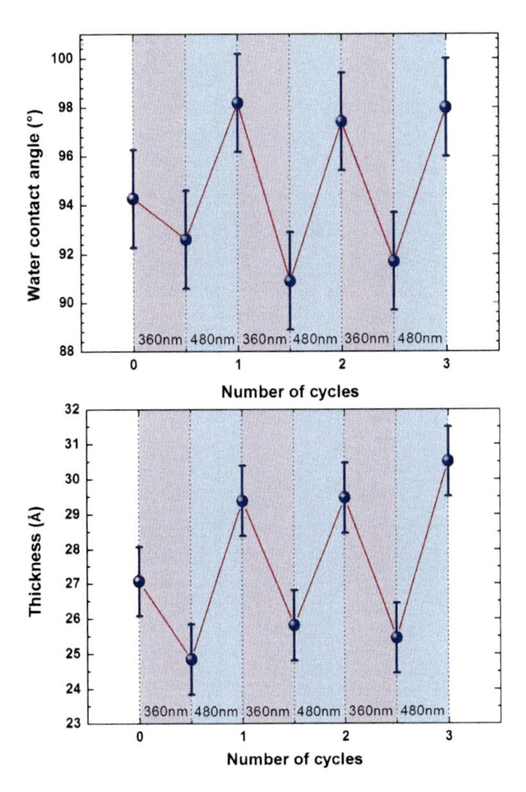

Figure 3.14 Evolution of the water contact angle (top) and thickness (bottom) of the SAM as a function of irradiation; after formation the SAM was exposed to three cycles composed of irradiations to UV light (360 nm) and blue light (480 nm). Copyright © 2010, American Chemical Society.

3.4.3 Electronic Transport Properties

The current–voltage curves measured by Conducting-AFM at the nanoscale (surface contact area of few 10 nm^2) and through the SAM showed a clear effect of the isomerization on the current. For these electrical characterizations the AFM tip was metalized with PtIr and the tip force in contact with the SAM controlled in the range 20–30 nN (see scheme in Fig. 3.15 and details in [62]).

In the typical current–voltage curves presented in the right side of the Fig. 3.15, the current measured on the pristine SAM is lower than the current measured on the SAM after irradiation by UV and formation of the *cis* isomer population. For example, at 1.5 V, the current increases from ~1 nA for the pristine SAM to ~40 nA for the SAM after UV irradiation (*cis* isomer). Similarly, after blue irradiation at 480 nm during 90 min and formation of the *trans* isomer population the current decreases to ~10 pA at 1.5 V. As observed for a majority of authors (see Section 3.1.1), the "ON" state of the AzBT SAM is associated to the *cis* isomer and the "OFF" state to the *trans* isomer. In the example presented here, the ratio of current ON/OFF is around 10^3 for higher bias ($|V| > 0.5$ V).

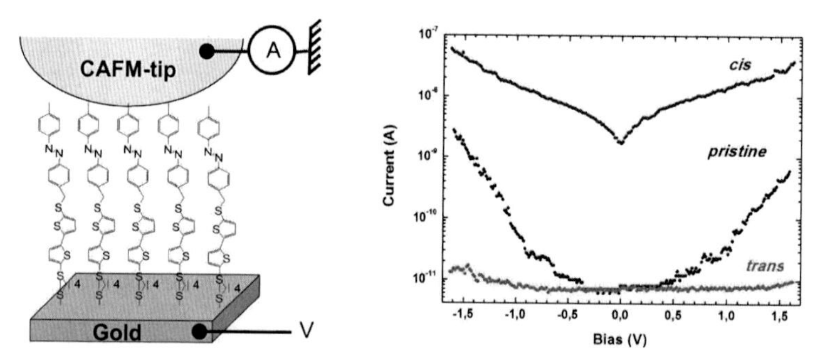

Figure 3.15 (Left) Scheme of the studied molecular junction: The AzBT SAM grafted on gold surface is connected with the metallized AFM tip. (Right) Typical *I*–*V* characteristics measured on the pristine SAM after formation (●), after irradiation by UV at 360 nm during 90 min (isomer *cis*) (●) and after irradiation by blue at 480 nm during 90 min (isomer *trans*) (●). Copyright © 2010, American Chemical Society.

For the totality of the current–voltage curves recorded (approximately 40 in each configuration), the histogram of the current at 1.5 V is plotted in a log scale (Fig. 3.16). Three distinguishable log-normal distributions of the current are observed associated to the three different configurations studied: the pristine SAM after formation, SAM after irradiation by UV at 360 nm during 90 min (isomer *cis*) and SAM after irradiation by blue at 480 nm during 90 min (isomer *trans*). This dispersion in the measured current in each distribution can be due to inhomogeneity in the SAM organization or variations of the tip force. Initially for the pristine SAM the peak is localized at $\sim 8.8 \times 10^{-8}$ A with a log-standard deviation of 1.5. After blue irradiation (*trans* isomer) the current decreases and the peak shift to lower values at $\sim 1.3 \times 10^{-11}$ A and a log standard deviation of 1.9. For the UV irradiation, current increases with a peak localized at $\sim 2.2 \times 10^{-8}$ A and a log standard deviation of 1.6. We can notice that the same behavior and same values were obtained for the histograms of current at –1.5 V [62], due to the fact that the current–voltage curves are almost symmetric. The intermediate current values for the pristine same suggests that just after its formation the SAM is composed of a mix of the two isomers, as already suggested before (Section 3.4.2). From the histograms, we deduce an average ON/OFF conductance ratio of 1.5×10^3 and a log standard deviation of 4.7. It means that 68% of the devices have an ON/OFF ratio comprised between 320 and 7×10^3. This high ON/OFF conductance ratio is one of the higher reported for azobenzene derivative (see Section 3.1.1).

Other current–voltage curves were obtained with the eGaIn technique (technique presented above in the left of Fig. 3.10). Globally the behavior observed was the same, with two distinguishable distributions localized at 3.4×10^{-7} A and 6×10^{-5} A for –0.6 V, respectively, associated to the *trans* and *cis* isomer. For this bias of –0.6 V, the average ON/OFF ratio is equal to 180 and increases at 750 for –1 V (for details [62]). We explain this reduction of the ratio compared to C-AFM by the increases of the surface contact area by eGaIn (area $\sim 10^{-4}$ cm^2 for eGaIn and ~ 10 nm^2 for C-AFM). Larger contact areas can include the contribution of some inactive switching domains or such defects in the organization of the SAM, which reduce the conductance ratio.

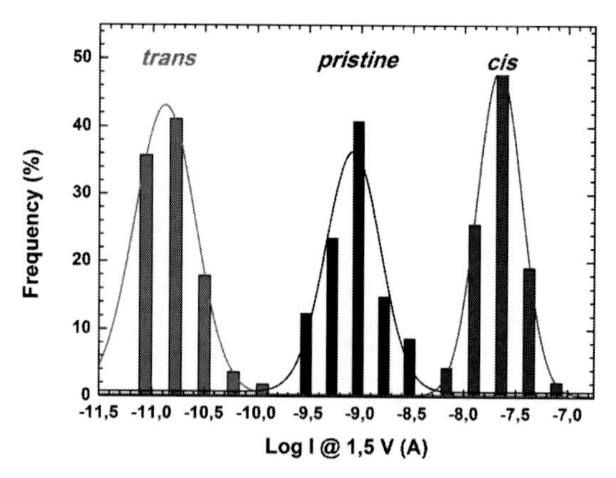

Figure 3.16 Histogram of the current in a log scale measured at 1.5 V for ~40 traces for each peaks and the adjustments with a Gaussian curve for: the pristine SAM after formation (—), after irradiation by UV at 360 nm during 90 min (isomer *cis*) (—), and after irradiation by blue at 480 nm during 90 min (isomer *trans*) (—). Copyright © 2010, American Chemical Society.

By this eGaIn technique (experimental details are exposed in [62]), the switching kinetics was investigated by following the time-dependent evolution of the current under blue and UV irradiation (Fig. 3.17). Under blue light exposure (*cis*-to-*trans* isomerization), we observed a main exponential rate equations with a characteristic time constant of 20 ± 1 min. For the UV light irradiation (*trans*-to-*cis* isomerization), we observed a main exponential rate equations with characteristic time constants of 90 ± 8 min. This difference of time constant between the two irradiations (ratio of about 4.5) can be explained by the difference of light power intensity (blue light power density is 3.6 higher than for UV light). From these kinetics curves, we estimated the values of the cross sections for the *trans*-to-*cis* and *cis*-to-*trans* photoisomerization, respectively, $\sigma_{TC} = 1.5 \times 10^{-18}$ cm^2 and $\sigma_{CT} = 5.4 \times 10^{-20}$ cm^2 for UV light irradiation, and $\sigma_{TC} = 3.9 \times 10^{-20}$ cm^2 and $\sigma_{CT} = 1.4 \times 10^{-18}$ cm^2 for the blue light irradiation. We obtain clear asymmetry behavior with $\sigma_{TC} > \sigma_{CT}$ under UV light and $\sigma_{CT} > \sigma_{TC}$ under blue light. These values are very close to those measured by sum-frequency generation vibrational spectroscopy on azobenzene derivative SAM [64].

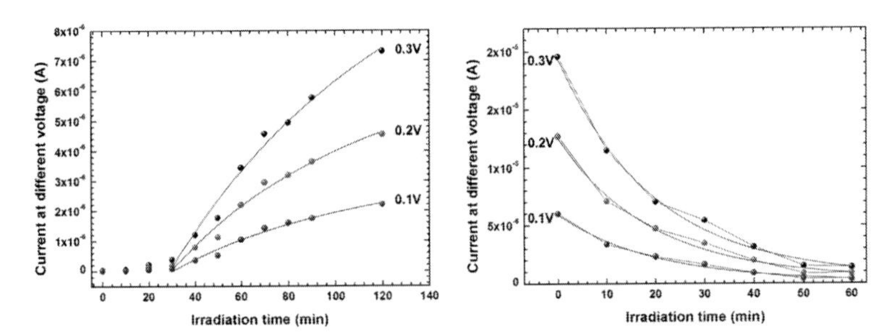

Figure 3.17 Evolution of the current measured at different bias 0.3 V, 0.2 V and 0.1 V in function of the time of irradiation by UV light for a *trans*-to-*cis* isomerization (left), and by blue light for a *cis*-to-*trans* isomerization (right). Copyright © 2010, American Chemical Society.

3.4.4 Origin of This High Conductance Switching Ratio

In order to understand the origin of this high conductance ratio measured in this system, we analyzed the current–voltage curves obtained by C-AFM with the TVS technique, already introduced in Section 3.3.2.

Histograms of the transition voltages at both positive (named V_{T+}) and negative biases (named V_{T-}) and also for the two isomers, deduced from the current–voltage characteristics measured by C-AFM, show four distinguish peaks obtained by adjustments with normal distributions (Fig. 3.18). These distributions are localized at $V_{T-} = -1.76 \pm 0.08$ V and $V_{T+} = 1.91 \pm 0.04$ V for the *trans* isomer, and $V_{T-} = -1.37 \pm 0.16$ V et $V_{T+} = 1.54 \pm 0.18$ V for the *cis* isomer. By using Bâldea's formulas [59] with these values of transition voltages (as in Section 3.3.2), we deduced an estimation of (i) the energy level of the molecular orbital involved in the electronic transport (with respect to the Fermi energy of the electrodes) called ε_0, and (ii) the degree of symmetry or asymmetry of the molecular orbitals in the junction. By this approach, the two isomers present a rather symmetric junction, i.e., with a factor of asymmetry close to zero. About, the energy level of the molecular orbital, a clear modification was observed between the *trans* and the *cis* isomer, respectively, calculated at $\varepsilon_0 = 1.59$ eV and $\varepsilon_0 = 1.26$ eV.

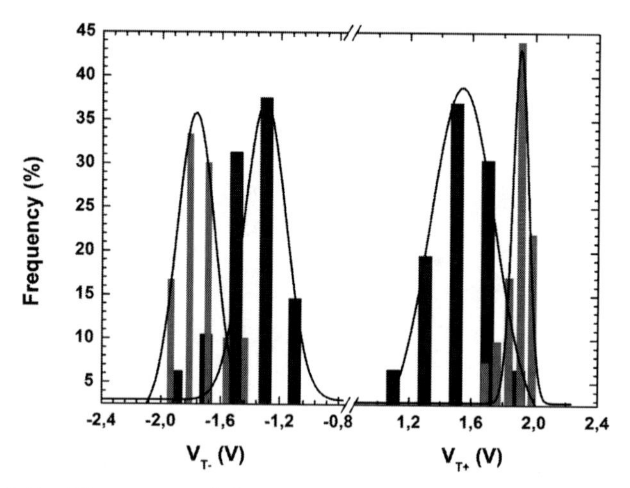

Figure 3.18 Histograms of the transition voltages V_T for positive (V_{T+}) and negative bias (V_{T-}) and for *cis* (■) and *trans* isomers (■). Copyright © 2010, American Chemical Society.

Until now this difference of conductance between isomers was explained by a change in the length of the molecule during the photoisomerization [10, 12]. Indeed, the ratio of conductance measured usually between 10 and 20 (see Section 3.1.1), can be explained (i) by using the classical expression of the tunnel conductance for a non-resonant tunneling transport:

$$G = G_0 \exp(-\beta d), \tag{3.1}$$

where G is the conductance of the junction, G_0 is an effective contact conductance, β the tunnel decay factor and d the tunnel barrier thickness [65, 66], (ii) by the variation of thickness between the *cis* and *trans* isomers comprised between 7 and 8 Å, and (iii) a constant value for β determined by electrical measurements [10]. Here for the important ratio measured (>10^3) this approach requires too high value for β. So this explanation is incomplete. In fact, β depends on the tunneling barrier height. And the variation of energy level of the molecular orbital (ε_0) observed before by TVS excludes the hypothesis of a constant β value for the two isomers. In fact, the β factor is related to the barrier height ε_0 by the following relation:

$$\beta = 2\sqrt{(2m_0\varepsilon_0)}/\hbar \tag{3.2}$$

with m_0 is the electron mass and $\hbar$ the reduced Planck constant [66]. Assuming the same value of G_0 for the two isomers, these two equations leads to an estimation of the ratio of conductance given by

$$\frac{G_{\text{cis}}}{G_{\text{trans}}} = \exp\left(\frac{2\sqrt{2m_0}}{\hbar}\left(e_{\text{trans}}\sqrt{\varepsilon_0^{\text{trans}}} - e_{\text{cis}}\sqrt{\varepsilon_0^{\text{cis}}}\right)\right) \quad (3.3)$$

with values of ε_0 determined thanks to Bâldea's formulas, and the thicknesses measured by ellipsometer for the two isomers (e_{cis} = 25 Å and e_{trans} = 29 Å; Section 3.4.2), this ratio is estimated to about 9.1×10^3. This value is in total agreement with the ratio measured by C-AFM up to 7.5×10^3 and 1.5×10^3 in average.

To conclude, this derivatized azobenzene molecule (AzBT) deposited in SAM exhibits clear reversible change of properties under light exposure, such as (i) the surface wettability, (ii) the thickness of the SAM, and (iii) the electronic properties via the modification of the molecular orbital energy level. These variations of thickness and energy levels explain the high conductance ratio measured by C-AFM and eGaIn between the two isomers, up to 7.5×10^3. This ratio is one of the higher, measured on azobenzene molecule to our knowledge. These results pave the way to the realization of solid-state molecular switch and memory devices with high ON/OFF conductance ratio.

3.5 Conclusion

We have shown that by an approach based on the use of SAMs, the orientation of the molecules on the surface can be controlled by optimizing the design of the molecule (for example, the number of anchor groups). For the first example of dynamic molecular junction, Pb^{2+} complexation by SAMs composed of molecules oriented parallel to the surface via two anchor groups, induces a clear increase up to 1.6×10^3 at −1.2 V, while only about one Pb^{2+} ion is captured per 7–8 molecules. The second example, based on SAM of derivative azobenzene molecules oriented perpendicular to the surface via a single anchor group exhibits clear reversible change of properties under light exposure. In particular the conductance can vary between two values separated by a ratio up to 7.5×10^3 and with an average value of 1.5×10^3.

These examples of dynamic molecular junction illustrate the great potential to use molecules with a function excitable by an external stimulus (here, for example, by presence of ions via complexation and by light via photoisomerization), in order to modulate the electronic properties of the junction. These effects of the stimulus can be viewed positively as an additional tool over and above the tuning of molecular structure in achieving control over molecular properties.

The possibilities of molecular systems for these dynamic molecular junctions are numerous, with a great variety of stimulus available (luminous, electronic, ionic, magnetic, thermic, mechanic, chemical, pH, biological, radioactive...). The quasi infinity of molecules potentially synthesizable by chemistry opens new opportunities for the demonstration of new dynamic molecular junctions. For example, recent molecular switches demonstrated in solution are isomerized by red light or near infrared light for biological applications (red and infrared light is more penetrating through tissues or cells compared to UV and visible light) [67, 68]. Another interest to develop molecular switches isomerizable by infrared light is for the development of molecular optoelectronic devices for applications in telecommunication, where infrared light is used to carry information [69]. Another recent example of a molecular switch: A dithienylethene oxazolidine hybrid system with height stable and reversible states was demonstrated in solution [70]. This octastate switch can change under external stimuli: light and/or chemicals (acid/base) and/or heat. This molecule is a good candidate to open the way to the demonstration of a multi-state molecular memory.

Moreover, by using only molecular switches, the three basic logic operations (AND, NOT and OR) and more complex logic functions have been reproduced at the molecular level [69]. According to the author, molecular switches can become the basic components of future logic devices.

Molecular electronic devices such as memory, switch, biosensor, sensor, and digital processor could be, in the future, integrated into current CMOS technology to create hybrid multifunctional integrated circuits as a significant step toward practical molecular electronics.

Acknowledgments

The author would like to thank all colleagues from the Nanostructures nanoComponents & Molecules Group in IEMN, in particular D. Vuillaume, D. Guérin, D. Deresmes, K. Smaali, and S. Godey, and colleagues J. Roncali, P. Blanchard, M. Oçafrain, T. K. Tran, S. Karpe from the Linear Conjugated Systems group at MOLTECH-Anjou for fruitful collaborations and discussions. The works done at IEMN were financially supported by ANR-PNANO under OPTOSAM project, CNRS, ministry of research, Région Nord-Pas de Calais.

References

1. Joachim, C., Gimzewski, J. K., Analysis of low-voltage I(V) characteristics of a single C_{60} molecule. *Europhys. Lett.*, 1995, **30**(7), 409–414.
2. *Molecular Switches* 2nd. Feringa, B. L., Browne, W. R., ed., Wiley-VCH, Weinheim, Germany, 2011.
3. Sauvage, J. P., *Molecular Machines and Motors.* Springer Berlin Heidelberg, 2001.
4. Samorì, P., Hecht, S., Gated systems for multifunctional optoelectronic devices. *Adv. Mater.*, 2013, **25**(3), 301–301; Weibel, N., Grunder, S., Mayor, M., Functional molecules in electronic circuits. *Org. Biomol. Chem.*, 2007, **5**(15), 2343–2353.
5. Kim, Y., Garcia-Lekue, A., Sysoiev, D., Frederiksen, T., Groth, U., Scheer, E., Charge transport in azobenzene-based single-molecule junctions. *Phys. Rev. Lett.*, 2012, **109**(22), 226801.
6. Cao, Y., Dong, S., Liu, S., Liu, Z., Guo, X., Toward functional molecular devices based on graphene–molecule junctions. *Angew. Chem. Int. Ed.*, 2013, **52**(14), 3906–3910.
7. Kumar, A. S., Ye, T., Takami, T., Yu, B.-C., Flatt, A. K., Tour, J. M., Weiss, P. S., Reversible photo-switching of single azobenzene molecules in controlled nanoscale environments. *Nano Lett.*, 2008, **8**(6), 1644–1648.
8. Comstock, M. J., Levy, N., Kirakosin, A., Cho, J., Lauterwasser, F., Harvey, J. H., Strubbe, D. A., Fréchet, J. M. J., Trauner, D., Louie, S. G., Crommie, M. F., Reversible photomechanical switching of individual engineered molecules at a metallic surface. *Phys. Rev. Lett.*, 2007, **99**, 038301.
9. Zhang, X., Wen, Y., Li, Y., Li, G., Du, S., Guo, H., Yang, L., Jiang, L., Gao, H., Song, Y., Molecularly controlled modulation of conductance on

azobenzene monolayer-modified silicon surfaces. *J. Phys. Chem. C*, 2008, **112**(22), 8288–8293.

10. Mativetsky, J. M., Pace, G., Elbing, M., Rampi, M. A., Mayor, M., Samori, P., Azobenzenes as light-controlled molecular electronic switches in nanoscale metal-molecule-metal junctions. *J. Am. Chem. Soc.*, 2008, **130**(29), 9192–9193.
11. Seo, S., Min, M., Lee, S. M., Lee, H., Photo-switchable molecular monolayer anchored between highly transparent and flexible graphene electrodes. *Nat. Commun.*, 2013, **4**, doi:10.1038/ncomms2937.
12. Ferri, V., Elbing, M., Pace, G., Dickey, M. D., Zharnikov, M., Samori, P., Mayor, M., Rampi, M. A., Light-powered electrical sitch based on cargo-lifting azobenzene monolayers. *Angew. Chem. Int. Ed. Engl.*, 2008, **47**, 3407–3409.
13. Ely, T., Das, S., Li, W. J., Kundu, P. K., Tirosh, E., Cahen, D., Vilan, A., Klajn, R., Photocontrol of electrical conductance with a nonsymmetrical azobenzene dithiol. *Synlett*, 2013, **24**(18), 2370–2374.
14. Yasuda, S., Nakamura, T., Matsumoto, M., Shigekawa, H., Phase switching of a single isomeric molecule and associated characteristic Rectification. *J. Am. Chem. Soc.*, 2003, **125**(52), 16430–16433.
15. Ferri, V., Elbing, M., Pace, G., Dickey, M. D., Zharnikov, M., Samorì, P., Mayor, M., Rampi, M. A., Light-powered electrical switch based on cargo-lifting azobenzene monolayers. *Angew. Chem. Int. Ed.*, 2008, **47**(18), 3407–3409.
16. Molen, S. J. V. D., Vegte, H. V. D., Kudernac, T., Amin, I., Feringa, B. L., Wees, B. J. V., Stochastic and photochromic switching of diarylethenes studied by scanning tunnelling microscopy. *Nanotechnology*, 2005, **17**(1), 310–314.
17. He, J., Chen, F., Liddell, P. A., Andréasson, J., Straight, S. D., Gust, D., Moore, T. A., Moore, A. L., Li, J., Sankey, O. F., Lindsay, S. M., Switching of a photochromic molecule on gold electrodes: Single-molecule measurements. *Nanotechnology*, 2005, **16**(6), 695–702.
18. Arramel, P. T. C., Kudernac, T., Katsonis, N., van der Maas, M., Feringa, B. L., van Wees, B. J., Reversible light induced conductance switching of asymmetric diarylethenes on gold: Surface and electronic studies. *Nanoscale*, 2013, **5**(19), 9277–9282.
19. Kim, Y., Hellmuth, T. J., Sysoiev, D., Pauly, F., Pietsch, T., Wolf, J., Erbe, A., Huhn, T., Groth, U., Steiner, U. E., Scheer, E., Charge transport characteristics of diarylethene photoswitching single-molecule junctions. *Nano Lett.*, 2012, **12**(7), 3736–3742.

20. Uchida, K., Yamanoi, Y., Yonezawa, T., Nishihara, H., Reversible On/Off conductance switching of single diarylethene immobilized on a silicon surface. *J. Am. Chem. Soc.*, 2011, **133**(24), 9239–9241.

21. Kronemeijer, A. J., Akkerman, H. B., Kudernac, T., van Wees, B. J., Feringa, B., Blom, P. W. M., de Boer, B., Reversible conductance switching in molecular devices. *Adv. Mater.*, 2008, **20**, 1467–1473.

22. Kim, D., Jeong, H., Lee, H., Hwang, W.-T., Wolf, J., Scheer, E., Huhn, T., Jeong, H., Lee, T., Flexible molecular-scale electronic devices composed of diarylethene photoswitching molecules. *Adv. Mater.*, 2014, **26**(23), 3968–3973.

23. Whalley, A. C., Steigerwald, M. L., Guo, X., Nuckolls, C., Reversible switching in molecular electronic devices. *J. Am. Chem. Soc.*, 2007, **129**(42), 12590–12591.

24. Dulic, D., van der Molen, S. J., Kudernac, T., Jonkman, H. T., de Jong, J. J. D., Bowden, T. N., van Esch, J., Feringa, B. L., van Wees, B. J., One-way optoelectronic switching of photochromic molecules on gold. *Phys. Rev. Lett.*, 2003, **91**(20), 207402.

25. Van der Molen, S., Liao, J., Kudernac, T., Agustsson, J. S., Bernard, L., Calame, M., Van Wees, B., Feringa, B., Schönenberger, C., Light-controlled conductance switching of ordered metal-molecule-metal devices. *Nano Lett.*, 2009, **9**(1), 76–80.

26. Roldan, D., Kaliginedi, V., Cobo, S., Kolivoska, V., Bucher, C., Hong, W., Royal, G., Wandlowski, T., Charge transport in photoswitchable dimethyldihydropyrene-type single-molecule junctions. *J. Am. Chem. Soc.*, 2013, **135**(16), 5974–5977.

27. Alemani, M., Peters, M. V., Hecht, S., Rieder, K.-H., Moresco, F., Grill, L., Electric field-induced isomerization of azobenzene by STM. *J. Am. Chem. Soc.*, 2006, **128**(45), 14446–14447.

28. Min, M., Seo, S., Lee, S. M., Lee, H., Voltage-controlled nonvolatile molecular memory of an azobenzene monolayer through solution-processed reduced graphene oxide contacts. *Adv. Mater.*, 2013, **25**(48), 7045–7050.

29. Lörtscher, E., Ciszek, J. W., Tour, J., Riel, H., Reversible and controllable switching of a single-molecule junction. *Small*, 2006, **2**(8–9), 973–977.

30. Chen, Y., Ohlberg, D. A. A., Li, X., Stewart, D. R., Williams, R. S., Jeppesen, J. O., Nielsen, K. A., Stoddart, J. F., Olynick, D. L., Anderson, E., Nanoscale molecular-switch devices fabricated by imprint lithography. *Appl. Phys. Lett.*, 2003, **82**(10), 1610–1612.

31. Chen, Y., Jung, G.-Y., Ohlberg, D. A. A., Li, X., Stewart, D. R., Jeppesen, J. O., Nielsen, K. A., Stoddart, J. F., Williams, R. S., Nanoscale molecular-switch crossbar circuits. *Nanotechnology*, 2003, **14**, 462–468.

32. Luo, Y., Collier, C. P., Jeppesen, J. O., Nielsen, K. A., Delonno, E., Ho, G., Perkins, J., Tseng, H. R., Yamamoto, T., Stoddart, J. F., Heath, J. R., Two-dimensional molecular electronics circuits. *Chemphyschem*, 2002, **3**(6), 519–525.

33. Green, J. E., Choi, J. W., Boukai, A., Bunimovich, Y., Johnston-Halperin, E., Delonno, E., Luo, Y., Sherrif, B. A., Xu, K., Shin, Y. S., Tseng, H.-R., Stoddart, J. F., Heath, J. R., A 160-kilobit molecular electronic memory patterned at 10^{11} bits per square centimeter. *Nature*, 2007, **445**, 414–417.

34. Stewart, D. R., Ohlberg, D. A. A., Beck, P. A., Chen, Y., Williams, R. S., Jeppesen, J. O., Nielsen, K. A., Stoddart, J. F., Molecule-independent electrical switching in Pt/organic monolayer/Ti devices. *Nano Lett.*, 2004, **4**(1), 133–136.

35. Williams, S., How we found the missing memristor. *IEEE Spectrum*, 2008, **45**(12), 24–31.

36. Chen, F., He, J., Nuckolls, C., Roberts, T., Klare, J. E., Lindsay, S., A Molecular switch based on potential-induced changes of oxidation state. *Nano Lett.*, 2005, **5**(3), 503–506.

37. Guo, X., Small, J. P., Klare, J. E., Wang, Y., Purewal, M. S., Tam, I. W., Hong, B. H., Caldwell, R., Huang, L., O'Brien, S., Yan, J., Breslow, R., Wind, S. J., Hone, J., Kim, P., Nuckolls, C., Covalently bridging gaps in single-walled carbon nanotubes with conducting molecules. *Science*, 2006, **311**, 356–359.

38. Guo, X., Single-molecule electrical biosensors based on single-walled carbon nanotubes. *Adv. Mater.*, 2013, **25**(25), 3397–3408.

39. Cao, Y., Dong, S., Liu, S., He, L., Gan, L., Yu, X., Steigerwald, M. L., Wu, X., Liu, Z., Guo, X., Building high-throughput molecular junctions using indented graphene point contacts. *Angew. Chem. Int. Ed.*, 2012, **51**(49), 12228–12232.

40. Liao, J., Agustsson, J. S., Wu, S., Schönenberger, C., Calame, M., Leroux, Y., Mayor, M., Jeannin, O., Ran, Y.-F., Liu, S.-X., Decurtins, S., Cyclic conductance switching in networks of redox-active molecular junctions. *Nano Lett.*, 2010, **10**(3), 759–764.

41. Quek, S. Y., Kamenetska, M., Steigerwald, M. L., Choi, H. J., Louie, S. G., Hybertsen, M. S., Neaton, J. B., Venkataraman, L., Mechanically controlled binary conductance switching of a single-molecule junction. *Nat. Nanotech.*, 2009, **4**(4), 230–234.

42. McCreery, R. L., Bergren, A. J., Progress with molecular electronic junctions: Meeting experimental challenges in design and fabrication. *Adv. Mater.*, 2009, **21**(43), 4303–4322.

43. Chen, S., Zhao, Z., Liu, H., Charge transport at the metal-organic interface. *Ann. Rev. Phys. Chem.*, 2013, **64**(1), 221–245.

44. Son, J. Y., Song, H., Molecular scale electronic devices using single molecules and molecular monolayers. *Curr. Appl. Phys.*, 2013, **13**(7), 1157–1171.

45. Ulman, A., *An Introduction to Ultrathin Organic Films: From Langmuir-Blodgett to Self-Assembly*. Academic press: Boston, 1991.

46. Tran, T. K., Qcafrain, M., Karpe, S., Blanchard, P., Roncali, J., Lenfant, S., Godey, S., Vuillaume, D., Structural control of the horizontal double fixation of oligothiophenes on gold. *Chem. Eur. J.*, 2008, **14**(20), 6237–6246.

47. Vance, A. L., Willey, T. M., van Buuren, T., Nelson, A. J., Bostedt, C., Fox, G. A., Terminello, L. J., XAS and XPS characterization of a surface-attached rotaxane. *Nano Lett.*, 2002, **3**(1), 81–84.

48. Rieley, H., Kendall, G. K., Zemicael, F. W., Smith, T. L., Yang, S., *Langmuir*, 1998, **14**, 5147–5153.

49. Laibinis, P. E., Whitesides, G. M., Allara, D. L., Tao, Y., Parikh, A. N., Nuzzo, R. G., Comparison of the structure and wetting properties of self-assembled monolayers of *n*-alkylthiols on the coinage metal surfaces, Cu, Ag, Au. *J. Am. Chem. Soc.*, 1991, **113**, 7152–7167.

50. Jousselme, B., Blanchard, P., Levillain, E., Delaunay, J., Allain, M., Richomme, P., Rondeau, D., Gallego-Planas, N., Roncali, J., Crown-annelated oligothiophenes as model compounds for molecular actuation. *J. Am. Chem. Soc.*, 2003, **125**(5), 1363–1370.

51. Demeter, D., Blanchard, P., Grosu, I., Roncali, J., Electropolymerization of crown-annelated bithiophenes. *Electrochem. Commun.*, 2007, **9**(7), 1587–1591.

52. Demeter, D., Blanchard, P., Grosu, I., Roncali, J., Poly(thiophenes) derivatized with linear and macrocyclic polyethers: From cation detection to molecular actuation. *J. Inclusion Phenom. Macrocyclic Chem.*, 2008, **61**(3–4), 227–239.

53. Tran, T. K., Smaali, K., Hardouin, M., Bricaud, Q., Ocafrain, M., Blanchard, P., Lenfant, S., Godey, S., Roncali, J., Vuillaume, D., A Crown-ether loop-derivatized oligothiophene doubly attached on gold surface as cation-binding switchable molecular junction. *Adv. Mater.*, 2013, **25**(3), 427–431.

54. Lenfant, S., Guerin, D., Van, F. T., Chevrot, C., Palacin, S., Bourgoin, J. P., Bouloussa, O., Rondelez, F., Vuillaume, D., Electron transport through rectifying self-assembled monolayer diodes on silicon: Fermi-level pinning at the molecule-metal interface. *J. Phys. Chem. B*, 2006, **110**(28), 13947–13958.

55. Chiechi, R. C., Weiss, E. A., Dickey, M. D., Whitesides, G. M., Eutectic gallium–indium (EGaIn): A moldable liquid metal for electrical characterization of self-assembled monolayers. *Angew. Chem.*, 2008, **120**(1), 148–150.

56. Huisman, E. H., Gueìdon, C. M., van Wees, B. J., van der Molen, S. J., Interpretation of transition voltage spectroscopy. *Nano Lett.*, 2009, **9**(11), 3909–3913.

57. Chen, J., Markussen, T., Thygesen, K. S., Quantifying transition voltage spectroscopy of molecular junctions: Ab initio calculations. *Phys. Rev. B*, 2010, **82**(12), 121412.

58. Araidai, M., Tsukada, M., Theoretical calculations of electron transport in molecular junctions: Inflection behavior in Fowler-Nordheim plot and its origin. *Phys. Rev. B*, 2010, **81**(23), 235114.

59. Bâldea, I., Ambipolar transition voltage spectroscopy: Analytical results and experimental agreement. *Phys. Rev. B*, 2012, **85**(3), 035442.

60. Ricoeur, G., Lenfant, S., Guerin, D., Vuillaume, D., Molecule/electrode interface energetics in molecular junction: A transition voltage spectroscopy study. *J. Phys. Chem. C*, 2012, **116**(39), 20722–20730.

61. Karpe, S., Ocafrain, M., Smaali, K., Lenfant, S., Vuillaume, D., Blanchard, P., Roncali, J., Oligothiophene-derivatized azobenzene as immobilized photoswitchable conjugated systems. *Chem. Commun.*, 2010, **46**(21), 3657–3659.

62. Smaali, K., Lenfant, S., Karpe, S., Ocafrain, M., Blanchard, P., Deresmes, D., Godey, S., Rochefort, A., Roncali, J., Vuillaume, D., High on-off conductance switching ratio in optically-driven self-assembled conjugated molecular systems. *Acs Nano*, 2010, **4**(4), 2411–2421.

63. Evans, S. D., Johnson, S. R., Ringsdorf, H., Williams, L. M., Wolf, H., Photoswitching of azobenzene derivatives formed on planar and colloidal gold surfaces. *Langmuir*, 1998, **14**(22), 6436–6440.

64. Wagner, S., Leyssner, F., Kördel, C., Zarwell, S., Schmidt, R., Weinelt, M., Rück-Braun, K., Wolf, M., Tegera, P., Reversible photoisomerization of an azobenzene-functionalized self-assembled monolayer probed by sum-frequency generation vibrational spectroscopy. *Phys. Chem. Chem. Phys.*, 2009, **11**, 6242–6248.

65. Holmlin, R. E., Haag, R., Chabinyc, M. L., Ismagilov, R. F., Cohen, A. E., Terfort, A., Rampi, M. A., Whitesides, G. M., Electron transport through thin organic films in metal–insulator–metal junctions based on self-assembled monolayers. *J. Am. Chem. Soc.*, 2001, **123**(21), 5075–5085.

66. Engelkes, V. B., Beebe, J. M., Frisbie, C. D., Length-dependent transport in molecular junctions based on SAMs of alkanethiols and alkanedithiols: Effects of metal work function and applied bias on tunneling efficiency and contact resistance. *J. Am. Chem. Soc.*, 2004, **126**(43), 14287–14296.

67. Yang, Y., Hughes, R. P., Aprahamian, I., Near-infrared light activated azo-BF2 switches. *J. Am. Chem. Soc.*, 2014, **136**(38), 13190–13193.

68. Samanta, S., Beharry, A. A., Sadovski, O., McCormick, T. M., Babalhavaeji, A., Tropepe, V., Woolley, G. A., Photoswitching azo compounds in vivo with red light. *J. Am. Chem. Soc.*, 2013, **135**(26), 9777–9784.

69. Raymo, F. M., Digital processing and communication with molecular switches. *Adv. Mater.*, 2002, **14**(6), 401–414.

70. Szalóki, G., Sevez, G., Berthet, J., Pozzo, J.-L., Delbaere, S., A simple molecule-based octastate switch. *J. Am. Chem. Soc.*, 2014, **136**(39), 13510–13513.

Chapter 4

Modulate and Control of Detailed Electron Transport of Single Molecule

Bingqian Xu, Kun Wang, Joseph Hamill, and Ryan Colvard

Single Molecule Study Laboratory,
College of Engineering and Nanoscale Science and Engineering Center,
University of Georgia, 507 Driftmier Engineering Center/168 Riverbend Research South, Athens, Georgia 30602, USA

bxu@engr.uga.edu

Molecular electronics has been frustrated by the lack of refined experimental control of the measured electron transport phenomena in molecular junctions. This chapter is intended to introduce the detailed experimental modulations and controls of the electron transport properties in single-molecule break junctions including the single-molecule conductance and current–voltage (*I*–*V*) characteristics. These experimental controls and modulations incorporate the mechanical modification of measurement process using single-molecule break junction technique, the manipulation of molecular structures and the molecule–electrode interfaces and the controls of measurement environment. Surrounding this core theme, some associated topics are also introduced including multiple experimental approaches prior to single-molecule break

Molecular Electronics: An Experimental and Theoretical Approach
Edited by Ioan Bâldea

ISBN 978-981-4613-90-3 (Hardcover), 978-981-4613-91-0 (eBook)
www.panstanford.com

junction technique used to determine single-molecule conductance, studied molecular species and the development of data analysis methods.

In general, we hope to offer readers a scope about how the electron transport properties of single-molecule break junction can be modulated and controlled in order to gain deep insight into the nature of molecular devices at the nanoscale level.

4.1 Introduction

4.1.1 Evolution of Experimental Determination of Single-Molecule Conductance

Since the 1970s, wiring single molecules to two electrodes has been proposed as the ultimate goal of downscaling active electronic components such as diodes and transistors to overcome the limit of Moore's prediction [1–3]. Molecular circuits built by this attractive idea could raise device density by a factor of $\sim 10^6$ compared to the current level of solid state devices [4]. Equally important is that the conceived nanometer long single-molecule device will add an overwhelming degree of functionality and structural flexibility compared to conventional semiconductors.

Back in 1974, Aviram and Ratner's theoretical discussion of electron transport through a single molecule performed as the catalytic agent towards experimentally building molecular rectifier [5]. This theoretical proposal involved unidirectional electron transport between donor and acceptor levels in a single molecule (Fig. 4.1). For the first time, it suggested the possibility of making and measuring a single-molecule circuit from the theoretical perspective. The original idea involved contacting a nanoscopic molecule with bulk material electrodes. Unlike the continuous and condensed energy-band dispersion of bulk materials, the energy levels, molecular orbitals (MOs), are quantized and discrete in a nanometer-scale molecule. This quantized feature is what distinguishes single-molecular devices from classical p–n junctions. The energy gap resides between the highest-occupied (HOMO) and the lowest-unoccupied molecular orbital (LUMO) is called the HOMO-LUMO gap (HLG). Just like the Fermi level of bulk electrodes varies from material to material, the HOMO and LUMO differ from molecule to molecule. Thus, bridging an individual molecule between two metallic electrodes results in the fusion of discrete states to continuous states. This unique trait is believed to lead to intriguing electrical and mechanical properties.

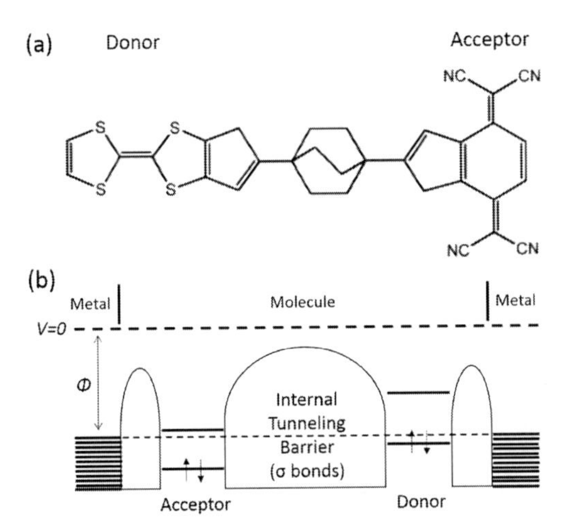

Figure 4.1 Molecular rectifier proposed by Aviram-Ratner in 1974. (a) Structure of the molecule with acceptor tetracyanoquinodimethane (TCNQ) and donor tetrathiolfulvalene (TTF) separated by a triple methylene bridge. (b) Energy diagram of the molecule (a) in contact with anode and cathode electrodes. Reprinted with permission from Ref. 5. Copyright (1974) by Elsevier.

On the other hand, with the initial excitement of downscaling electronic component size by orders via using molecular devices, real experimental attempt has to start with successfully connecting a single molecule to metallic electrodes or attaching electrodes to an individual molecule. However, in the decade since the concept was conceived, no significant progress was achieved by experimentalists. Attaching the molecule to the electrodes was greatly challenging back then. Experimental efforts were made employing increasingly clever ways: using mercury drops as electrodes [6], using Lorentz force to cross metallic wires [7], and trapping molecules in a nanopore [8, 9]. However, the experimental results turned out to be not reliable enough. Other techniques involving sandwiching robust molecules between electrode layers were proved to be too destructive for small molecules [10–12]. Achieving a reliable contact of the molecule and electrodes was hardly reachable and remained a major obstacle that hindered the progress of science towards single-molecule device for more than a decade [13].

However, in the 1990s, it was the development of scanning tunneling microscopy (STM) and later the atomic force microscopy (AFM) that brought us the most significant breakthrough in molecular electronics [14]. It rapidly turned out that these scanning probe microscopy (SPM) techniques could realize the measurement of electrical signals, like conductance and *I–V* characteristics, at the single-molecule level. The first experimental attempt was conducted by Mark Reed's group at Yale University, collaborating with James Tour's group, then at the University of South Carolina [15]. They successfully bridged benzene-1,4′-dithiol molecules between two facing gold electrodes and tested the current–voltage characteristics, which built the most fundamental step for the emerging area of molecular electronics. Then a review paper by Aviram et al. in 2000 summarized the proposed molecular structures of the day that were promising for building electronic rectifier, switches and storages, and also pictured the possible architecture of hybrid molecular circuits (Fig. 4.2) [16]. A series of papers later on in the early 2000s shed light on how such measurements could be made and advanced the understanding of electron transport properties through molecules. Meanwhile, worldwide interest was sparked due to their success. This point in time was marked as the true beginning of molecular electronics [14].

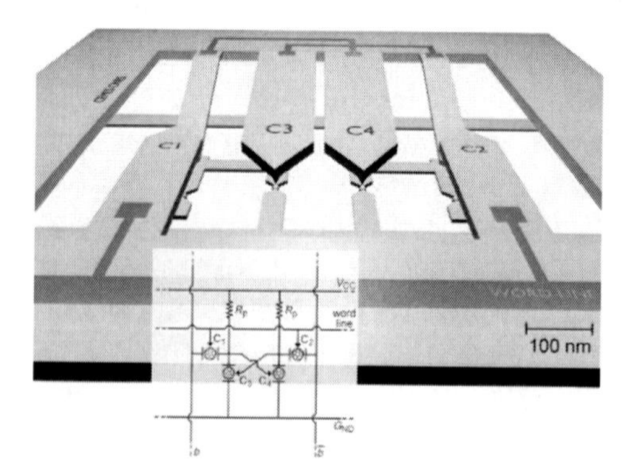

Figure 4.2 Example architecture of a hybrid molecular electronic device proposed in Aviram's review paper in 2000. Reprinted with permission from Ref. 16. Copyright (2000) by Nature Publishing Group.

Unfortunately, large fluctuations in the data sets of the conductance measurements in the early 2000s appeared to be the major barrier that hampered experimentalists from understanding the real mechanism of the electron migration process through molecules. Many discrepancies emerged not only between experimental data and simulated results, but also among data sets collected from different labs [17–22]. This later turned out to be caused by the lack of appropriate controls of experimental conditions. For example, the contact interfaces that connect the molecule and electrodes could vary significantly during the measurement process, which strongly influenced the resulting data. Furthermore, thermal fluctuations of molecular conformations were found to be able to result in dramatic current–voltage response, which had not been well considered. Overall, it was way too early to discuss agreement with theory at this stage. Hence, what was in urgent need was a repeatable and reliable experimental platform that could at least minimize the fluctuations in experimental data, and raise confidence for further investigation. Meeting with the imperative requirement of the field, Xu and Tao's work in 2003 realized a repeatable STM "break junction" (STMBJ) method [23]. This method involved using a metallic STM tip as one electrode and a metallic substrate as the other electrode (Fig. 4.3a). Via piezoelectric control, the STM tip was forced to engage into the substrate covered with sample molecules. When the tip was retracted, molecules were incorporated into the space between the tip and substrate, and as the tip was pulled away further, the molecules broke one by one. During the tip retracting process, conductance versus time (distance) traces could be recorded. As the tip retracted far enough, there would be no molecule sandwiched between the tip and substrate, and then the tip was driven to engage to the substrate again to repeat the entire process. The schematic illustration of the whole process is shown in Fig. 4.3b. As the tip was pulled away from the substrate, the conductance traces usually show step-wise features (Figs. 4.4a,c). The individual steps differ from the next by a unique value of conductance. For atomic thin metal junctions, the quantized steps were predicted, and experimentally illustrated, to have a conductance of $N \times G_0$ ($G_0 = 2e^2/h$, where N is an integer, e is the electronic charge, and h is Planck's constant) [24]. These steps correspond to multiple integers of single metal (Au) atomic contact conductance, as exemplified in Fig. 4.4a. After the last metallic junction, conductance steps with conductance values much smaller than G_0 corresponds to integer numbers of molecules in real molecular junctions. This process is exemplified as the transition from Fig. 4.4a to 4.4c. Thus, the last significant step before the conductance falls to zero is the conductance across an individual molecule. In real experiments, defective traces that carry unknown steps and nonspecific features may mislead researchers. Therefore, by repeating this engaging-retracting cycle thousands of times, statistical analysis of thousands of similar traces could reveal peaks at the positions where steps most frequently appear in a conductance histogram. This repeating process averaged out the randomly placed features of the traces generated from factors irrelevant to molecular junctions. The first prominent peak in the histogram was believed to be the conductance when only one individual molecule was left in the break junction, namely, the single-molecule conductance. This method washed out the contact variations to a certain extent, and eliminated most discrepancies in earlier measurements. Now it is probably the most widely used method for studying electron transport in single-molecule junctions. An analogy to STMBJ is a conducting-AFM break junction (CAFMBJ). CAFMBJ uses an AFM tip as the tip electrode and involves a laser-reflection controlled force signal detector, which enables measurements of conductance and force in parallel. CAFMBJ works the same way STMBJ does, but adds one more detectable parameter, force, to the system. Another method, mechanically controlled break junction (MCBJ) uses notched metal wire fixed on elastic substrate [15, 25, 26]. A diagram of MCBJ is shown in Fig. 4.3c. The substrate is usually covered with an insulator, and the metal wire is mechanically broken by bending the substrate. A single-molecule break junction (SMBJ) is formed when only one molecule is left in the gap between two terminals of the broken metal wire. These methods (STMBJ, CAFMBJ and MCBJ) collecting signals by sandwiching a single molecule in a scheme of break junctions are called SMBJ techniques. With these recently developed SMBJ techniques, experimentalists have obtained extensive data, and phenomena of great interest have been observed in the past decade, which have extended our physical understanding of charge transport through molecules [27–36].

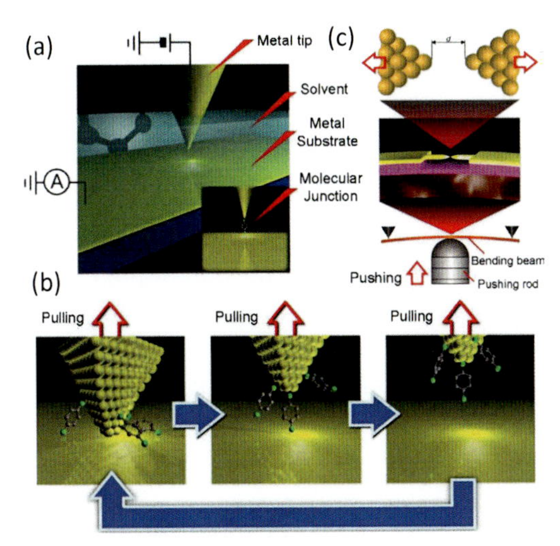

Figure 4.3 (a) Schematic of SPMBJ technique and (b) SPMBJ working principle. (c) Schematic of MCBJ technique.

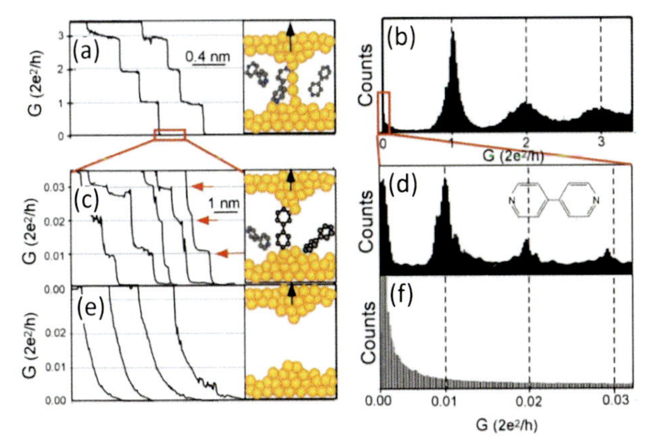

Figure 4.4 Single-molecule conductance measurement conducted by Xu and Tao. (a) Conductance trace of a gold contact between gold tip and gold substrate with quantum steps near G_0. (b) Conductance histograms constructed from 1000 traces shown in (a). (c) Conductance traces and junction schematic for 4,4′-bypiridine molecular junctions after the contact in (a) completely breaks. (d) Corresponding conductance histogram constructed from traces shown in (c). (e) and (f) show the conductance traces and histograms respectively, when there is no molecule in solution. Reprinted with permission from Ref. 23. Copyright (2003) by the American Association for the Advancement of Science.

Other techniques employing the emerging STM technique were also developed at the same time. The I(s) technique introduced by Haiss et al. in 2003 also used a STM tip to form molecular junctions [37]. However, the core difference is the junction formation method. The $I(s)$ method avoids the contact between two metal electrodes. The STM tip approaches the surface of analyte molecules, then retracts away while the tunneling current is measured. Schematics of the $I(s)$ technique and the corresponding conductance signal are exemplified in Fig. 4.5a. A further development using STM employs similar ideas but focuses on the time domain and has been referred to as the $I(t)$ method, as exemplified in Fig. 4.5b [38]. The STM tip is placed at a constant distance from the substrate. This distance is usually set to be less than the length of a fully extended molecule. Then characteristic vibrations in current can be monitored which behave like telegraphic noise signals (rightmost panel in Fig. 4.5b). Current jumps have been attributed to the attachment or detachment of the molecule to or from the STM tip. The conductance can then be determined by calculating the peak value of the current jump and the applied bias [39].

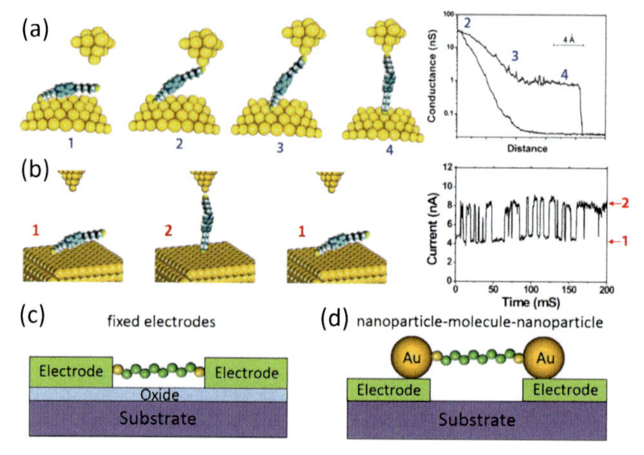

Figure 4.5 (a) Schematic illustration (left) Schematic of $I(s)$, and example conductance signal (right). (b) Schematic of $I(t)$ technique (left), and example current signal. (c) Structure of fixed electrode junction. (d) Structure of nanoparticle-molecule-nanoparticle junction. *Note*: (a and b) Reprinted with permission from Ref. 30. Copyright (2008) by the Royal Society of Chemistry.

Unlike the SPM technique, another conceptual method to measure the single-molecule conductance involves fabricating a pair of facing electrodes on a solid substrate. Molecules are then embedded in the nanoscale-gap between the electrodes using proper anchoring groups (Fig. 4.5c). Building such structures is technically challenging. To fabricate the molecular-scale slit, fabrication techniques, such as electromigration [40, 41], electrochemical etching and deposition [42–45], have been developed. Another strategy of this method employs synthesizing a dimer structure that consists of two Au nanoparticles connected by an individual molecule, and trapping the structure between two metal electrodes (Fig. 4.5d) [46]. These methods involving fixed electrodes ensure the stability of the electrode, which is ideal for studying electron transport molecules. Another advantage of this method is that one can use the substrate as a gate electrode to manipulate the electron transport through molecules. This approach has been used to study intriguing physical phenomena, such as Kondo effects and electron charging at low temperature [40, 41, 47–49]. However, the fabrication yield is usually rather low, and it is difficult to determine how many molecules are bridged across the electrodes and how many contributed to the measured results, making it difficult for statistical analysis [44, 50, 51].

4.1.2 Statistical Analysis

Even though multiple conductance measurement techniques have been developed, neither how many molecules form the junction for one experiment nor the exact situation of the molecule–metal interaction is known. Statistically analyzing the measured data turned out to be the most appropriate way. In 2001, Cui et al. applied a statistical analysis to a monothiol matrix containing single or a few alkanedithiol molecules [52]. This work for the first time proved that electrical measurements could be made on single molecules with the help of statistical analysis. They found that current–voltage curves were quantized as integer multiples of one another. By plotting histograms, sharp peaks at integer multiples of the single-unit peak could be seen. This single-unit peak was used to identify single-molecule junctions. Now one-dimensional (1D) histograms are widely used to analyze current (conductance) versus distance traces produced by SPMBJ, MCBJ, $I(s)$, and $I(t)$ techniques.

The strength and sharpness of the conductance peaks in a 1D histogram are determined by the mechanical stability of the junction conformation as well as the robustness of the conductance. For instance, the alteration of molecular conformation or molecule–metal bonding geometries could change the conductance of a particular molecular junction in certain ways, and statistical histograms of the conductance traces generated from thousands of such measurements in turn would broaden the distribution of one peak or even split one into multiple peaks since it is formed by adding conductance traces from different junction conformations. Therefore, for a conductance strongly dominated by the molecular-electrode contact or particular molecular conformation, one might expect a rather wide distribution of conductance peaks or even multiple peaks. Oppositely, a conductance weakly depending on the molecular-electrode bonding geometry and molecular torsional angle would contribute to a relatively narrow peak.

However, analyses based on the 1D histogram ignore a great deal of information hidden in the data. As the shape of the histograms reveal, it is a complex ensemble average of many molecular junction conformations [53]. They are not able to display the particular time history of individual trajectories of junction formation and breaking. This somehow hinders further investigation into the latent yet significant subtleties in the data. To overcome this, researchers have been making efforts to advance the data analysis methods used. This issue will be discussed in detail in Section 4.4.

4.1.3 Measured Molecule

In terms of molecular species, both simple saturated molecules, mainly alkane molecules, and intricate conjugated molecules, have been extensively examined [52, 54–57]. In order to make contact with metal electrodes, molecules are usually modified with chemical linkers at the two terminals. Chemical linkers including amines ($-NH_2$), thiols (–SH), carboxyls (–COOH), dimethyl phosphines ($-PMe_2$), and recently selenols (–SeH) have been tested for various purposes [39, 58–63]. Materials of the electrodes involve the most commonly used Au and others like Pt, Ag and Pd [30, 64–66].

Through the chemical linkers, the hybridization of the electrode energy levels and molecular orbitals occurs. This involves the broadening of the discrete molecular orbitals to quasi-continuum density of states and thus increases the complexity when considering the alignment of frontier molecular orbitals with Fermi levels of the electrodes, which is the decisive factor of electron transport properties of a SMBJ. The detailed electronic properties of some particular molecular junctions will be introduced in later sections.

4.2 Controversy in SMBJ Measurements

A dilemma that had puzzled researchers for a long time in those initial single-molecule conductance measurements was the differing values of conductance measured for the same molecule. The controversy in the conductance measurement results occur not only on simple carbon-based alkane molecules, but also on π-conjugated molecules.

Using SPMBJ, Xu et al. measured a conductance of $2.5 \times 10^{-4}G_0$ for octanedithiol (C8DT), while Haiss et al. reported a value of around $1.2 \times 10^{-5}G_0$ for the same molecule [23, 38]. Following measurements in different labs similarly reported multiple conductance values between these two [30, 60, 67–69]. Measurements using different techniques revealed even larger differences (Fig. 4.6b). Interestingly, all the methods contacted the molecule with gold electrodes. Other than C8DT, experimental investigations on alkanedithiol molecules containing different numbers of carbon units (CH_2) revealed consistent diversity of single-molecule conductance value (Fig. 4.6a) [67]. More detailed studies sorted these different conductance sets into three major groups: low, medium, and high (Fig. 4.7a). Comprehensive measurements of molecular junctions with identical alkane molecular cores but different anchoring groups yielded large variation in conductance as well (Fig. 4.6a) [70, 71]. The multiple order differing in conductance values of these molecules was surprising for methods of seeming similarity, and it stimulated both experimentalists and theorists to probe the nature of the just emerged SMBJ and the unknown secret in it.

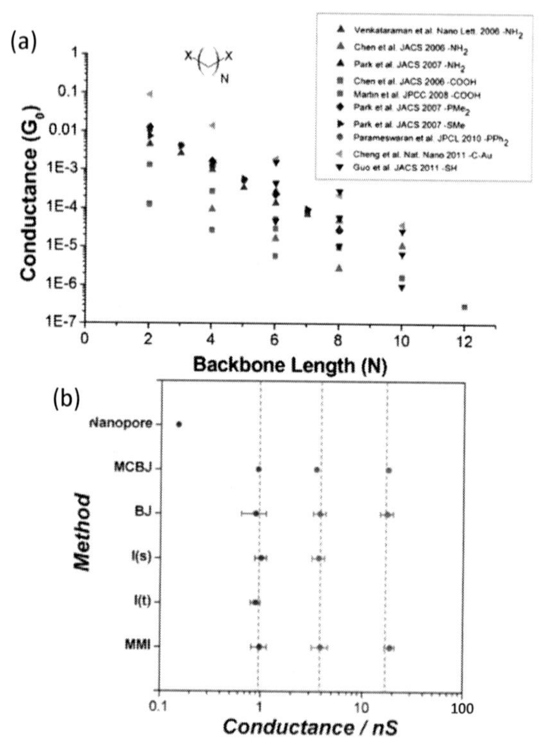

Figure 4.6 (a) Conductance values for alkane chains of different lengths with different end groups. (b) Conductance values for octanedithiol between gold contacts by different methods. Note the orders of variation in conductance values measured by different groups or different methods. *Note*: (a) Reprinted with permission from Ref. 70. Copyright (2014) by the Pan-Stanford Publishing; (b) Reprinted with permission from Ref. 39. Copyright (2010) by the Royal Society of Chemistry.

The first tentative explanation ascribed the conductance diversity to be related to different molecule–electrode contact configurations, namely the difference in Au–sulfur (Au–S) bonding morphologies [38]. In-depth simulations later on showed various Au–S contact geometries concerning the origin of these different conductance groups. There have been some disagreements among them. These modeled Au–S contact configurations mainly differed in the manner in which the sulfur head groups attached to the Au clusters, and hence they were given different names, including "atop," "bridge," "hollow" and "gauge" [30, 67, 68, 72–75].

Configuration "atop" describes the case in which sulfur atom only interacts with one Au atom that resides at the top of a pyramidal shaped Au cluster, or other analogous situations, as shown in the middle panel of Fig. 4.7b. Configuration "bridge" describes the situation where the head group sulfur atom directly interacts with two Au atoms, as illustrated in the right panel of Fig. 4.7b. In "bridge," the sulfur atom locates on the median line of two Au atoms and has similar bonding distance to two nearest Au atoms. Configuration "hollow" describes the occasion where the sulfur atom directly couple with three Au atoms. In "hollow," the sulfur atom sits above the center of an equilateral triangle-shaped Au cluster with three Au atoms at the apex of the triangle. Configuration "gauge" represents structural defects, which could lower the junction conductance up to a factor of 10 [30]. One example of the "gauge" effect with both terminal sulfur atom coordinated in the "atop" position is given in the left panel of Fig. 4.7b. Given that at least two molecule–electrode contact interfaces are involved in a SMBJ, the combination of these proposed contact configurations could form various molecular junction conformations. These different Au–S coupling scenarios will greatly influence the coupling strength of the molecule–electrode interfaces. The coupling strength, in turn, determines how easily the electrons can migrate across the molecule–electrode interface. Hence, the contact geometry undoubtedly plays an essential role in determining the electronic behavior of a SMBJ.

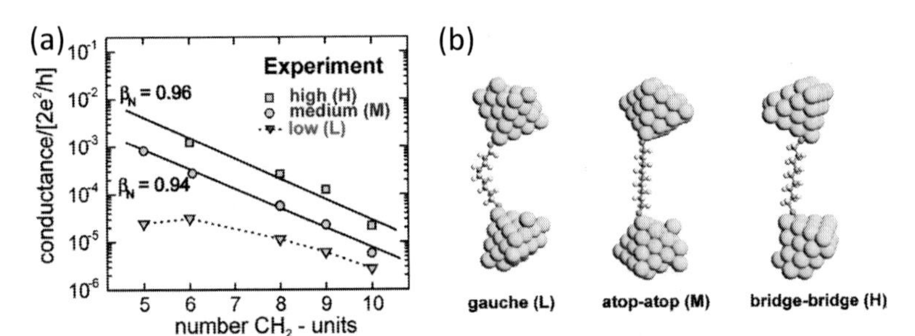

Figure 4.7 (a) Chain-length dependence of single-molecule conductance of alkanedithiol in contact with gold electrodes. Three conductance sets (high, medium and low) were revealed for each chain length. (b) Three contact configurations of Au–C9DT–Au junction. Reprinted with permission from Ref. 30. Copyright (2008) by the American Chemical Society.

Conductance discrepancies also exist for intricate conjugated molecules. For these molecular species, the π–π orbital coupling is the dominating factor that determines the electronic properties of a molecular junction. In the first STMBJ experiment, 4,4′-bypiridine was measured to have a conductance of around $0.01G_0$ [23]. Nevertheless, following measurements on the same molecule revealed different conductance sets, and the studies focused on the individual current versus distance traces illustrated obvious conductance jumps and drops on a conductance step [76]. These phenomena suggested the change of junction conformation as the junction evolved from formation to breaking. Simulations imitating the exact evolution of a molecular junction have been of immense help, because as of yet the direct observation of atomic scale event has been extremely difficult. The simulation results showed obvious orientation change of the 4,4′-bypiridine molecule as the junction separation increases, and the molecule could have two or more probable orientations within a junction, each with a unique conductance signature, as shown in Fig. 4.8 [74].

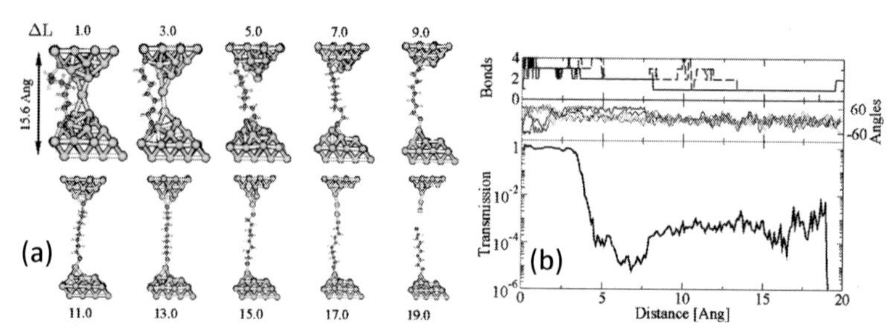

Figure 4.8 (a) Snapshots of structure evolution of an Au–C8DT–Au junction as the junction is being stretched. (b) (bottom) Calculated electron transmission probability as a function of stretching distance. The dihedral angle (middle) for S-C8-S chain and the number (top) of Au–S bonds are also shown. Reprinted with permission from Ref. 74. Copyright (2009) by the American Chemical Society.

A recent study by Tao et al. reported counterintuitive experimental features on single-molecule plateaus of an Au–benzenedithiol(BDT)–Au junction [77]. The experiments were conducted at 4 K under high-vacuum conditions, and some conductance plateaus showed a severe increase in elevation by more than an order of magnitude during stretching, and then decreased again as the junction was compressed (Fig. 4.9a). The large increase in conductance is counterintuitive because increasing the separation distance between the two electrodes is expected to weaken molecule–electrode coupling and increase tunneling distance, which one would expect to lead to a decrease in conductance rather than an increase. The reason of the conductance increase during stretching was attributed to the HOMO lifting of BDT caused by the elongation of the junction. Theoretical simulations by Ratner et al. reported a similar phenomenon for another π-conjugated molecule [78]. In their study, a junction-pulling model revealed obvious conductance blinking for a π-stacked molecule sandwiched between gold electrodes. The conductance switching mechanism was reported to predominantly depend on the unfolding of the π-stacking conformation, an event that could result in 3–4 orders of conductance drop. Furthermore, the resulting switch was reversible and robust, causing the observed conductance blinking. Their follow-up simulation on a different π-stacking molecular junction also illustrated an unintuitive current increase with the forced junction elongation (Fig 4.9b) [79]. A thorough calculation analysis suggested that the inverted trend in the transport was determined by the dual role played by H-bonds in both stabilizing π-stacking for particular extensions and introducing extra electronic couplings between the complementary aromatic rings that also enhance the electron transport across the molecule. As Tao's and Ratner's work elucidated, the conductance changes by orders while the junction is being stretched showed the extent to which electronic characteristics are sensitive to molecular conformations. The combining effect of the complexity of the molecule–electrode interfaces and internal molecular orientations requires a rather delicate data interpretation of the experimental findings.

As discussed above, the in-depth probing into the experimental contradictions and counter-intuitions with the cooperation of experimentalists and theorists have discovered abundant results and advanced understanding. To date, although great efforts have been devoted to it, the experimental results and the simulation system are still not fully connected. It is still rather hard to imitate experimental conditions using computer simulations. And some simulation predictions, for instance, Ratner's work, have not yet been experimentally confirmed. A great deal of information already discovered or still hidden in the data is yet under explanation. The experimental data using existing techniques only reflects a complicated ensemble of contributions from many factors that get involved in the SMBJ system. It is necessary to elaborately define each element before reaching any conclusion. Thereby, more refined experimental controls of the molecular junction that can single out individual factors for particular study, more powerful data analysis approaches that can reveal latent yet decisive information from the data, and more advanced simulations that enable the imitation of real experimental situations and the modeling of intricate molecular systems are highly in demand.

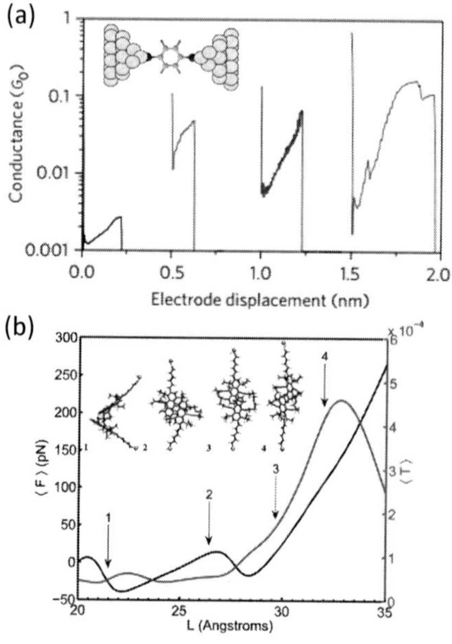

Figure 4.9 (a) Example conductance traces of Au–1,4′-benzenedithiol–Au junction (structure shown in the inset) showing conductance increase as electrode separation increases. (b) shows the average force (black) and transmission (red) as the function of junction extension. The inset shows the snapshots of structures during pulling encountered with the positions arrows point at. Note the increase of transmission with the pulling proceeds. *Note*: (a) Reprinted with permission from Ref. 77. Copyright (2012) by the American Chemical Society; (b) Reprinted with permission from Ref. 79. Copyright (2011) by the American Chemical Society.

4.3 Experimental Modulations and Controls of SMBJ System

To clear the mist above the field and gain deep insight into the SMBJ system, innovations have been triggered to advance the measurement techniques and modulate the measurement process using existing techniques. In this section, the recently developed experimental modulations and controls of SMBJ system will be discussed.

4.3.1 Modulations of SPMBJ Measurement Process

The major issue that has been puzzling researchers lies in the controversies in measured conductance data using SPMBJ, MCBJ and other techniques. In terms of manipulating the measurement process, SPMBJ serves as a better candidate due to its working principle. The core principle of SPMBJ involves taking advantage of the PZT-driven movement of a metal tip. The formation and break of a molecular junction is realized by approaching the tip to the metal substrate and then pulling it away from the substrate. This unique feature of the SPM (STM/CAFM) experimental setup offers great flexibility to experimentalists. Given that the data gathering/collecting is usually carried out during the tip retraction process, the process in which SPM tip approaches and contacts with the metal substrate is trial to the conductance measurements. This tip engaging process is usually defined by a threshold current of several nA or a distance of a few nm to ensure the physical contact of the metal tip with the metal substrate. Once the contact is made, there is no need to apply further controls. The essence of manipulating the SPMBJ measurement concentrates on the mechanical controls of the tip retraction process in which the tip is pulled away from the substrate following a certain manner. The piezoelectric transducer (PZT) provides precise displacement of the SPM tip, allowing for the separation between the tip and substrate to be adjusted in multiple modes. Example PZT modulation signals are shown in Fig. 4.10. Furthermore, SPM allows for an external bias voltage to be applied between the tip and substrate. This external bias is also tunable. These variables, such as the displacement of the tip, tip retracting speed, external bias and so on, can be controlled by a computer program and a feedback mechanism (see the SPM system in Fig. 4.10). The controlling of these variables could bring experimentalists one step closer to the nature of SMBJ, especially the molecule–electrode interfaces.

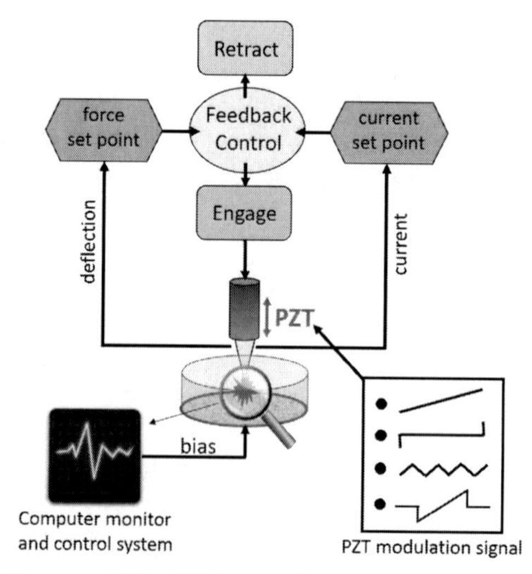

Figure 4.10 Diagram of the SPM control system and possible modulations of PZT signals.

Conventional SPMBJ measurements usually apply a constant tip retracting speed throughout the whole measurements, and the PZT signal appears to be a straight line as shown in Fig. 4.11a. Here we name this measurement procedure as the "continuous-stretch mode" SPMBJ. The histograms resulting from these continuously stretched conductance traces often illustrate conductance peaks with a rather broad distribution and low intensity, which misses significant details important for understanding the SMBJ systems it is measuring. The junctions are also short lived in continuous-stretch mode SPMBJ. As discussed earlier, the peak shape is predominantly determined by the stability of the molecular junction conformation. Thus, the ability to stabilize the molecular conformation while it is measured is necessary. In Zhou et al.'s report, this issue was resolved by changing the tip retraction mode from continuous-stretch to a stretch-hold modification [69]. The stretch-hold modification involves stair-stepping the retraction process so that the system pauses or holds momentarily and allows for the junction to settle into a quasi-relaxed state. As shown in Fig. 4.11b, the conductance step became much more distinct and well defined, and the junction had a much longer lifetime. This modification of the tip retraction was proven to eliminate, or at least minimize, the variations of experimental conditions, such as the fluctuation of the junction conformation. This stretch-hold-stretch was controlled by coding the Labview program that prescribes the displacement of the PZT controlling the SPM tip. Using this modified SPMBJ, the conductance histograms of C8DT molecules revealed much finer peaks and extra conductance sets (Fig. 4.11c). The conductance sets at around $7.0 \times 10^{-5}G_0$ and $9.0 \times 10^{-5}G_0$ were between the previously reported medium ($5 \times 10^{-5}G_0$) and high ($2.5 \times 10^{-4}G_0$) conductance values for the Au–C8DT–Au molecular junction (Fig. 4.11d). Thereby, the stretch-hold modification of the tip retraction process successfully discovered those less populated conductance sets of the molecular junctions. These conductance groups should be contributed to by different junction conformations. However, for alkanedithol molecules, the molecular core can be considered as a rigid structure because the C–C covalent bond has been studied to be much stronger than the Au–SH and Au–Au bonds at the molecule–electrode interfaces [80, 81]. Thus, the conductance change mainly came from the contact parts of the molecular junction. In this study, the source of the different conductance sets was the variation of the different Au–S bonding geometries. As introduced earlier, various Au–S bonding configurations have been proposed to be responsible for the differing conductance values experimentally observed. Therefore, the discovered less populated conductance sets not only illustrated the coexistence of more preferential contact conformations but also suggested the high sensitivity of these contact conformations to the measurement procedures. It also provided constructive hints for detailed theoretical simulations.

Other than the stretch-hold modification, the tip can also be modulated to cycle through "saw-tooth" displacements while it is retracted [82, 83]. This could be achieved by applying a triangular shaped ac signal to the SPM PZT movement (Fig. 4.12a). Modulations in this manner could influence the orientation of specific contact geometries by regularly compressing and elongating the molecular junction. The CAFMBJ allows one to measure force simultaneously with conductance, and when this saw-tooth modulation is applied to CAFMBJ, the effects of contact configurations can be illustrated in a more prominent way.

The first paper using this saw-tooth modulation measured the conductance of C8DT in a STMBJ system. Compared with continuous-stretch mode conductance traces, the saw-tooth modulated conductance traces displayed distinctively responsive features to the PZT signal. In Fig. 4.12a, when the saw-tooth modulation was applied on the continuous-stretch mode STMBJ, the typical conductance traces revealed regular fluctuations with a 180° phase shift with respect to the PZT signal (see the inset of Fig. 4.12a). It is easy to understand that when the PZT signal was at its peak, the junction distance was the longest and the electron had to tunnel through a longer barrier, which for sure will result in a conductance decrease. And the case in which the PZT signal was at its valley was the other way around. The 1D histogram of the ensemble of many such modulated traces illustrated a wide distribution of multiple peaks, suggesting that the conductance of C8DT molecular junction can be any value because of the sensitive dependence of the microscopic details on the molecular-electrode interfaces [82].

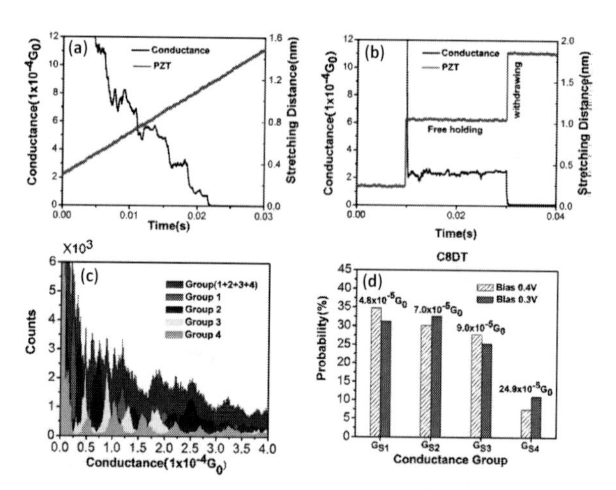

Figure 4.11 Example continuous-stretch mode (a) and stretch-hold mode (b) conductance trace (black) and corresponding Piezo movement signal (blue). (c) Conductance histograms of Au–octanediamine(C8DA)–Au junction with stretch-hold modulation applied. (d) Four conductance sets extracted from the conductance histograms of Au–octanedithiol (C8DT)–Au junction with stretch-hold modulation applied. Reprinted with permission from Ref. 69. Copyright (2009) by the American Chemical Society.

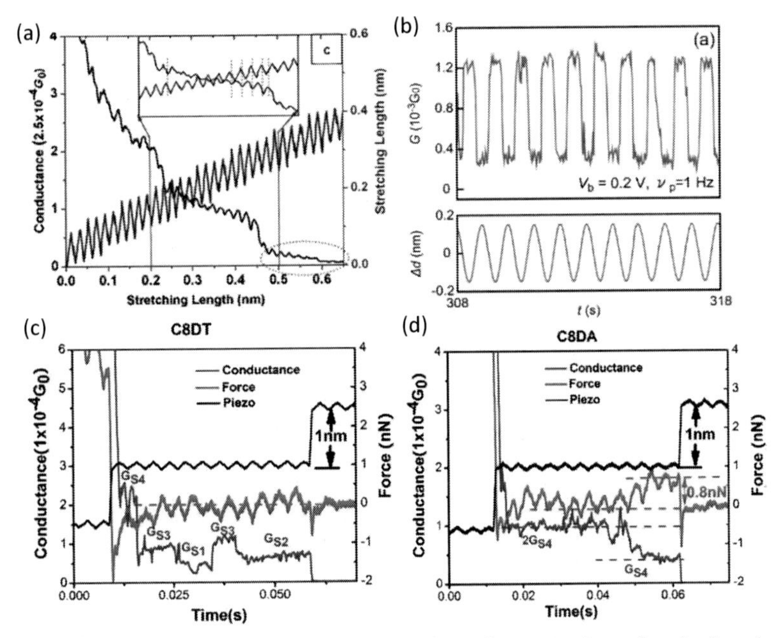

Figure 4.12 Experimental modulations based on single-molecule break junction technique. (a) Representative continuous-stretch mode conductance trace under saw-tooth modulation for Au–C8DT–Au junction. The inset shows 180° phase shift between the PZT signal with the conductance fluctuation. (b) Conductance switching (upper) between G_H and G_L induced by mechanical oscillations (lower). (c) Example stretch-hold mode force and conductance traces under saw-tooth modulation for Au–C8DT–Au junction, showing obvious conductance switch among different conductance sets. (d) Example stretch-hold mode force and conductance traces under saw-tooth modulation for Au–C8DA–Au junction, showing conductance change with correspondent force change. *Note*: (a) Reprinted with permission from Ref. 82. Copyright (2007) by the American Chemical Society; (b) Reprinted with permission from Ref. 84. Copyright (2010) by the Royal Society of Chemistry; (c and d) Reprinted with permission from Ref. 83. Copyright (2010) by the American Chemical Society.

However, this saw-tooth modulation applied on the continuous-stretch mode SPMBJ performs while the junction was constantly moving, which is still not an ideal platform to study the detailed behavior of a SMBJ. This is like if one aims to study the functionality of the leg muscle of a panther, one cannot capture the detailed movements of muscle segments while the panther is running, unless the movement can be snap-shot. The combination of the stretch-hold modification and saw-tooth modulation allowed experimentalists to snap-shot the modulated molecular junctions, which served as a well-defined platform to explore the detailed electron transport properties of the SMBJs. Using the combination of these two modulations, the regular compression and elongation of the PZT signal could be applied on the stabilized molecular junctions while the tip was free-held. Such experimental manipulation carried out on the alkane molecules isolated the contact parts to investigate the detailed relationship between force and conductance. Zhou et al. reported their measurement results utilizing these two modulations in a CAFMBJ system, and individual conductance traces revealed intriguing phenomena never observed before (Fig. 4.12c,d) [83]. For both C8DT and C8DA, they observed the perfect correspondence of force and simultaneously measured conductance. Interestingly, some conductance traces show conductance switching between different conductance sets that were reported in previous studies without an obvious force change (Fig. 4.12c) [67–69]. This phenomenon was attributed to the switching among different adsorption sites of Au atoms induced by the mechanical modulation. While for C8DA, conductance traces that displayed a conductance switch accompanied with a force change close to the Au–NH_2 binding force was discovered (Fig. 4.12d), strongly suggesting that the conductance variations are a combined effect of the switching between different adsorption sites and the dissociation of the molecule from the Au electrode with the bond breaking. These mechanical modulations of the measurement process in a SPMBJ system helped experimentalists to monitor the behavior of individual molecular junctions, providing a powerful tool to tune the electron transport properties and learn the real mechanisms of molecular junctions. However, the relationship between force and conductance observed on individual traces could not elucidate the average effect of many such behaviors. As many force-conductance trace pairs were obtained in the experiments, the approach to analyze the average correlation between force and conductance changes induced by the mechanical modulations is not available yet. Hence, more powerful data analysis methods that can distinguish the significant information that is latent in the data sets are also required.

Another experimental modulation involved using nanofabricated MCBJ approach to achieve the transition between the bistable molecule–electrode contact configurations in an Au–hexanedithiol (HDT)–Au system [84]. The experiment was conducted at room temperature in high vacuum condition. After a repeated mechanical elongation and compression cycle was imposed on the MCBJ system, an immediate conductance switch from G_H ($1.3 \times 10^{-3} G_0$) to G_L ($0.4 \times 10^{-3} G_0$) and then back to G_H was observed (Fig. 4.12b). The multiple conductance states were attributed to the conformational distortions in alkyl chains and hollow-to-atop configuration transitions of Au–S bonding [85, 86]. A detailed inelastic electron tunneling spectroscopy (IETS) probed the atomic-scale vibration modes of the molecular junctions and provided evidence that the two-state conductance switching occurred via repeated deformation of Au–S bonding between hollow and atop motifs. This study also noted that this conductance switching occurred only within a certain frequency and amplitude range of the mechanical modulations, indicating the high specificity of the contact configuration to the junction distance. The accomplishment of the controllable conductance switching of molecular junctions thereby offered a hint for further developing electronic switching devices.

4.3.2 Tuning Molecular Structures

A series of experimental modulations embedded in the traditional SPMBJ have illustrated abundant features. These features are mainly contributed by variation at the molecule–electrode interfaces. However, another focus of SMBJ study rests with how molecules behave in a molecular junction and the relation between molecular structures and the resultant electronic behavior. Efforts towards this direction require close collaboration of the chemical synthesizing of characteristic molecules and comprehensive measurements of electron transport properties of SMBJs. In this section, representative works concentrating on the molecular conformation effect will be introduced.

An elegant experimental study by Venkataraman et al. explored the influence of the dihedral angle of ring structures on electron transport properties of molecular junctions (Fig. 4.13) [87]. In their work, the amine group (–NH_2) was chosen as the anchoring group because of the relatively lower flexibility of the Au–NH_2 bonding geometries at the molecule–electrode contacts [63, 88]. This reduced the impact of variations rooted from the contact interfaces, and thereby the change in measured results could primarily be contributed by the structural variations of the molecular core. The fundamental molecule their study was based on was a biphenyl—a molecule with two phenyl rings linked by C–C bond. A series of seven biphenyl molecules with different ring substitutions that differed in twist angles were tested. The substituents forced a range of dihedral angles between the phenyl rings [89, 90]. As shown in Fig. 4.13b, the conductance decreased by a factor of 20 when the dihedral angle increased from 0° to 88°, which was attributed to the reduced conjugation degree between the phenyl rings for larger angles. This experimental feature was noted to be consistent with a cosine-squared relation theoretically predicted for the transport through π-conjugated biphenyl systems [91]. A series of following studies by Venkataraman's group have advanced the understanding of how carbon-ring structures could perturb the conductance of measured molecular junctions [89, 92, 93].

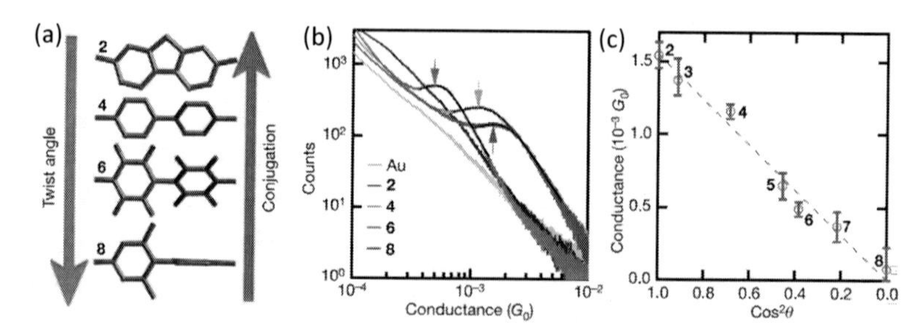

Figure 4.13 Single-molecule conductance of diaminobiphenyl derivatives as a function of twist angle. (a) Relation between twist angle and conjugation for four structures. (b) Example conductance histograms (log-scaled). (c) The dependence of conductance on twist angle. Reprinted with permission from Ref. 87. Copyright (2006) by Nature Publishing Group.

The electron transport through the π-stacked systems has been of remarkable interest due to its critical role in biological systems, polymers, and materials sciences [94–96]. Recent improvement of synthetic techniques allows us to arrange the π molecules in a controllable manner [97, 98]. Kiguchi et al. have achieved the investigation of electron transport through a π-stacked system using π-stacked molecules [99]. In their study, the conductance measurements were performed for π-stacked aromatics where the π-stack was sequentially increased from four to six stacked aromatic molecules. A schematic of their π-stacked molecule is shown in Fig. 4.14a. Schneebeli et al. also succeeded in characterizing π-stacked system using stacked benzene rings held together in an eclipsed fashion via a paracyclophane scaffold (Fig. 4.14b) [90]. Overall, the comparison of a vital parameter, decay constant β, which determines the decay rate of electron tunneling process indicated that the β of single π-stacked molecule in Kiguchi's work is 0.1 Å^{-1} and is much smaller than that of alkane chains (β = 0.7 ~ 0.9 Å^{-1}), and comparable to that of the π-conjugated organic molecules (β = 0.05 ~ 0.2 Å^{-1}) [100]. This small β for the caged π-stacked molecules indicates a relatively high efficiency of electron transport through π-stacked systems. In the case of the π-stacked benzene rings, the β (0.63 Å^{-1}) was smaller than that of alkane molecules, but greater than that of the π-stacked molecules. Electron transport through π-stacked system was also probed by other experimental methods and theoretical simulations, and the observed phenomena offer promising guidance for developing more intricate and functional molecular structures [101–104].

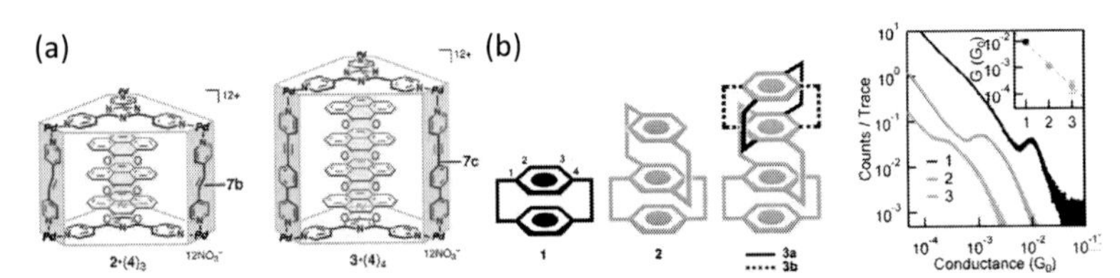

Figure 4.14 (a) Structure of π-stacked molecules in the work of Kiguchi et al., (b) structure (left panel) of π-stacked Benzene rings in the work of Schneebeli et al. and the conductance measurement results (right panel). *Note*: (a) Reprinted with permission from Ref. 99. Copyright (2011) by the American Chemical Society; (b) Reprinted with permission from Ref. 90. Copyright (2011) by the American Chemical Society.

Another direction of manipulating the molecular conformation involves tuning the structure of the terminal groups of sample molecules, and investigating the impact of the terminal groups on the electron transport properties of SMBJs. Research in this direction could not only probe the effect of molecular structure, but could also search for terminal groups that couple to the electrodes more efficiently. The simplest case is shown in Fig. 4.15a, for BDA molecules with anchoring groups ($-NH_2$) placed at different sites along a benzene ring [105]. Figure 4.15b shows the conductance histograms of 1,2-BDA, 1,3-BDA and 1,4-BDA. For 1,2-BDA, the molecules adsorbed only on one electrode and would not form a complete molecular junction with the molecule wired to two electrodes. Therefore, the conductance histogram revealed no obvious peak for 1,2-BDA. The single-molecule conductance of 1,4-BDA is larger than that of 1,3-BDA, which was also reproduced by theoretical calculations [105, 106]. To explain the observed results, it was highlighted that the change of anchoring position from 1,4 to 1,3 in benzene ring deformed the HOMO level and provided an energy shift that led to the reduction in the conductance of the single molecule. Recently, intricate carbon-based structures, such as carbon nanotubes (CNT) and C60, have been used as electrodes and anchoring groups [107, 108]. Martin et al. reported their conductance measurement results using C60 as the anchoring group that weakly interacts with metal electrodes (Fig. 4.15c). The conductance histogram of C60-anchored molecular junctions showed a small spread in conductance values, compared to a thiol anchoring unit [108]. A recent work by Wandlowski's group studied the anchoring group's dependence of conductance through a series of oligoyne molecular junctions (Fig. 4.15d) [109]. They synthesized five different anchoring structures, labeled as PY, NH2, CN, SH, and BT. Interestingly, two conductance sets, high and low, were determined using both STMBJ and MCBJ techniques for all the five terminal structures. The most probable single-molecule conductance of different anchoring structures varied by more than two orders for both high and low conductance sets. The experimental β for high conductance sets range between 1.7 nm^{-1} (CN) and 3.2 nm^{-1} (SH) and show the trend of $\beta_H(CN) < \beta_H(NH_2) < \beta_H(BT) < \beta_H(PY) \approx \beta_H(SH)$. They also noted that the low conductance states could be attributed to two possible scenarios: (i) single-molecule junctions involving the pulling out of gold atoms from the electrodes or (ii) multimolecular junctions involving π–π stacking.

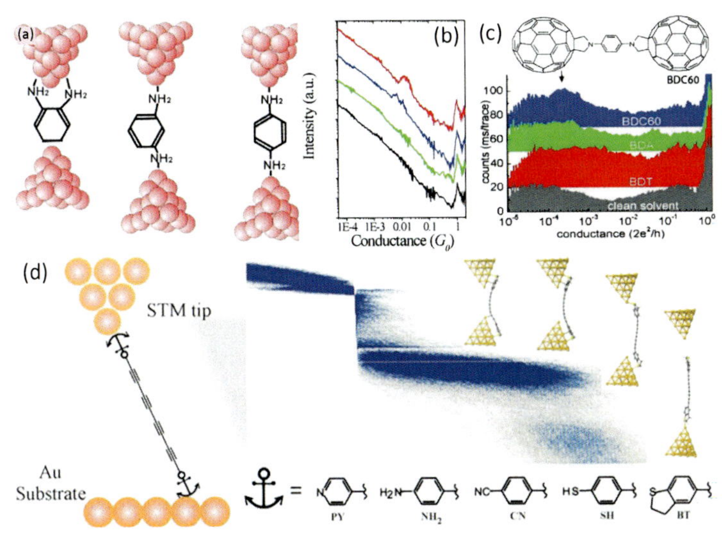

Figure 4.15 (a) Schematics of the single-molecule break junction of 1,2-BDA (left), 1,3-BDA (middle) and 1,4-BDA (right). (b) Conductance histograms measured in solution without molecules (black) and with 1,2-BDA (green), 1,3-BDA (blue) and 1,4-BDA (red). (c) Schematic (upper panel) of 1,4-bis(fullero(c)pyrrolidin-1-yl)benzene (BDC60), and conductance histograms (bottom panel) of BDC60 and other reference molecules. (d) Schematics of functionalized oligoyenes molecular junctions with different anchoring moieties (PY, NH2, CN, SH, and BT as labeled in ref. 104). *Note*: (A-B) Reprinted with permission from Ref. 105. Copyright (2010) by the American Chemical Society; (c) Reprinted with permission from Ref. 108. Copyright (2008) by the American Chemical Society; (d) Reprinted with permission from Ref. 109. Copyright (2013) by the American Chemical Society.

Vazquez et al. recently investigated the conductance superposition law for parallel components in single-molecule circuits [110]. They synthesized a series of molecular systems that contained either one or two backbones in parallel, bonded together cofacially by a common linker on each end (Fig. 4.16). The conductance measurements showed that the conductance of a double-backbone molecular junction can be more than twice that of a single-backbone junction. This intriguing phenomenon is not consistent with Kirchhoff's laws in which the net conductance of two parallel components is the sum of the individual conductances.

This inconsistency was attributed to constructive quantum interference when circuit dimensions are comparable to the electronic phase coherence length [111–113].

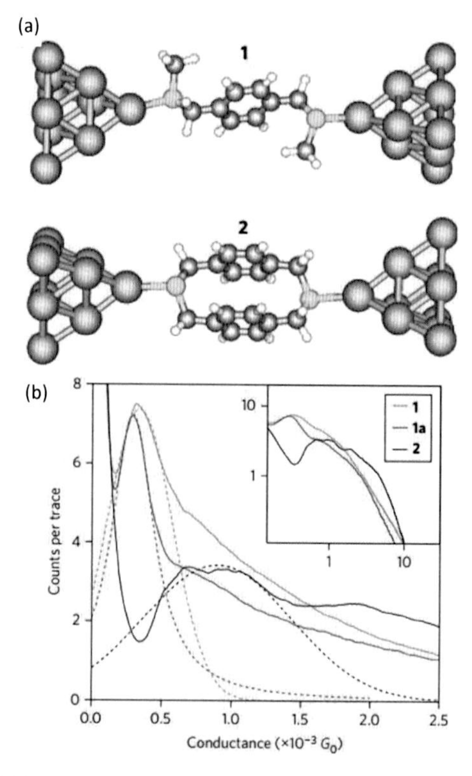

Figure 4.16 (a) Chemical structure of molecules with single- and double-backbone molecules studied in ref. 105. (b) Conductance histograms for each molecule: 1, 1a, and 2. *Note*: 1a is another two-backbone molecule with one side conjugated (CH_2–benzene–CH_2) the other side saturated (C_4H_8), which is not shown in (a). Dashed lines represent Lorentzian fits to the conductance peaks. Reprinted with permission from Ref. 110. Copyright (2012) by Nature Publishing Group.

Various manipulations of molecular structures discussed above have provided a great deal of experimental evidence and promoted our physical understanding of molecular structures. The combination of experimental modulations and control of molecular structures would serve as the fundamental steps towards the ultimate goal of molecular devices.

4.3.3 Environmental Controls of SMBJ System

Besides a molecular junction itself, the environment where single-molecule measurements are conducted have been of great interest as well. Electrical measurements are usually carried out in appropriate buffer solutions. More importantly, the ions in the solution have been proven to play a vital role in the electron transport properties of molecular junctions [95, 114, 115]. To date, the dramatic development in junction fabrication allows more precise measurements, but also begs further investigation into the non-negligible contribution from ionic transport in the solution [13]. Recently, Doi et al. reported the transient electrical response in the vicinity of the biased electrodes (Fig. 4.17a) [115]. Their results showed a rapid response of ions to strong fields near the electrode surface after the turning on of an applied voltage. Ions gradually translocated in the weak electric field, and slowly relaxed within the diffusion layer, implying a significant influence of ionic current in the solution. The interaction between the molecules and surrounding ions has been highlighted in studies focusing on DNA molecules as well [116–118]. For example, the conductance of a single DNA molecule measured in solution was reported to be one order of magnitude greater than the conductance of the same DNA molecule in dry condition [119]. DNA molecules have been proven to exhibit surprising conformational versatility, such as right handed B- and A-DNA and left-handed Z-DNA, which strongly depend on ionic concentrations in the solution where DNA molecules stay [120, 121]. Therefore, the ionic environment has a significant influence on the electrical measurements for DNA-based molecular junctions. Wang et al. measured the conductance of a poly $d(GC)_4$ DNA duplex in solutions containing different $MgCl_2$ concentrations [122]. Their experimental results showed that the increase of ionic concentration induced a secondary structural transition from the B to the Z form of DNA molecules, and the conformational change reduced the DNA conductance by two orders of magnitude.

Temperature also plays an important role in metal–molecule–metal junctions [123, 124]. Tsutsui et al. reported conductance measurements of BDT at 77 K [124]. The temperature dependence of a BDT molecular junction lifetime was revealed as shown in the left panel of Fig. 4.17b. The junction lifetime can be explained in terms of gold single-atom contact stability. The experimental results suggested that the molecular junction lifetime at 77 K started to become shorter than its lifetime at room temperature (293 K) under low-strain-rate conditions, where differences in the effects of thermo fluctuations on gold single-atom contact stability became pronounced.

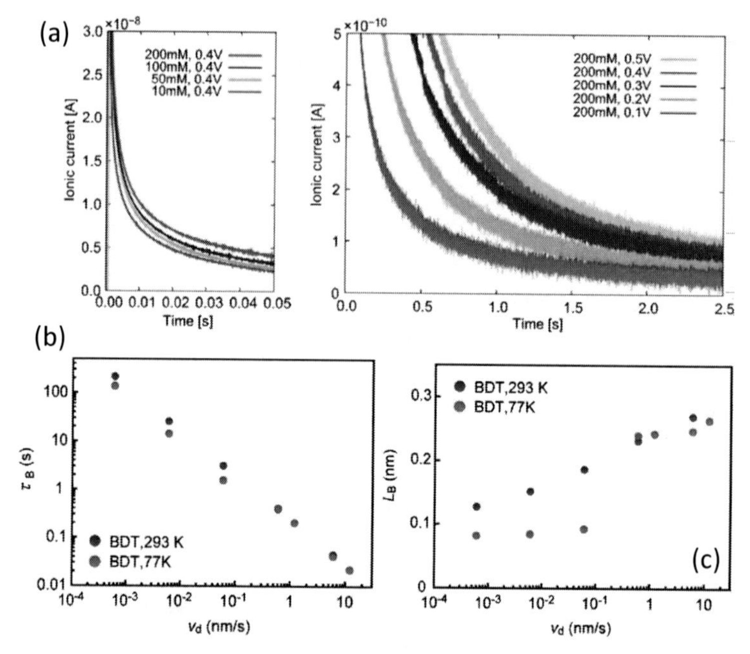

Figure 4.17 (a) Transient current response of NaCl solution under an applied bias of 0.4 V for several salt concentrations (left); current response for a concentration of 200 mM NaCl at various bias voltages (right). (b) Junction lifetime τ_B versus junction-stretching rate v_d (left), and junction breakdown length L_B versus junction-stretching rate v_d (right) for benzenedithiol (BDT) single-molecule junction measured at 77 K (red) and 293 K (blue). *Note*: (a) Reprinted with permission from Ref. 115. Copyright (2014) by the American Chemical Society; (b) Reprinted with permission from Ref. 124. Copyright (2009) by the American Chemical Society

In addition, local heating, a common issue in current carrying devices, is of great influence, too. Another of Tsutsui's reports studied the local heating effect of Au–BDT–Au system at room temperature [123]. Contact destabilization was observed when subjected to a high electric field. They found that the junction lifetime decayed exponentially with the intensity of bias, and the junction local temperature was raised to 463 K at 1 V. By performing room temperature IETS, the severe local heating was attributed to phonon mismatch at the molecule–electrode interfaces. Also, Huang et al. studied the local heating effect for Au–C8DT–Au molecular junctions by measuring the average force needed to break down the molecule–electrode bond using CAFMBJ technique [125]. The measurement results showed that the temperature was raised by 30 K above ambient room temperature at 1 V, and the junctions became increasingly unstable when bias went beyond 1 V. It is notable that the degree of local heating is different for different molecular junctions, and therefore local heating has to be carefully considered when discussing the resulting electronic properties.

4.3.4 Characteristic Current–Voltage Behavior

The original considerations of molecular electronics were to search for molecular candidates with functional current–voltage properties, such as rectifying effect and negative differential resistance (NDR) [5]. This asymmetric *I–V* behavior can be experimentally achieved by breaking the symmetry of the molecular junction conformation. To date, rectification can be observed either from asymmetric molecules [16, 29] or with inconsistent molecule–electrode interfaces [126, 127]. However, most rectifications have been observed from small organic molecules with the structure containing an electron donor and an acceptor group separated by an insulating group [5, 16, 128–130]. Example donor–insulator–acceptor molecules are shown in Fig. 4.18a. In these donor–insulator–acceptor molecules, the origin of the rectification would involve charge transfer from one electrode to the acceptor, to the donor, and finally to the second electrode at forward bias. However, at the reverse bias, charge has to migrate through the junction in the opposite direction, which requires a larger potential to align the energy levels of the donor and acceptor with that of the electrodes. Hence, the molecule would produce much larger current under forward bias, which is the signature of rectification. In terms of asymmetric molecule–electrode interface, it involves two strategies: (i) asymmetric anchoring groups and (ii) asymmetric electrodes. In these cases, the asymmetry occurs at the contacts. The frontier molecular orbital couples with one electrode stronger than it couples with the other electrode, which weakens the charge transport ability when charge transfers from the electrode through the weakly coupled molecule–electrode interface under the reverse bias. Example *I–V* behavior of molecular junctions based on donor–insulator–acceptor molecules and inconsistent anchoring groups are shown in Figs. 4.18b,c, respectively.

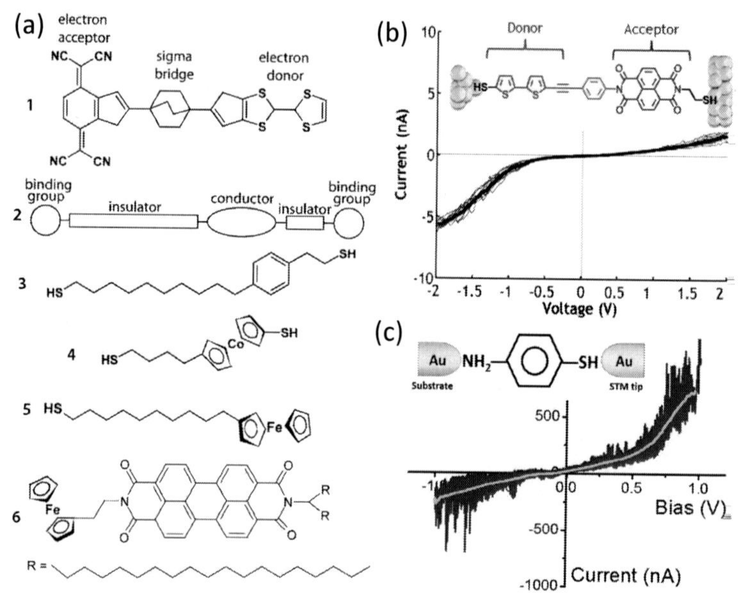

Figure 4.18 (a) Example donor–insulator–acceptor molecules, (b) Example *I–V* curves (raw data in blue and average curve in black) of an inverse rectification of a donor-acceptor typed molecular junction. (c) Example rectifying *I–V* curves (raw data in blue and average curve in cyan) of a molecular junction with asymmetric anchoring groups. *Note*: (a) Reprinted with permission from Ref. 126. Copyright (2010) by the American Chemical Society; (b) Reprinted with permission from Ref. 130. Copyright (2011) by the American Chemical Society; (c) Reprinted with permission from Ref. 127. Copyright (2014) by the Royal Society of Chemistry.

To understand the mechanisms causing the rectification behavior of various molecular junctions, multiple theoretical models have been developed. The groups of Williams and Baranger et al. proposed that molecular junctions with a single conducting molecular orbital that is slightly shifted from the Fermi levels of the electrodes—either HOMO or LUMO—and asymmetrically coupled to one of the electrodes (i.e., in closer spatial proximity to one electrode than the other) can rectify [126, 128, 129]. Figures 4.19a,b outline the schematic paradigm of energy levels of the proposed molecular rectifier junction by Wiliams and Baranger, respectively. These molecular rectifiers have a "conductive" HOMO or LUMO level that is energetically positioned just above the Fermi levels of the electrodes (a small difference in energy between the Fermi levels of the electrodes and the conducting molecular orbital ensures a low switch-on bias of the molecular rectifier). The conductive molecular orbital follows the potential of the nearer electrode, and thus participates in charge transport more easily at one polarity of bias. Zhao et al. recently proposed a model for the molecular rectifiers with asymmetric molecule–electrode contacts which regarded the stronger coupling at one molecular-electrode interface as a closer affinity between the conducting molecular orbital and the Fermi level of the electrode (Fig. 4.19c) [131]. In this model, the conducting molecular orbital (HOMO or LUMO) tends to shift with the Fermi level of the strongly coupled electrode as a whole, when a bias is applied. The conducting molecular orbital can lean towards the Fermi level of the strongly coupled electrode and is dragged into the conduction window under forward bias. Under a reverse bias, the conducting orbital is out of the conduction window, which generates little current.

However, these theoretical results suggest that molecular rectifiers in the proposed models based on one conducting molecular orbital or double-barrier cannot achieve rectification ratios (RR) exceeding ~22 [132, 126]. Calculation results on different types of molecular rectifiers, including the donor–insulator–acceptor model, showed that molecular rectifiers in tunneling regime cannot have RR greater than ~20 [133]. The upper limits for RRs of molecular diodes are far smaller than the values commercially achieved with semiconductor diodes (RR = $10^6 \sim 10^8$), but still higher than the low values actually observed in many experiments (RR = $1 \sim 10$).

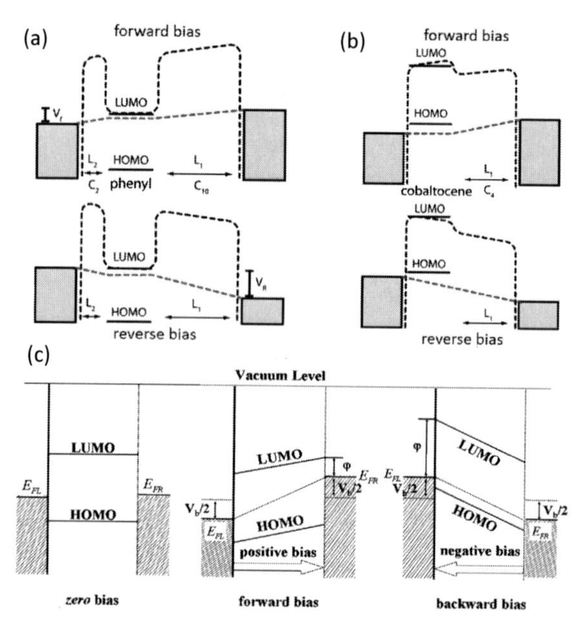

Figure 4.19 Mechanisms of rectification proposed by Williams (a), Baranger (b) and Zhao (c). *Note*: (a and b) Reprinted with permission from Ref. 126. Copyright (2010) by the American Chemical Society; (c) Reprinted with permission from Ref. 131. Copyright (2010) by the American Chemical Society.

Another intriguing current–voltage characteristic is the NDR effect. NDR effect was defined as a non-monotonic dependence of current on the bias voltage (i.e., the current decreases with increasing bias). This effect has been experimentally observed for various molecular junctions [8, 134–136]. An example *I–V* curve with a NDR feature is shown as the red curve in Fig. 4.20a. However, the source of the NDR is still in controversy. Efforts have been devoted by theorists to explore the cause using theoretical simulations. The main cause for NDR has been attributed to many reasons: (i) the bias-dependent electron-phonon interactions [137, 138], (ii) a potential-drop-induced shifting of molecular orbitals [139, 140], (iii) image charge effect [141], (iv) the change in effective contact coupling [142], and (v) the narrower density of state features of the tip apex [143]. Dubi et al. recently reported another mechanism that the coulomb interaction induced electron-hole binding across the molecule–electrode interface, resulting in a renormalized and enhanced molecule–electrode coupling (Fig. 4.20b) [144]. They showed that the effective coupling is non-monotonic in bias voltage, leading to the NDR effect. Figures 4.19c,d exemplify a mechanism of the NDR effect that involves a bias-dependent transmission as the function of injected energy (applied bias). Detailed understanding of the NDR effect requires deeper investigation into the cause of this phenomenon, and the efforts of simulations and experimental studies are both needed.

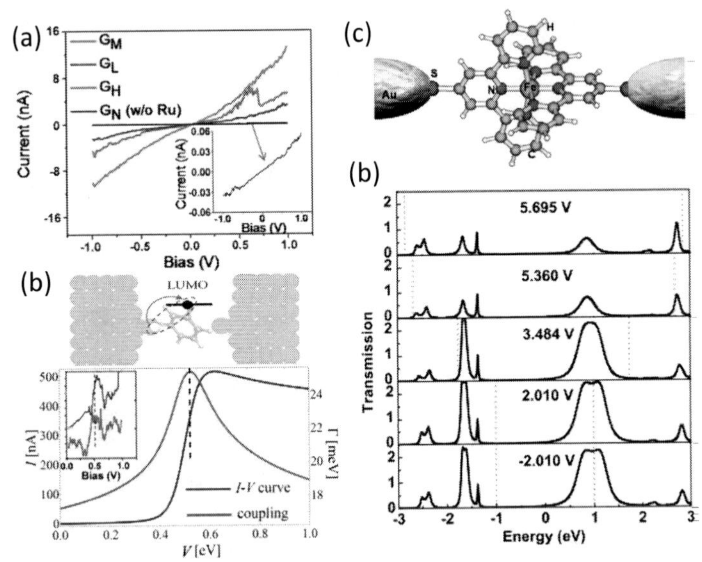

Figure 4.20 (a) Contact specific NDR effect (red) for Ru(tpy-SH)2 molecular junctions, (b) Schematic of Coulomb interaction induced electron-hole binding across the molecule–electrode interface (upper), and calculated *I–V* curve (blue) and effective coupling (red) as a function of bias voltage (bottom). (c) Structure of a strongly coupled Fe-terpyridine (FETP) molecular junction (top), (d) bias dependent transmission as a function of injection energy. Fermi energy is set to zero in the energy scale, and the dotted line in each panel represents the chemical potential window. *Note*: (a) Reprinted with permission from Ref. 136. Copyright (2013) by the American Chemical Society; (b) Reprinted with permission from Ref. 144. Copyright (2013) by the AIP Publishing; (c-d) Reprinted with permission from Ref. 142 as follows: Pati, R., McClain, M., and Bandyopadhyay, A., *Phys. Rev. Lett.*, **100**, 246801. Copyright (2008) by the American Physical Society.

4.4 The Development of Data Analysis Approaches

As introduced in previous sections, the single-molecule conductance can be determined using a conventional one-dimensional (1D) conductance histogram. However, the 1D conductance histogram only involves one significant parameter: conductance, which could not reveal other detailed features hidden in the conductance versus distance traces (i.e., the junction evolution distance and correlation between two traces). To unravel this hidden yet vital information in the collected data sets, researchers have developed advanced data analysis methods. For example, two-dimensional (2D) conductance versus distance plots have been used to reveal not only the conductance value but also the junction evolution features [78, 92]. In Fig. 4.21a, the 2D conductance versus distance plots of different molecular junctions revealed the slight increase in junction elongation distance prior to junction breaking when target molecule alternated from 1 to 3. In order to unravel the relationship between specific signals in individual traces, such as the relation between two conductance values, Makk et al. developed a conductance 2D auto-correlation histogram(C-2DACH) by introducing a time series analysis [76]. Figure 4.21d shows the C-2DACH for Au–4,4′-bipyridine–Au junctions. A clear blue negatively or anti-correlated region is where high and low conductance sets of the junction cross were revealed. A careful check of the histograms made of traces (Fig. 4.21e) with a long plateau at either high or low conductance values proved that the observed blue region is an anti-correlation in the length of the plateaus instead of an anti-correlation in the existence of two junction configurations. To explore further, Hamill et al. recently expanded the auto-correlation to a cross-correlation by adding another variable, force, to the calculations [145]. This enabled the correlation analysis of simultaneously measured force and conductance signals using the CAFMBJ technique. Figure 4.21f shows the force-conductance 2D cross-correlation histogram (FC-2DCCH) for Au–4,4′-bipyridine–Au junctions by clipping the traces to the single-molecule plateaus. The black circle highlights the strong correlation between a force of 0.1 nN and a conductance of $6.0 \times 10^{-4} G_0$, which could be attributed to the molecular twist during the initial stages of a break junction. The black square represents an anti-correlation between a force of −0.4 nN and a conductance of 1.2 × $10^{-4}G_0$, which was suggested to be related to the slipping of molecule–electrode bonding from one site to another. These advances in the analysis method could discover significant subtleties that help to gain insights into the mechanical and electrical properties of SMBJs.

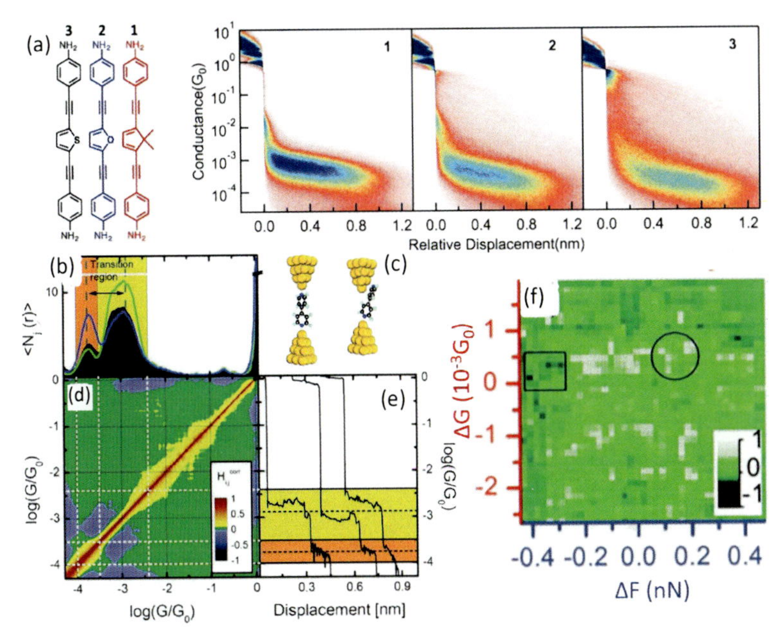

Figure 4.21 (a) Structure the three molecular wires (left) and their 2D conductance versus distance plots (right). Auto-correlation analysis of Au–4,4′-bipyridine–Au junctions are illustrated in b, c, d and e. (b) Conductance histogram using equidistant bins at log-scale. (c) Two contact binding configurations. (d) 2DACH of Au–4,4′-bipyridine–Au junctions. (e) Sample conductance traces. (f) shows the force-conductance cross-correlation analysis for Au–4,4′-bipyridine–Au junctions with conductance and force traces clipped to single-molecule plateaus. *Note*: (a) Reprinted with permission from Ref. 92. Copyright (2014) by the American Chemical Society; (b–e) Reprinted with permission from Ref. 76. Copyright (2012) by the American Chemical Society; (f) Reprinted with permission from Ref. 145. Copyright (2014) by the Royal Society of Chemistry.

Probably the most important information researchers want to acquire from the measured I–V curves of molecular junctions is the energy gap between the Fermi level of the electrode and the frontier molecular orbital of the molecule. To extract such important parameters, data analysis methods to interpret the measured I–V characteristics are also well-developed. An I–V curve can be fit to a Landauer formula using a Levenberg–Marquardt least-squares-fitting algorithm [146, 147]. The fitting results in three parameters: the energy gap ε, and the degree of coupling, Γ_R and Γ_L, to each electrode separately. Particularly, for asymmetric I–V curves, the asymmetry in the coupling strength at the two contact interfaces can be interpreted by comparing the resulted fitting parameters Γ_R and Γ_L. Symmetric I–V curves can also be fitted to the Simmons model, which could extract additional parameters, notably the tunneling barrier height Φ and the barrier decay constant β [148, 149]. Recent developments of the Simmons model allow one to fit the rectification ratio curves derived from the asymmetric experimental I–V curves, which could acquire the additional decay constant $\Delta\beta$ under the reverse bias [127, 150]. These analysis methods fitting experimental data to theoretical models have served as a powerful tool to study the electron transport properties of molecular junctions. Furthermore, Beebe et al. proposed another widely used method, transition voltage spectroscopy (TVS), to interpret the I–V characteristics [151]. In TVS, an I–V curve can be derived to a Fowler–Nordheim (F–N) plot, which describes a relationship of ln (I/V^2) versus $1/V$. The yielded minima in an F–N plot signifies the transition voltage where a trapezoid-shaped tunneling barrier turns to a triangle-shaped barrier or direct tunneling converts to field emission. Figure 4.22 shows an example F–N plot and the schematic of tunneling barrier change (inset). Usually asymmetric I–V curves yield transition voltage at different bias values under opposite bias polarities, which indicates that a trapezoid tunneling barrier converts to a triangle tunneling barrier at a smaller bias under forward bias than it does under the reverse bias. In real experiments, an F–N plot may not always produce minima. The non-minima F–N plot implies that the transition voltage for the measured molecular junction is not reached within the bias range applied in the experiments.

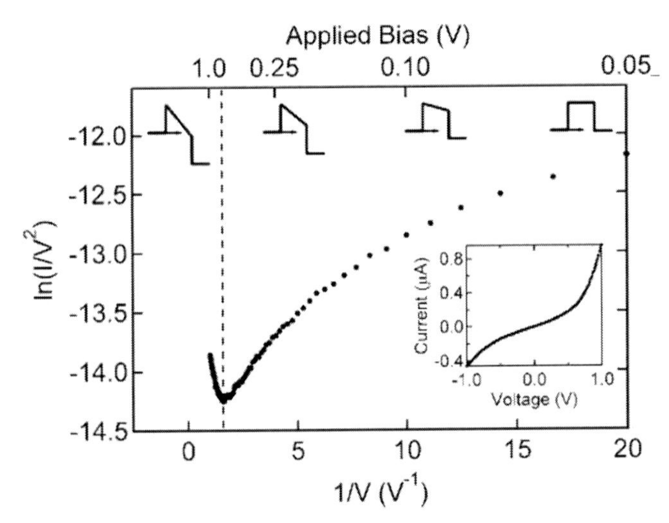

Figure 4.22 Example F–N plot derived from *I–V* curve (inset) and schematics of tunneling barrier as a function of applied bias. Reprinted with permission from Ref. 151 as follows: Beebe, J. M., Kim, B., Gadzuk, J. W., Frisbie, C. D., and Kushmerick, J. G., *Phys. Rev. Lett.*, **97**, 026801. Copyright (2006) by the American Physical Society.

4.5 Summary and Outlook

In this chapter, recent developments of experimental modulations based on the traditional SMBJ technique, including mechanical modulations of the measurement process, modifications of various molecular structures, and environmental controls have been discussed. In order to extract the detailed features hidden in the data sets, data analysis methods, such as two-dimensional signal versus distance plots, two-dimensional auto-correlation analysis, and two-dimensional cross-correlation analysis, have been introduced. To interpret characteristic current–voltage features, fitting methods, such as the Landauer formula and Simmons model, and transition voltage spectroscopy have been developed. These experimental modulations and data analysis methods have proven to be powerful in discovering the subtle yet significant information that determines the electron transport properties of single-molecule break junctions.

As revealing as these developments are, there is still plenty of space for improvement in the techniques involved in SMBJs. For example, adding a reliable gating electrode into the metal–molecule–metal system is still challenging. Experimentalists have already made attempts towards this direction, and many more attempts are needed [152]. Experimental manipulations involving optical effects, thermoelectric effects, spintronics and quantum interference could be highly inspiring but have not yet gained enough credit [153]. Along the same lines, new methods for analyzing multivariate data sets are just emerging, with great potential for expansion and application. It is beneficial to incorporate multivariate correlation to study magnetic phenomena, such as Kondo effects [154]. Additionally, theoretical methods enabling the simulation of complicated junction systems, such as DNA molecular junctions with ionic effect considered, will be rather helpful but are not yet available.

Molecular electronics is still in its fledgling stage, though it can be traced back to 1971. Since the invention of the STM technique, developments throughout the past two decades have been extraordinary. With some fundamental problems overcome recently, it is now the time for researchers to move beyond simple descriptions of charge transport and explore numerous intrinsic features of molecules [14].

Acknowledgment

The authors thank the U.S. National Science Foundation for funding this work (Grant Nos. ECCS 0823849 and ECCS 1231967).

References

1. Feynman, R. P. (1960). There's plenty of room at the bottom, *Eng. Sci.*, **23**, 22–36.
2. Moore, G. E. (1965). Cramming more components onto integrated circuits, *Electronics*, **38**, 114–117.
3. Vonhippel, A. (1956). Molecular engineering, *Science*, **123**, 315–317.
4. McCreery, R. L., Yan, H., and Bergren, A. J. (2013). A critical perspective on molecular electronic junctions: There is plenty of room in the middle, *Phys. Chem. Chem. Phys.*, **15**, 1065–1081.

5. Aviram, A., and Ratner, M. A. (1974). Molecular rectifiers, *Chem. Phys. Lett.*, **29**, 277–283.
6. Slowinski, K., Chamberlain, R. V., Miller, C. J., and Majda, M. (1997). Through-bond and chain-to-chain coupling. Two pathways in electron tunneling through liquid alkanethiol monolayers on mercury electrodes, *J. Am. Chem. Soc.*, **119**, 11910–11919.
7. Gregory, S. (1990). Inelastic tunneling spectroscopy and single-electron tunneling in an adjustable microscopic tunnel junction, *Phys. Rev. Lett.*, **64**, 689–692.
8. Chen, J., Reed, M. A., Rawlett, A. M., and Tour, J. M. (1999). Large on-off ratios and negative differential resistance in a molecular electronic device, *Science*, **286**, 1550–1552.
9. Zhou, C., Deshpande, M. R., Reed, M. A., Jones, L., and Tour, J. M. (1997). Nanoscale metal self-assembled monolayer metal heterostructures, *Appl. Phys. Lett.*, **71**, 611–613.
10. Fischer, C. M., Burghard, M., Roth, S., and Vonklitzing, K. (1995). Microstructured gold/langmuir-blodgett film/gold tunneling junctions, *Appl. Phys. Lett.*, **66**, 3331–3333.
11. McCreery, R. L., and Bergren, A. J. (2009). Progress with molecular electronic junctions: Meeting experimental challenges in design and fabrication, *Adv. Mater.*, **21**, 4303–4322.
12. Zhong, Z. H., Wang, D. L., Cui, Y., Bockrath, M. W., and Lieber, C. M. (2003). Nanowire crossbar arrays as address decoders for integrated nanosystems, *Science*, **302**, 1377–1379.
13. Hamill, J., Wang, K., and Xu, B. (2014). Characterizing molecular junctions through the mechanically controlled break-junction approach, *Rep. Electrochem.*, **4**, 1–11.
14. Ratner, M. (2013). A brief history of molecular electronics, *Nat. Nanotech.*, **8**, 378–381.
15. Reed, M. A., Zhou, C., Muller, C. J., Burgin, T. P., and Tour, J. M. (1997). Conductance of a molecular junction, *Science*, **278**, 252–254.
16. Joachim, C., Gimzewski, J. K., and Aviram, A. (2000). Electronics using hybrid-molecular and mono-molecular devices, *Nature*, **408**, 541–548.
17. Dunlap, D. D., Garcia, R., Schabtach, E., and Bustamante, C. (1993). Masking generates contiguous segments of metal-coated and bare DNA for scanning tunneling microscope imaging, *Proc. Natl. Acad. Sci. U.S.A.*, **90**, 7652–7655.

18. Fink, H. W., and Schonenberger, C. (1999). Electrical conduction through DNA molecules, *Nature*, **398**, 407–410.
19. Kasumov, A. Y., Kociak, M., Gueron, S., Reulet, B., Volkov, V. T., Klinov, D. V., and Bouchiat, H. (2001). Proximity-induced superconductivity in DNA, *Science*, **291**, 280–282.
20. Lindsay, S. M., and Ratner, M. A. (2007). Molecular transport junctions: Clearing mists, *Adv. Mater.*, **19**, 23–31.
21. Porath, D., Bezryadin, A., de Vries, S., and Dekker, C. (2000). Direct measurement of electrical transport through DNA molecules, *Nature*, **403**, 635–638.
22. Salomon, A., Cahen, D., Lindsay, S., Tomfohr, J., Engelkes, V. B., and Frisbie, C. D. (2003). Comparison of electronic transport measurements on organic molecules, *Adv. Mater.*, **15**, 1881–1890.
23. Xu, B. Q., and Tao, N. J. (2003). Measurement of single-molecule resistance by repeated formation of molecular junctions, *Science*, **301**, 1221–1223.
24. Landauer, R. (1957). Spatial variation of currents and fields due to localized scatterers in metallic conduction, *IBM J. Res. Dev.*, **1**, 223–231.
25. Agrait, N., Yeyati, A. L., and van Ruitenbeek, J. M. (2003). Quantum properties of atomic-sized conductors, *Phys. Rep.*, **377**, 81–279.
26. Muller, C. J., Vanruitenbeek, J. M., and Dejongh, L. J. (1992). Experimental observation of the transition from weak link to tunnel junction, *Phys. C*, **191**, 485–504.
27. Chen, F., He, J., Nuckolls, C., Roberts, T., Klare, J. E., and Lindsay, S. (2005). A molecular switch based on potential-induced changes of oxidation state, *Nano Lett.*, **5**, 503–506.
28. Dalgleish, H., and Kirczenow, G. (2006). A new approach to the realization and control of negative differential resistance in single-molecule nanoelectronic devices: Designer transition metal-thiol interface states, *Nano Lett.*, **6**, 1274–1278.
29. Diez-Perez, I., Hihath, J., Lee, Y., Yu, L., Adamska, L., Kozhushner, M. A., Oleynik, I. I., and Tao, N. (2009). Rectification and stability of a single molecular diode with controlled orientation, *Nat. Chem.*, **1**, 635–641.
30. Li, C., Pobelov, I., Wandlowski, T., Bagrets, A., Arnold, A., and Evers, F. (2008). Charge transport in single au vertical bar alkanedithiol vertical bar au junctions: Coordination geometries and conformational degrees of freedom, *J. Am. Chem. Soc.*, **130**, 318–326.

31. Love, J. C., Estroff, L. A., Kriebel, J. K., Nuzzo, R. G., and Whitesides, G. M. (2005). Self-assembled monolayers of thiolates on metals as a form of nanotechnology, *Chem. Rev.*, **105**, 1103–1169.

32. Nitzan, A., and Ratner, M. A. (2003). Electron transport in molecular wire junctions, *Science*, **300**, 1384–1389.

33. Taniguchi, M., Tsutsui, M., Mogi, R., Sugawara, T., Tsuji, Y., Yoshizawa, K., and Kawai, T. (2011). Dependence of single-molecule conductance on molecule junction symmetry, *J. Am. Chem. Soc.*, **133**, 11426–11429.

34. Tao, N. J. (2006). Electron transport in molecular junctions, *Nat. Nanotech.*, **1**, 173–181.

35. Vazquez, H., Skouta, R., Schneebeli, S., Kamenetska, M., Breslow, R., Venkataraman, L., and Hybertsen, M. S. (2012). Probing the conductance superposition law in single-molecule circuits with parallel paths, *Nat. Nanotech.*, **7**, 663–667.

36. Wang, K., Hamill, J. M., Zhou, J., Guo, C., and Xu, B. (2014). Measurement and control of detailed electronic transport properties of single molecule junctions, *Faraday Discuss.*, DOI: 10.1039/C4FD00080C.

37. Haiss, W., van Zalinge, H., Higgins, S. J., Bethell, D., Hobenreich, H., Schiffrin, D. J., and Nichols, R. J. (2003). Redox state dependence of single molecule conductivity, *J. Am. Chem. Soc.*, **125**, 15294–15295.

38. Haiss, W., Nichols, R. J., van Zalinge, H., Higgins, S. J., Bethell, D., and Schiffrin, D. J. (2004). Measurement of single molecule conductivity using the spontaneous formation of molecular wires, *Phys. Chem. Chem. Phys.*, **6**, 4330–4337.

39. Nichols, R. J., Haiss, W., Higgins, S. J., Leary, E., Martin, S., and Bethell, D. (2010). The experimental determination of the conductance of single molecules, *Phys. Chem. Chem. Phys.*, **12**, 2801–2815.

40. Liang, W. J., Shores, M. P., Bockrath, M., Long, J. R., and Park, H. (2002). Kondo resonance in a single-molecule transistor, *Nature*, **417**, 725–729.

41. Park, J., Pasupathy, A. N., Goldsmith, J. I., Chang, C., Yaish, Y., Petta, J. R., Rinkoski, M., Sethna, J. P., Abruna, H. D., McEuen, P. L., and Ralph, D. C. (2002). Coulomb blockade and the kondo effect in single-atom transistors, *Nature*, **417**, 722–725.

42. Kervennic, Y. V., Thijssen, J. M., Vanmaekelbergh, D., Dabirian, R., Jenneskens, L. W., van Walree, C. A., and van der Zant, H. S. J. (2006). Charge transport in three-terminal molecular junctions incorporating sulfur-end-functionalized tercyclohexylidene spacers, *Angew. Chem. Int. Ed. Engl.*, **45**, 2540–2542.

43. Li, C. Z., Bogozi, A., Huang, W., and Tao, N. J. (1999). Fabrication of stable metallic nanowires with quantized conductance, *Nanotechnology*, **10**, 221–223.

44. Li, X. L., He, H. X., Xu, B. Q., Xiao, X. Y., Nagahara, L. A., Amlani, I., Tsui, R., and Tao, N. J. (2004). Measurement of electron transport properties of molecular junctions fabricated by electrochemical and mechanical methods, *Surf. Sci.*, **573**, 1–10.

45. Morpurgo, A. F., Marcus, C. M., and Robinson, D. B. (1999). Controlled fabrication of metallic electrodes with atomic separation, *Appl. Phys. Lett.*, **74**, 2084–2086.

46. Dadosh, T., Gordin, Y., Krahne, R., Khivrich, I., Mahalu, D., Frydman, V., Sperling, J., Yacoby, A., and Bar-Joseph, I. (2005). Measurement of the conductance of single conjugated molecules, *Nature*, **436**, 677–680.

47. Chae, D. H., Berry, J. F., Jung, S., Cotton, F. A., Murillo, C. A., and Yao, Z. (2006). Vibrational excitations in single trimetal-molecule transistors, *Nano Lett.*, **6**, 165–168.

48. Kubatkin, S., Danilov, A., Hjort, M., Cornil, J., Bredas, J. L., Stuhr-Hansen, N., Hedegard, P., and Bjornholm, T. (2003). Single-electron transistor of a single organic molecule with access to several redox states, *Nature*, **425**, 698–701.

49. Yu, L. H., Keane, Z. K., Ciszek, J. W., Cheng, L., Tour, J. M., Baruah, T., Pederson, M. R., and Natelson, D. (2005). Kondo resonances and anomalous gate dependence in the electrical conductivity of single-molecule transistors, *Phys. Rev. Lett.*, **95**, 256803.

50. de Picciotto, A., Klare, J. E., Nuckolls, C., Baldwin, K., Erbe, A., and Willett, R. (2005). Prevalence of coulomb blockade in electro-migrated junctions with conjugated and non-conjugated molecules, *Nanotechnology*, **16**, 3110–3114.

51. Lee, J. O., Lientschnig, G., Wiertz, F., Struijk, M., Janssen, R. A. J., Egberink, R., Reinhoudt, D. N., Hadley, P., and Dekker, C. (2003). Absence of strong gate effects in electrical measurements on phenylene-based conjugated molecules, *Nano Lett.*, **3**, 113–117.

52. Cui, X. D., Primak, A., Zarate, X., Tomfohr, J., Sankey, O. F., Moore, A. L., Moore, T. A., Gust, D., Harris, G., and Lindsay, S. M. (2001). Reproducible measurement of single-molecule conductivity, *Science*, **294**, 571–574.

53. Natelson, D. (2012). Mechanical break junctions: Enormous information in a nanoscale package, *ACS Nano*, **6**, 2871–2876.

54. Choi, S. H., Kim, B., and Frisbie, C. D. (2008). Electrical resistance of long conjugated molecular wires, *Science*, **320**, 1482–1486.
55. Engelkes, V. B., Beebe, J. M., and Frisbie, C. D. (2004). Length-dependent transport in molecular junctions based on sams of alkanethiols and alkanedithiols: Effect of metal work function and applied bias on tunneling efficiency and contact resistance, *J. Am. Chem. Soc.*, **126**, 14287–14296.
56. Holmlin, R. E., Haag, R., Chabinyc, M. L., Ismagilov, R. F., Cohen, A. E., Terfort, A., Rampi, M. A., and Whitesides, G. M. (2001). Electron transport through thin organic films in metal-insulator-metal junctions based on self-assembled monolayers, *J. Am. Chem. Soc.*, **123**, 5075–5085.
57. Kelley, S. O., Jackson, N. M., Hill, M. G., and Barton, J. K. (1999). Long-range electron transfer through DNA films, *Angew. Chem. Int. Ed.*, **38**, 941–945.
58. Adaligil, E., Shon, Y.-S., and Slowinski, K. (2009). Effect of headgroup on electrical conductivity of self-assembled monolayers on mercury: N-alkanethiols versus n-alkaneselenols, *Langmuir*, **26**, 1570–1573.
59. Chen, F., Li, X., Hihath, J., Huang, Z., and Tao, N. (2006). Effect of anchoring groups on single-molecule conductance: Comparative study of thiol-, amine-, and carboxylic-acid-terminated molecules, *J. Am. Chem. Soc.*, **128**, 15874–15881.
60. Haiss, W., Martin, S., Leary, E., van Zalinge, H., Higgins, S. J., Bouffier, L., and Nichols, R. J. (2009). Impact of junction formation method and surface roughness on single molecule conductance, *J. Phys. Chem. C*, **113**, 5823–5833.
61. Kamenetska, M., koentopp, M., Whalley, A. C., Park, Y. S., Steigerwald, M. L., Nuckolls, C., Hybertsen, M. S., and Venkataraman, L. (2009). Formation and evolution of single-molecule junctions, *Phys. Rev. Lett.*, **102**, 126803.
62. Park, Y. S., Whalley, A. C., Kamenetska, M., Steigerwald, M. L., Hybertsen, M. S., Nuckolls, C., and Venkataraman, L. (2007). Contact chemistry and single-molecule conductance: A comparison of phosphines, methyl sulfides, and amines, *J. Am. Chem. Soc.*, **129**, 15768–15769.
63. Venkataraman, L., Klare, J. E., Tam, I. W., Nuckolls, C., Hybertsen, M. S., and Steigerwald, M. L. (2006). Single-molecule circuits with well-defined molecular conductance, *Nano Lett.*, **6**, 458–462.
64. Beebe, J. M., Kim, B., Frisbie, C. D., and Kushmerick, J. G. (2008). Measuring relative barrier heights in molecular electronic junctions with transition voltage spectroscopy, *ACS Nano*, **2**, 827–832.

65. Kiguchi, M., Miura, S., Takahashi, T., Hara, K., Sawamura, M., and Murakoshi, K. (2008). Conductance of single 1,4-benzenediamine molecule bridging between Au and Pt electrodes, *J. Phys. Chem. C*, **112**, 13349–13352.

66. Kim, C. M., and Bechhoefer, J. (2013). Conductive probe AFM study of Pt-thiol and Au-thiol contacts in metal-molecule-metal systems, *J. Chem. Phys.*, **138**, 014707.

67. Guo, S., Hihath, J., Diez-Perez, I., and Tao, N. (2011). Measurement and statistical analysis of single-molecule current-voltage characteristics, transition voltage spectroscopy, and tunneling barrier height, *J. Am. Chem. Soc.*, **133**, 19189–19197.

68. Li, X., He, J., Hihath, J., Xu, B., Lindsay, S. M., and Tao, N. (2006) Conductance of single alkanedithiols: Conduction mechanism and effect of molecule–electrode contacts, *J. Am. Chem. Soc.*, **128**, 2135–2141.

69. Zhou, J., Chen, F., and Xu, B. (2009). Fabrication and electronic characterization of single molecular junction devices: A comprehensive approach, *J. Am. Chem. Soc.*, **131**, 10439–10446.

70. Hihath, J. (2014). Controling the Molecule-Electrode Contact in Single-Molecule Devices, 1st ed. Moth-Poulsen, K., ed., Singapore: Pan-Stanford Publishing, at press.

71. Joshua, H., and Nongjian, T. (2014). The role of molecule-electrode contact in single-molecule electronics, *Semicond. Sci. Technol.*, **29**, 054007.

72. Demir, F., and Kirczenow, G. (2012). Identification of the atomic scale structures of the gold-thiol interfaces of molecular nanowires by inelastic tunneling spectroscopy, *J. Chem. Phys.*, **136**, 014703.

73. Kaun, C.-C., and Seideman, T. (2008). Conductance, contacts, and interface states in single alkanedithiol molecular junctions, *Phys. Rev. B*, **77**, 033414.

74. Paulsson, M., Krag, C., Frederiksen, T., and Brandbyge, M. (2009). Conductance of alkanedithiol single-molecule junctions: A molecular dynamics study, *Nano Lett.*, **9**, 117–121.

75. Tachibana, M., Yoshizawa, K., Ogawa, A., Fujimoto, H., and Hoffmann, R. (2002). Sulfur-gold orbital interactions which determine the structure of alkanethiolate/Au(111) self-assembled monolayer systems, *J. Phys. Chem. B*, **106**, 12727–12736.

76. Makk, P., Tomaszewski, D., Martinek, J., Balogh, Z., Csonka, S., Wawrzyniak, M., Frei, M., Venkataraman, L., and Halbritter, A. (2012). Correlation analysis of atomic and single-molecule junction conductance, *ACS Nano*, **6**, 3411–3423.

77. Bruot, C., Hihath, J., and Tao, N. (2012). Mechanically controlled molecular orbital alignment in single molecule junctions, *Nat. Nanotech.*, **7**, 35–40.

78. Franco, I., George, C. B., Solomon, G. C., Schatz, G. C., and Ratner, M. A. (2011). Mechanically activated molecular switch through single-molecule pulling, *J. Am. Chem. Soc.*, **133**, 2242–2249.

79. Franco, I., Solomon, G. C., Schatz, G. C., and Ratner, M. A. (2011). Tunneling currents that increase with molecular elongation, *J. Am. Chem. Soc.*, **133**, 15714–15720.

80. Rubio-Bollinger, G., Bahn, S. R., Agrait, N., Jacobsen, K. W., and Vieira, S. (2001). Mechanical properties and formation mechanisms of a wire of single gold atoms, *Phys. Rev. Lett.*, **87**, 026101.

81. Xu, B. Q., Xiao, X. Y., and Tao, N. J. (2003). Measurements of single-molecule electromechanical properties, *J. Am. Chem. Soc.*, **125**, 16164–16165.

82. Xu, B. (2007). Modulating the conductance of a Au-octanedithiol-Au molecular junction, *Small*, **3**, 2061–2065.

83. Zhou, J., Chen, G., and Xu, B. (2010). Probing the molecule-electrode interface of single-molecule junctions by controllable mechanical modulations, *J. Phys. Chem. C*, **114**, 8587–8592.

84. Taniguchi, M., Tsutsui, M., Yokota, K., and Kawai, T. (2010). Mechanically-controllable single molecule switch based on configuration specific electrical conductivity of metal-molecule-metal junctions, *Chem. Sci.*, **1**, 247–253.

85. Fujihira, M., Suzuki, M., Fujii, S., and Nishikawa, A. (2006). Currents through single molecular junction of Au/hexanedithiolate/Au measured by repeated formation of break junction in STM under UHV: Effects of conformational change in an alkylene chain from gauche to trans and binding sites of thiolates on gold, *Phys. Chem. Chem. Phys.*, **8**, 3876–3884.

86. He, J., Sankey, O., Lee, M., Tao, N. J., Li, X. L., and Lindsay, S. (2006). Measuring single molecule conductance with break junctions, *Faraday Discuss.*, **131**, 145–154.

87. Venkataraman, L., Klare, J. E., Nuckolls, C., Hybertsen, M. S., and Steigerwald, M. L. (2006). Dependence of single-molecule junction conductance on molecular conformation, *Nature*, **442**, 904–907.

88. Quek, S. Y., Venkataraman, L., Choi, H. J., Louie, S. G., Hybertsen, M. S., and Neaton, J. B. (2007). Amine-gold linked single-molecule circuits: Experiment and theory, *Nano Lett.*, **7**, 3477–3482.

89. Chen, W., Widawsky, J. R., Vázquez, H., Schneebeli, S. T., Hybertsen, M. S., Breslow, R., and Venkataraman, L. (2011). Highly conducting π-conjugated molecular junctions covalently bonded to gold electrodes, *J. Am. Chem. Soc.*, **133**, 17160–17163.

90. Schneebeli, S. T., Kamenetska, M., Cheng, Z., Skouta, R., Friesner, R. A., Venkataraman, L., and Breslow, R. (2011). Single-molecule conductance through multiple π–π-stacked benzene rings determined with direct electrode-to-benzene ring connections, *J. Am. Chem. Soc.*, **133**, 2136–2139.

91. Woitellier, S., Launay, J. P., and Joachim, C. (1989). The possibility of molecular switching: Theoretical study of [(nh3)5ru-4,4′-bipy-ru(nh3)5]5+, *Chem. Phys.*, **131**, 481–488.

92. Chen, W., Li, H., Widawsky, J. R., Appayee, C., Venkataraman, L., and Breslow, R. (2014). Aromaticity decreases single-molecule junction conductance, *J. Am. Chem. Soc.*, **136**, 918–920.

93. Dell, E. J., Capozzi, B., DuBay, K. H., Berkelbach, T. C., Moreno, J. R., Reichman, D. R., Venkataraman, L., and Campos, L. M. (2013). Impact of molecular symmetry on single-molecule conductance, *J. Am. Chem. Soc.*, **135**, 11724–11727.

94. Bendikov, M., Wudl, F., and Perepichka, D. F. (2004). Tetrathiafulvalenes, oligoacenenes, and their buckminsterfullerene derivatives: The brick and mortar of organic electronics, *Chem. Rev.*, **104**, 4891–4946.

95. Genereux, J. C., and Barton, J. K. (2010). Mechanisms for DNA charge transport, *Chem. Rev.*, **110**, 1642–1662.

96. Wu, J., Pisula, W., and Müllen, K. (2007). Graphenes as potential material for electronics, *Chem. Rev.*, **107**, 718–747.

97. Kiguchi, M., Tal, O., Wohlthat, S., Pauly, F., Krieger, M., Djukic, D., Cuevas, J. C., and van Ruitenbeek, J. M. (2008). Highly conductive molecular junctions based on direct binding of benzene to platinum electrodes, *Phys. Rev. Lett.*, **101**, 046801.

98. Yamauchi, Y., Yoshizawa, M., Akita, M., and Fujita, M. (2009). Engineering double to quintuple stacks of a polarized aromatic in confined cavities, *J. Am. Chem. Soc.*, **132**, 960–966.

99. Kiguchi, M., Takahashi, T., Takahashi, Y., Yamauchi, Y., Murase, T., Fujita, M., Tada, T., and Watanabe, S. (2011). Electron transport through single molecules comprising aromatic stacks enclosed in self-assembled cages, *Angew. Chem. Int. Ed.*, **50**, 5707–5710.

100. Kiguchi, M., and Kaneko, S. (2013). Single molecule bridging between metal electrodes, *Phys. Chem. Chem. Phys.*, **15**, 2253–2267.

101. Martín, S., Grace, I., Bryce, M. R., Wang, C., Jitchati, R., Batsanov, A. S., Higgins, S. J., Lambert, C. J., and Nichols, R. J. (2010). Identifying diversity in nanoscale electrical break junctions, *J. Am. Chem. Soc.*, **132**, 9157–9164.

102. Pontes, R. B., Novaes, F. D., Fazzio, A., and da Silva, A. J. R. (2006). Adsorption of benzene-1,4-dithiol on the Au(111) surface and its possible role in molecular conductance, *J. Am. Chem. Soc.*, **128**, 8996–8997.

103. Solomon, G. C., Herrmann, C., Vura-Weis, J., Wasielewski, M. R., and Ratner, M. A. (2010). The chameleonic nature of electron transport through π-stacked systems, *J. Am. Chem. Soc.*, **132**, 7887–7889.

104. Wu, S., Gonzalez, M. T., Huber, R., Grunder, S., Mayor, M., Schoenenberger, C., and Calame, M. (2008). Molecular junctions based on aromatic coupling, *Nat. Nanotech.*, **3**, 569–574.

105. Kiguchi, M., Nakamura, H., Takahashi, Y., Takahashi, T., and Ohto, T. (2010). Effect of anchoring group position on formation and conductance of a single disubstituted benzene molecule bridging Au electrodes: Change of conductive molecular orbital and electron pathway, *J. Phys. Chem. C*, **114**, 22254–22261.

106. Ke, S.-H., Yang, W., and Baranger, H. U. (2008). Quantum-interference-controlled molecular electronics, *Nano Lett.*, **8**, 3257–3261.

107. Feldman, A. K., Steigerwald, M. L., Guo, X., and Nuckolls, C. (2008). Molecular electronic devices based on single-walled carbon nanotube electrodes, *Acc. Chem. Res.*, **41**, 1731–1741.

108. Martin, C. A., Ding, D., Sørensen, J. K., Bjørnholm, T., van Ruitenbeek, J. M., and van der Zant, H. S. J. (2008). Fullerene-based anchoring groups for molecular electronics, *J. Am. Chem. Soc.*, **130**, 13198–13199.

109. Moreno-García, P., Gulcur, M., Manrique, D. Z., Pope, T., Hong, W., Kaliginedi, V., Huang, C., Batsanov, A. S., Bryce, M. R., Lambert, C., and Wandlowski, T. (2013). Single-molecule conductance of functionalized oligoynes: Length dependence and junction evolution, *J. Am. Chem. Soc.*, **135**, 12228–12240.

110. Vazquez, H., Skouta, R., Schneebeli, S., Kamenetska, M., Breslow, R., Venkataraman, L., and Hybertsen, M. S. (2012). Probing the conductance superposition law in single-molecule circuits with parallel paths, *Nat. Nanotech.*, **7**, 663–667.

111. Aharonov, Y., and Bohm, D. (1959). Significance of electromagnetic potentials in the quantum theory, *Phys. Rev.*, **115**, 485–491.

112. Beenakker, C. W. J., and Vanhouten, H. (1991). Quantum transport in semiconductor nanostructures, *Solid State Phys.*, **44**, 1–228.

113. Webb, R. A., Washburn, S., Umbach, C. P., and Laibowitz, R. B. (1985). Observation of h/e aharonov–bohm oscillations in normal-metal rings, *Phys. Rev. Lett.*, **54**, 2696–2699.

114. Barnett, R. N., Cleveland, C. L., Joy, A., Landman, U., and Schuster, G. B. (2001). Charge migration in DNA: Ion-gated transport, *Science*, **294**, 567–571.

115. Doi, K., Tsutsui, M., Ohshiro, T., Chien, C.-C., Zwolak, M., Taniguchi, M., Kawai, T., Kawano, S., and Di Ventra, M. (2014). Nonequilibrium ionic response of biased mechanically controllable break junction (MCBJ) electrodes, *J. Phys. Chem. C*, **118**, 3758–3765.

116. Mukherjee, S., and Bhattacharyya, D. (2013). Influence of divalent magnesium ion on DNA: Molecular dynamics simulation studies, *J. Biomol. Struct. Dyn.*, **31**, 896–912.

117. Patel, D. J., Canuel, L. L., and Pohl, F. M. (1979). Alternating β-DNA conformation for the oligo(dg-dc) duplex in high-salt solution., *Proc. Natl. Acad. Sci. U.S.A.*, **76**, 2508–2511.

118. Xu, M. S., Endres, R. G., Tsukamoto, S., Kitamura, M., Ishida, S., and Arakawa, Y. (2005). Conformation and local environment dependent conductance of DNA molecules, *Small*, **1**, 1168–1172.

119. Tran, P., Alavi, B., and Gruner, G. (2000). Charge transport along the λ-DNA double helix, *Phys. Rev. Lett.*, **85**, 1564.

120. Hamori, E., and Jovin, T. M. (1987). The b-z conformational transition and aggregation of poly d(g-c) induced by moderate concentrations of $Mg(ClO_4)_2$, *Biophys. Chem.*, **26**, 375–383.

121. Jovin, T. M., Soumpasis, D. M., and McIntosh, L. P. (1987). The transition between β-DNA and z-DNA, *Ann. Rev. Phys. Chem.*, **38**, 521–560.

122. Wang, K., Hamill, J. M., Wang, B., Guo, C., Jiang, S., Huang, Z., and Xu, B. (2014). Structure determined charge transport in single DNA molecule break junctions, *Chem. Sci.*, **5**, 3425–3431.

123. Tsutsui, M., Taniguchi, M., and Kawai, T. (2008). Local heating in metal–molecule–metal junctions, *Nano Lett.*, **8**, 3293–3297.

124. Tsutsui, M., Taniguchi, M., and Kawai, T. (2009). Atomistic mechanics and formation mechanism of metal–molecule–metal junctions, *Nano Lett.*, **9**, 2433–2439.

125. Huang, Z., Xu, B., Ventra, M. D., and Tao, N. J. (2006). Measurement of current-induced local heating in a single molecule junction, *Nano Lett.*, **6**, 1240–1244.

126. Nijhuis, C. A., Reus, W. F., and Whitesides, G. M. (2010). Mechanism of rectification in tunneling junctions based on molecules with asymmetric potential drops, *J. Am. Chem. Soc.*, **132**, 18386–18401.

127. Wang, K., Zhou, J., Hamill, J. M., and Xu, B. (2014). Measurement and understanding of single-molecule break junction rectification caused by asymmetric contacts, *J. Chem. Phys.*, **141**, 054712.

128. Kornilovitch, P. E., Bratkovsky, A. M., and Williams, R. S. (2002). Current rectification by molecules with asymmetric tunneling barriers, *Phys. Rev. B*, **66**, 165436.

129. Liu, R., Ke, S. H., Yang, W. T., and Baranger, H. U. (2006). Organometallic molecular rectification, *J. Chem. Phys.*, **124**, 024718.

130. Yee, S. K., Sun, J., Darancet, P., Tilley, T. D., Majumdar, A., Neaton, J. B., and Segalman, R. A. (2011). Inverse rectification in donor-acceptor molecular heterojunctions, *ACS Nano*, **5**, 9256–9263.

131. Zhao, J., Yu, C., Wang, N., and Liu, H. (2010). Molecular rectification based on asymmetrical molecule-electrode contact, *J. Phys. Chem. C*, **114**, 4135–4141.

132. Armstrong, N., Hoft, R. C., McDonagh, A., Cortie, M. B., and Ford, M. J. (2007). Exploring the performance of molecular rectifiers: Limitations and factors affecting molecular rectification, *Nano Lett.*, **7**, 3018–3022.

133. Stadler, R., Geskin, V., and Cornil, J. (2008). A theoretical view of unimolecular rectification, *J. Phys. Cond. Matter*, **20**, 374105.

134. Buerkle, M., Viljas, J. K., Vonlanthen, D., Mishchenko, A., Schoen, G., Mayor, M., Wandlowski, T., and Pauly, F. (2012). Conduction mechanisms in biphenyl dithiol single-molecule junctions, *Phys. Rev. B*, **85**, 075417.

135. Migliore, A., and Nitzan, A. (2011). Nonlinear charge transport in redox molecular junctions: A marcus perspective, *ACS Nano*, **5**, 6669–6685.

136. Zhou, J., Samanta, S., Guo, C., Locklin, J., and Xu, B. (2013). Measurements of contact specific low-bias negative differential resistance of single metalorganic molecular junctions, *Nanoscale*, **5**, 5715–5719.

137. Galperin, M., Ratner, M. A., and Nitzan, A. (2004). Hysteresis, switching, and negative differential resistance in molecular junctions: A polaron model, *Nano Lett.*, **5**, 125–130.

138. Härtle, R., and Thoss, M. (2011). Resonant electron transport in single-molecule junctions: Vibrational excitation, rectification, negative differential resistance, and local cooling, *Phys. Rev. B*, **83**, p. 115414.

139. Thygesen, K. S. (2008). Impact of exchange-correlation effects on the iv characteristics of a molecular junction, *Phys. Rev. Lett.*, **100**, 166804.

140. Wang, Y., and Cheng, H.-P. (2012). Electronic and transport properties of azobenzene monolayer junctions as molecular switches, *Phys. Rev. B*, **86**, 035444.

141. Kaasbjerg, K., and Flensberg, K. (2011). Image charge effects in single-molecule junctions: Breaking of symmetries and negative-differential resistance in a benzene single-electron transistor, *Phys. Rev. B*, **84**, 115457.

142. Pati, R., McClain, M., and Bandyopadhyay, A. (2008). Origin of negative differential resistance in a strongly coupled single molecule-metal junction device, *Phys. Rev. Lett.*, **100**, 246801.

143. Grobis, M., Wachowiak, A., Yamachika, R., and Crommie, M. F. (2005). Tuning negative differential resistance in a molecular film, *Appl. Phys. Lett.*, **86**, 204102.

144. Dubi, Y. (2013). Dynamical coupling and negative differential resistance from interactions across the molecule-electrode interface in molecular junctions, *J. Chem. Phys.*, **139**, 154710.

145. Hamill, J. M., Wang, K., and Xu, B. (2014). Force and conductance molecular break junctions with time series crosscorrelation, *Nanoscale*, **6**, 5657–5661.

146. Briechle, B. M., Kim, Y., Ehrenreich, P., Erbe, A., Sysoiev, D., Huhn, T., Groth, U., and Scheer, E. (2012). Current-voltage characteristics of single-molecule diarylethene junctions measured with adjustable gold electrodes in solution, *Beilstein J. Nanotech.*, **3**, 798–808.

147. Büttiker, M., Imry, Y., Landauer, R., and Pinhas, S. (1985). Generalized many-channel conductance formula with application to small rings, *Phys. Rev. B*, **31**, 6207.

148. Akkerman, H. B., Naber, R. C. G., Jongbloed, B., van Hal, P. A., Blom, P. W. M., de Leeuw, D. M., and de Boer, B. (2007). Electron tunneling through alkanedithiol self-assembled monolayers in large-area molecular junctions, *Proc. Nat. Acad. Sci. U.S.A.*, **104**, 11161–11166.

149. Simmons, J. G. (1963). Generalized formula for the electric tunnel effect between similar electrodes separated by a thin insulating film, *J. Appl. Phys.*, **34**, 1793.

150. Cui, B., Xu, Y., Ji, G., Wang, H., Zhao, W., Zhai, Y., Li, D., and Liu, D. (2014). A single-molecule diode with significant rectification and negative differential resistance behavior, *Org. Electron.*, **15**, 484–490.

151. Beebe, J. M., Kim, B., Gadzuk, J. W., Frisbie, C. D., and Kushmerick, J. G. (2006). Transition from direct tunneling to field emission in metal-molecule-metal junctions, *Phys. Rev. Lett.*, **97**, 026801.

152. Capozzi, B., Chen, Q., Darancet, P., Kotiuga, M., Buzzeo, M., Neaton, J. B., Nuckolls, C., and Venkataraman, L. (2014). Tunable charge transport in single-molecule junctions via electrolytic gating, *Nano Lett.*, **14**, 1400–1404.

153. Aradhya, S. V., and Venkataraman, L. (2013). Single-molecule junctions beyond electronic transport, *Nat. Nanotech.*, **8**, 399–410.

154. Parks, J. J., Champagne, A. R., Hutchison, G. R., Flores-Torres, S., Abruña, H. D., and Ralph, D. C. (2007). Tuning the kondo effect with a mechanically controllable break junction, *Phys. Rev. Lett.*, **99**, 026601.

Chapter 5

Vibronic Effects in Electron Transport through Single-Molecule Junctions

Rainer Härtle[a] **and Michael Thoss**[b]

[a]*Institute for Theoretical Physics,*
Georg-August-Universität Göttingen,
Friedrich-Hund-Platz 1, 37077 Göttingen, Germany
[b]*Institute for Theoretical Physics and Interdisciplinary Center for Molecular Materials, Friedrich-Alexander-Universität Erlangen-Nürnberg,*
Staudtstr. 7/B2, 91058 Erlangen, Germany

Rainer.Haertle@theorie.physik.uni-goettingen.de, michael.thoss@physik.uni-erlangen.de

5.1 Introduction

The study and understanding of quantum transport processes in single-molecule junctions, i.e., molecules chemically bound to metal or semiconductor electrodes, has been of great interest recently (Reed et al., 1997; Nitzan, 2001; Cuniberti et al., 2005; Cuevas and Scheer, 2010). These systems combine the possibility to study fundamental aspects of non-equilibrium many-body quantum physics at the nanoscale with the perspective for technological applications in nanoelectronic devices. Recent experimental studies of transport in molecular junctions have revealed a wealth of interesting transport phenomena such as Coulomb blockade,

Molecular Electronics: An Experimental and Theoretical Approach
Edited by Ioan Bâldea

ISBN 978-981-4613-90-3 (Hardcover), 978-981-4613-91-0 (eBook)
www.panstanford.com

Kondo effects, negative differential resistance, transistor- or diode-like behavior, as well as switching and hysteresis. (Gaudioso et al., 2000; Liang et al., 2002; Park et al., 2002; Elbing et al., 2005; Blum et al., 2005; Lörtscher et al., 2006; Choi et al., 2006; Osorio et al., 2010; van der Molen and Liljeroth, 2010).

An important aspect that distinguishes nanoscale molecular conductors from mesoscopic semiconductor devices is the influence of the nuclear degrees of freedom of the molecular bridge on transport properties (Galperin et al., 2007b; Härtle and Thoss, 2011a; Härtle et al., 2013c). Due to the small size and mass of molecules, the charging of the molecular bridge is often accompanied by significant changes of the nuclear geometry, indicating strong electronic-vibrational (vibronic) coupling. This coupling manifests itself in vibronic structures in the transport characteristics and may result in a multitude of non-equilibrium phenomena such as current-induced local heating and cooling, multistability, switching and hysteresis, as well as decoherence (Galperin et al., 2005, 2007b; de Leon et al., 2008; Ioffe et al., 2008; Ward et al., 2008; Härtle et al., 2009; Hüttel et al., 2009; Repp et al., 2010; Hihath et al., 2010; Ballmann et al., 2010; Osorio et al., 2010; Arroyo et al., 2010; Secker et al., 2011; Härtle and Thoss, 2011a,b; Kim et al., 2011; Ward et al., 2011; Ballmann et al., 2012; Albrecht et al., 2012; Wilner et al., 2013; Ballmann et al., 2013; Härtle et al., 2013a,c). Vibrational signatures indicating strong vibronic coupling as well as strong excitation of vibrational modes were observed in experiments for a number of molecular junctions (Wu et al., 2004; LeRoy et al., 2004; Yu et al., 2004; Pasupathy et al., 2005; Sapmaz et al., 2006; Thijssen et al., 2006; Parks et al., 2007; Böhler et al., 2007; de Leon et al., 2008; Hüttel et al., 2009; Hihath et al., 2010; Ballmann et al., 2010; Jewell et al., 2010; Osorio et al., 2010). Novel experimental techniques based on measuring the force needed to break a junction (Huang et al., 2006) or employing Raman spectroscopy (Ward et al., 2008; Ioffe et al., 2008) allow, furthermore, a characterization of the current-induced vibrational non-equilibrium state of a single-molecule junction and, thus, complement the information carried by the respective current–voltage characteristics.

Various theoretical approaches have been employed to describe vibrationally coupled electron transport through single molecules (for an earlier overview see Galperin et al. (2007b)).

While scattering theory approaches (Cizek et al., 2004; Toroker and Peskin, 2007; Zimbovskaya and Kuklja, 2009; Jorn and Seidemann, 2009) can be used to address the regime of strong molecule–lead coupling, non-equilibrium Green's function (NEGF) approaches (Flensberg, 2003; Mitra et al., 2004; Galperin et al., 2006c; Ryndyk et al., 2006; Frederiksen et al., 2007a; Tahir and MacKinnon, 2008; Härtle et al., 2008; Bergfield and Stafford, 2009; Härtle et al., 2009) additionally allow a non-perturbative description of the associated non-equilibrium state of such a junction, especially with respect to the vibrational degrees of freedom. Numerically exact methods, based on path integrals (Mühlbacher and Rabani, 2008; Weiss et al., 2008; Hützen et al., 2012; Simine and Segal, 2013), multi-configurational wave-function methods (Wang and Thoss, 2009), the scattering state numerical renormalization group approach (Jovchev and Anders, 2013), or a combination of reduced density matrix techniques and impurity solvers (Cohen and Rabani, 2011; Wilner et al., 2013) provide valuable insights and benchmarks for model systems and problems that may not be addressed by perturbation theory or other approximative schemes. Master equation approaches (May, 2002; Mitra et al., 2004; Lehmann et al., 2004; Pedersen and Wacker, 2005; Harbola et al., 2006; Zazunov et al., 2006; Siddiqui et al., 2007; Timm, 2008; May and Kühn, 2008a,b; Härtle et al., 2009; Leijnse and Wegewijs, 2008; Esposito and Galperin, 2009, 2010; Härtle et al., 2010; Härtle and Thoss, 2011a), although perturbative with respect to the coupling between the molecule and the leads, have been proven to be very efficient and useful.

In this chapter, we give an overview on fundamental mechanisms of vibrationally coupled electron transport and vibronic effects in molecular junctions. This includes effects in the resonant and the off-resonant transport regimes. We discuss specifically non-equilibrium effects related to current-induced vibrational excitation, effects due to multiple electronic states and the role of electron–hole pair creation processes. We also give a brief overview on the role of these effects for technological applications of single-molecule junctions. The results discussed in this chapter have been obtained using a NEGF method (Galperin et al., 2006c; Härtle et al., 2008, 2009; Volkovich et al., 2011), which is reviewed in some detail. A survey of other theoretical methods is also provided.

5.2 Theoretical Description of Vibrationally Coupled Electron Transport

The theoretical description of electron transport in a molecular junction requires the combination of an electronic structure method with transport theory. While for purely electronic transport, i.e., neglecting electronic-vibrational coupling, a well-established first principles methodology exists, which combines density-functional theory (DFT) (or higher level electronic structure methods) with NEGF theory in a self-consistent scheme (Brandbyge et al., 2002; Evers et al., 2004; Ke et al., 2004; Cuevas and Scheer, 2010), the description of vibrationally coupled electron transport is in many cases based on model Hamiltonians with parameters, which are determined by first principles electronic structure calculations. In this chapter, we review the methodology we have used in this context in the recent years.

5.2.1 Model Hamiltonian

To obtain a suitable model Hamiltonian for vibrationally coupled electron transport in a molecular junction, we follow the scheme outlined in Cederbaum and Domcke (1974, 1976), Kondov et al. (2007), Benesch et al. (2006, 2008, 2009), and Benesch (2009). In general, the Hamiltonian of the (isolated) molecular bridge, H_{M}, can be written as

$$H_{\mathrm{M}} = H_{\mathrm{nuc}} + H_{\mathrm{el}}, \tag{5.1}$$

$$H_{\mathrm{nuc}} = -\sum_{\mathrm{a}} \frac{1}{2M_{\mathrm{a}}} \Delta_{\mathrm{a}} + V_{\mathrm{nuc}}(\mathbf{R}), \tag{5.2}$$

$$V_{\mathrm{nuc}}(\mathbf{R}) = \sum_{\mathrm{a<b}} \frac{Z_{\mathrm{a}} Z_{\mathrm{b}} e^2}{|\mathbf{R}_{\mathrm{a}} - \mathbf{R}_{\mathrm{b}}|}, \tag{5.3}$$

$$H_{\mathrm{el}}(\mathbf{R}) = -\sum_{i} \frac{1}{2m_{\mathrm{e}}} \Delta_i + \sum_{i<j} \frac{e^2}{|\mathbf{r}_i - \mathbf{r}_j|} - \sum_{\mathrm{a}i} \frac{Z_{\mathrm{a}} e^2}{|\mathbf{R}_{\mathrm{a}} - \mathbf{r}_i|}, \tag{5.4}$$

where H_{nuc} describes the nuclear and $H_{\mathrm{el}}(\mathbf{R})$ the electronic degrees of freedom of the molecule[1]. The nuclear part of the Hamiltonian,

[1]In this work, we give energy values in units of electron-volts (eV), charges in Coulomb (C), charge electrical currents in Ampere (A) and voltages in Volt (V). In all the equations and formulas we set $\hbar = 1$ for notational convenience. The charge of an electron is denoted by $-e = -1.6022 \times 10^{-19}$ C.

H_{nuc}, includes the kinetic energy of the nuclei and the Coulomb repulsion between them, $V_{nuc}(\mathbf{R})$. Similarly, $H_{el}(\mathbf{R})$ represents the kinetic energy of the electrons, the Coulomb repulsion between them, and the Coulomb attraction between the electrons and the nuclei. Thereby, the vector $\mathbf{R}$ summarizes the position vectors $\mathbf{R}_a$ of the nuclei and $\mathbf{r}_i$ denotes the position vector of the i-th electron. The charge and the mass of the nuclei are given by eZ_a and M_a, respectively, while the mass of an electron is denoted by m_e.

Properties of a molecule, such as for example its vibrational level structure, the frequency of the vibrational modes etc., are crucially dependent on the specific electronic state of the molecule. A natural choice to determine these properties, in particular when ab initio calculations are employed (Xue and Ratner, 2003a,b; Evers et al., 2004; Benesch et al., 2006; Frederiksen et al., 2007a,b; Troisi and Ratner, 2006; Gagliardi et al., 2007; Wheeler and Dahnovsky, 2008; Benesch et al., 2008, 2009; Chen et al., 2009; Monturet et al., 2010; Cuevas and Scheer, 2010), is the electronic ground state of the uncharged molecule, which fulfills

$$H_{el}(\mathbf{R})\,\Psi_{ref}(\mathbf{r};\mathbf{R}) = E_{ref,el}(\mathbf{R})\,\Psi_{ref}(\mathbf{r};\mathbf{R}). \tag{5.5}$$

Here, the semicolon denotes a parametric dependence of $\Psi_{ref}(\mathbf{r};\mathbf{R})$ on the nuclear coordinates $\mathbf{R}$ and $\mathbf{r}$ subsumes the position vectors of the electrons $\mathbf{r}_i$. Employing an effective single particle description (such as, e.g., Hartree–Fock theory or DFT), the electronic ground state is given by a single Slater determinant,

$$|\Psi_{ref}(\mathbf{R})\rangle = \prod_{m\in\{\text{occ.}\}} c_m^\dagger(\mathbf{R})|0\rangle. \tag{5.6}$$

This determinant involves single-particle states or molecular orbitals $\psi_m(\mathbf{r}_j;\mathbf{R})$, which are occupied ($m \in \{\text{occ.}\}$) in the reference state $\Psi_{ref}(\mathbf{r};\mathbf{R})$. Using the creation operators $c_m^\dagger(\mathbf{R})$ and the corresponding destruction operators $c_m(\mathbf{R})$ for these single-particle states, the electronic part of the Hamiltonian can be rewritten in the occupation number representation as follows:

$$\begin{aligned} H_{el}(\mathbf{R}) = E_{ref,el}(\mathbf{R}) &+ \sum_m \varepsilon_m(\mathbf{R})(c_m^\dagger(\mathbf{R})c_m(\mathbf{R}) - \delta_m) \\ &+ \sum_{m<n} U_{mn}(\mathbf{R})(c_m^\dagger(\mathbf{R})c_m(\mathbf{R}) - \delta_m)(c_n^\dagger(\mathbf{R})c_n(\mathbf{R}) - \delta_n), \end{aligned} \tag{5.7}$$

where the parameters δ_m are defined as

$$\delta_m = \begin{cases} 1, m \in \{\text{occ.}\} \\ 0, m \notin \{\text{occ.}\} \end{cases} \tag{5.8}$$

and $\varepsilon_m(\mathbf{R})$ denotes orbital energies. Here, we simplify the electron–electron interaction terms by taking $U_{mn}(\mathbf{R}) = U_{mnmn}(\mathbf{R}) - U_{mnnm}(\mathbf{R})$, using the original matrix elements (Cederbaum and Domcke, 1974, 1976)

$$U_{mnop}(\mathbf{R}) \approx U_{mnop}(\mathbf{R})(\delta_{mo}\delta_{np} - \delta_{mp}\delta_{no}). \tag{5.9}$$

Thus, we account for an additional charging energy $U_{mn}(\text{R})$, if the molecular bridge is doubly charged, or different electron–electron interaction strengths if the molecular bridge is in a neutral excited state. A direct coupling of the single-particle states due to electron–electron interactions is not included in the Hamiltonian (5.7) (vide infra).

Next, we characterize the nuclear or vibrational degrees of freedom of the molecular bridge. Within the Born–Oppenheimer approximation, the motion of the nuclei is determined by the adiabatic potential energy surface (PES):

$$V_{\text{ad}}(\mathbf{R}) = V_{\text{nuc}}(\mathbf{R}) + E_{\text{ref,el}}(\mathbf{R}). \tag{5.10}$$

Employing a harmonic expansion of the PES around the equilibrium geometry of the nuclei, $\mathbf{R}_{\text{eq}}$, we obtain

$$H_{\text{nuc}} + E_{\text{ref,el}}(\mathbf{R}) \approx \sum_\alpha \Omega_\alpha \left(a_\alpha^\dagger a_\alpha + \frac{1}{2} \right) + E_{\text{ref,el}}(\mathbf{R}_{\text{eq}}). \tag{5.11}$$

Thereby, we have introduced the normal modes α of the junction with frequencies Ω_α and corresponding ladder operators $a_\alpha^\dagger / a_\alpha$.

In line with the harmonic approximation for the PES, we expand the orbital and correlation energies, $\varepsilon_m(\mathbf{R})$ and $U_{mn}(\mathbf{R})$, to first order in the vibrational coordinates $Q_\alpha = a_\alpha^\dagger + a_\alpha$ and obtain

$$\varepsilon_m(\mathbf{R}) \approx \varepsilon_m(\mathbf{R}_{\text{eq}}) + \sum_\alpha Q_\alpha \frac{\partial \varepsilon_m(\mathbf{R}_{\text{eq}})}{\partial Q_\alpha} \equiv \varepsilon_m + \sum_\alpha Q_\alpha \lambda_{m\alpha}, \tag{5.12}$$

$$U_{mn}(\mathbf{R}) \approx U_{mn}(\mathbf{R}_{\text{eq}}) + \sum_{\alpha} Q_{\alpha} \frac{\partial U_{mn}(\mathbf{R}_{\text{eq}})}{\partial Q_{\alpha}} \equiv U_{mn} + \sum_{\alpha} Q_{\alpha} W_{mn\alpha}, \quad (5.13)$$

where $\lambda_{m\alpha}$ and $W_{mn\alpha}$ denote coupling strengths between the vibrational and the electronic degrees of freedom of the molecular bridge. Moreover, we drop the parametric dependence of the single-particle operators $c_m(\mathbf{R})$ on the nuclear coordinates and define $c_m(\mathbf{R}_{\text{eq}}) \equiv c_m$. Neglecting this dependence, we discard, in particular, off-diagonal non-adiabatic coupling terms, $\sim c_m^\dagger c_n$ and $\sim c_m^\dagger c_n c_o^\dagger c_p$ with $m \neq n, p$. These terms are relevant if level crossings become significant (Repp et al., 2010) (for example in the vicinity of conical intersections (Domcke et al., 2004). Dropping an irrelevant constant, we thus summarize the expansion of $H_{\text{nuc}} + H_{\text{el}}(\mathbf{R})$ in Q_α as

$$\begin{aligned} H_{\text{M}} = & \sum_{\alpha} \Omega_{\alpha} a_{\alpha}^{\dagger} a_{\alpha} + \sum_{m} \varepsilon_m (c_m^{\dagger} c_m - \delta_m) \\ & + \sum_{m<n} U_{mn} (c_m^{\dagger} c_m - \delta_m)(c_n^{\dagger} c_n - \delta_n) \\ & + \sum_{m\alpha} \lambda_{m\alpha} Q_{\alpha} (c_m^{\dagger} c_m - \delta_m) \\ & + \sum_{m<n,\alpha} W_{mn\alpha} Q_{\alpha} (c_m^{\dagger} c_m - \delta_m)(c_n^{\dagger} c_n - \delta_n). \end{aligned} \quad (5.14)$$

Note that, due to the parameters δ_m, there is no electronic-vibrational coupling in the reference state $\Psi_{\text{ref}}(\mathbf{r}; \mathbf{R})$. In the following, we do not consider vibrationally dependent electron–electron interactions ($W_{mn\alpha} = 0$).

Next, we consider the modeling of the electrodes. They provide macroscopic reservoirs of electrons, which can be represented by a continuum of non-interacting electronic quasiparticle states. The Hamiltonian of the left (L)/right (R) lead can thus be written as

$$H_{\text{L/R}} = \sum_{k \in \text{L/R}} \varepsilon_k c_k^{\dagger} c_k \quad (5.15)$$

with energies ε_k and corresponding creation/annihilation operators $c_k^\dagger/c_k$. Since the electrodes are macroscopic, they are not affected by the coupling to the molecule. The density operator describing the state of the left/right lead is therefore given by

$$\rho_{\mathrm{L/R}} = \frac{1}{\mathrm{Tr}\{e^{-(H_{\mathrm{L/R}}-\mu_{\mathrm{L/R}}N_{\mathrm{L/R}})/(k_{\mathrm{B}}T)}\}} e^{-(H_{\mathrm{L/R}}-\mu_{\mathrm{L/R}}N_{\mathrm{L/R}})/(k_{\mathrm{B}}T)} \tag{5.16}$$

with $N_{\mathrm{L/R}} = \sum_{k\in \mathrm{L/R}} c_k^\dagger c_k$, T the temperature in the leads and k_{B} the Boltzmann constant. At zero bias, $e\Phi = \mu_{\mathrm{L}} - \mu_{\mathrm{R}} = 0$, the chemical potentials in the left lead μ_{L} and the right lead μ_{R} are at the same level as the Fermi energy of the system $\varepsilon_{\mathrm{F}} = 0$, while at $\Phi \neq 0$ they are different. To model this difference, we use the concept of a voltage division factor η. Thereby, according to the discussion given in Refs. Datta et al., 1997; Mujica et al. (2002), we assume the potential bias to drop symmetrically at the contacts, that is we use $\mu_{\mathrm{L/R}} = \eta e\Phi = e\Phi/2$ and $\mu_{\mathrm{L/R}} = -(1 - \eta)e\Phi = -e\Phi/2$, even for an inherently asymmetric molecular junction. Note that a more detailed description of the potential profile in a molecular junction can be obtained for models, where the specific (contact) geometry is known. In that case, the electrostatic potential profile can be obtained employing the Poisson equation (Xue and Ratner, 2003a; Elbing et al., 2005; Cuniberti et al., 2005). Another strategy to describe the potential drop in a molecular junction is based on an equivalent circuit model of a molecular junction (Hanna and Tinkham, 1991; Ingold and Nazarov, 1992; Hanke et al., 1995; Galperin et al., 2006a).

The coupling between the molecular bridge and the leads can be expressed in terms of interaction matrix elements V_{mk} by

$$H_{\mathrm{ML/MR}} = \sum_{k\in \mathrm{L/R},\, m\in \mathrm{M}} V_{mk} c_k^\dagger c_m + V_{mk}^* c_m^\dagger c_k. \tag{5.17}$$

Due to the coupling to the electrodes, electrons acquire a finite lifetime on the molecule, or in other words, the corresponding electronic states become broadened. This broadening can be quantified by the level-width functions

$$\Gamma_{\mathrm{L/R},mn}(\varepsilon) = 2\pi \sum_{k\in \mathrm{L/R}} V_{mk}^* V_{nk} \delta(\varepsilon - \varepsilon_k). \tag{5.18}$$

The Hamiltonian of the overall molecular junction, H, is thus given by

$$H = H_{\mathrm{M}} + H_{\mathrm{L}} + H_{\mathrm{R}} + H_{\mathrm{ML}} + H_{\mathrm{MR}}. \tag{5.19}$$

The parameters of this Hamiltonian can be determined, for example, employing first-principles methods (cf. Refs. Xue and Ratner, 2003a,b; Evers et al., 2004; Benesch et al., 2006; Frederiksen et al., 2007a,b; Troisi and Ratner, 2006; Gagliardi et al., 2007; Wheeler and Dahnovsky, 2008; Benesch et al., 2008, 2009; Chen et al., 2009; Monturet et al., 2010), or fitted to experimental data (Böhler et al., 2004; Elbing et al., 2005; Sapmaz et al., 2005; Tao, 2006; Keane et al., 2006; Osorio et al., 2010; Martin et al., 2010; Ballmann et al., 2010; Choi et al., 2008; Huisman et al., 2009; Trouwborst et al., 2011; Secker et al., 2011; Pump et al., 2008; Lafferentz et al., 2009; Brumme et al., 2011). In this chapter, we will consider parametrized models to discuss the fundamental transport mechanisms.

5.2.2 Survey of Theories for Vibrationally Coupled Charge Transport

During the past decade, a number of methods have been developed to solve the transport problem that is posed by the Hamiltonian (5.19). This includes approximate as well as (numerically) exact methods. In general, exact methods include all relevant physical mechanisms. They are, however, also very demanding such that often only a limited range of parameters can be considered. On the other hand, approximate methods may miss essential physical phenomena, but are typically efficient enough to scan systematically through large sectors of the phase space of the problem. A comparison of methods often allows us to reveal the physical mechanism at work (Galperin et al., 2007b; Eckel et al., 2010; Volkovich et al., 2011; Wang et al., 2011; Härtle et al., 2013b).

Numerically exact approaches employed to study vibrationally coupled electron transport include numerical path-integral approaches (Mühlbacher and Rabani, 2008; Weiss et al., 2008; Hützen et al., 2012; Simine and Segal, 2013), the multilayer multiconfiguration time-dependent Hartree (ML-MCTDH) method (Wang and Thoss, 2009; Wang et al., 2011), the scattering state numerical renormalization group approach (Jovchev and Anders, 2013), and a combination of reduced density matrix techniques and impurity solvers (Wilner et al., 2013, 2014).

Approximate methods used are, for example, based on scattering theory (Nitzan, 2001; Cizek et al., 2004, 2005; Toroker and Peskin, 2007; Zimbovskaya and Kuklja, 2009; Jorn and Seidemann, 2009; Bedkihal et al., 2013), where the non-equilibrium transport problem is treated on the same footing as an electron–molecule scattering experiment. Thereby, the molecular bridge is described as a scattering region, where an electron incoming from one of the leads scatters elastically or inelastically on its way to the other lead. These methods are designed to describe the scattering process on a single-electron level, which means that the electrical current is described by these methods as a sum of independent electron scattering events. Thus, changes of the state of the molecular bridge due to inelastic processes are not taken into account. Scattering theory is therefore very useful, if these state changes do not influence the next electron tunneling through the junction, for example, due to fast relaxation processes. In addition, scattering theory yields the exact result for coherent transport through a non-interacting molecular bridge.

Vibrationally coupled electron transport through single molecules has been successfully described employing density matrix theory (May, 2002; Mitra et al., 2004; Lehmann et al., 2004; Koch et al., 2006a; Siddiqui et al., 2007; Timm, 2008; Leijnse and Wegewijs, 2008; Esposito and Galperin, 2009) often based on master equations for the density matrix of the molecular bridge. Many master equation approaches are invoke the Born–Markov approximation (Mitra et al., 2004; May and Kühn, 2008a,b; Schultz and von Oppen, 2009; Härtle et al., 2010; Härtle and Thoss, 2011a; Pshenichnyuk and Cizek, 2011; Volkovich and Peskin, 2011; Härtle and Thoss, 2011b; Volkovich et al., 2011). It employs a second-order expansion of the exact equation of motion of the molecular density matrix, which, in general, is referred to as the Nakajima–Zwanzig equation (Nakajima, 1958; Zwanzig, 1960), in the coupling of the molecular bridge to the leads. Other approaches to solve the Nakajima–Zwanzig equation are based on higher-order expansions (May, 2002; Pedersen and Wacker, 2005; Koch et al., 2006b; Leijnse and Wegewijs, 2008), full-counting statistics (Koch et al., 2005; Flindt et al., 2008) or other advanced approximation schemes (Schultz and von Oppen, 2009; Esposito and Galperin, 2009). Besides the stationary transport regime of a molecular junction, time-dependent phenomena, such as transient currents, the effect of AC voltages or optical phenomena, are also considered employing master equation methodologies (Lehmann et al., 2004; May and Kühn, 2008a,b; Volkovich et al., 2008; Volkovich and Peskin, 2011).

Alternatively, non-equilibrium Green's function theory (NEGF) (Haug and Jauho, 1996; Cuevas and Scheer, 2010) can be used to describe transport through a single-molecule junction. These approaches are based on perturbation theory (Mitra et al., 2004; Galperin et al., 2004; Frederiksen et al., 2004; Ryndyk et al., 2006; Frederiksen et al., 2007a; Entin-Wohlman et al., 2010), advanced equation of motion techniques (Galperin et al., 2007a), projection operator techniques (Flensberg, 2003; Braig and Flensberg, 2003; Galperin et al., 2008), full-counting statistics (Avriller and Levy Yeyati, 2009; Schmidt and Komnik, 2009; Haupt et al., 2009), or other non-perturbative schemes (Kral, 1997; Galperin et al., 2006c; Härtle et al., 2008, 2009; Koch et al., 2010). While most of these methods address the stationary state of a molecular junction, there are also a number of approaches that describe transient behavior, optical phenomena and/or the effect of AC voltages (Galperin and Tretiak, 2008; Tahir and MacKinnon, 2008; Wilner et al., 2014). A comparison of results obtained from NEGF to other methods can be found, for example, in Refs. Härtle et al., 2009; Eckel et al., 2010; Härtle et al., 2010; Volkovich et al. (2011).

Other approaches, where the electronic and the vibrational degrees of freedom are treated on different theoretical levels, have also been developed. For example, approaches based on Langevin equations (Koch et al., 2006a; Shen et al., 2007; Dzhioev and Kosov, 2011), Navier–Stokes equations (D'Agosta et al., 2006) or molecular dynamics simulations (McEniry et al., 2007, 2008, 2009) indeed account for the quantum mechanical nature of the electronic degrees of freedom but consider vibrational motion on a classical level. These approaches often address the adiabatic regime of vibrationally coupled electron transport, where $\Omega << \Gamma$ (cf. the discussion given in the following Eq. (5.26)). Another strategy towards a classical treatment of nuclear dynamics in molecular transport junctions is based on mapping the fermionic electronic degrees of freedom to bosonic degrees of freedom, which have a clear classical analog, and thus allows a treatment of all degrees of freedom on the same (classical) footing (Swenson et al., 2012; Li et al., 2013, 2014).

5.2.3 A Non-Equilibrium Green's Function Approach

All results presented in this work have been obtained by the NEGF approach, which is outlined in the following. This method, which, in contrast to many other NEGF schemes, is non-perturbative with respect to electronic-vibrational coupling, was originally proposed by Galperin et al. (Galperin et al., (2006c)). It is based on a factorization of electronic and vibrational time scales in a molecular junction and employs self-consistent second-order perturbation theory with respect to the molecule–lead coupling. We extended this method to account for multiple vibrational modes and multiple electronic states (Härtle et al., 2008, 2009, 2010, 2011), including a non-perturbative description of electron–electron interactions in terms of the elastic co-tunneling approximation (Groshev et al., 1991; Averin and Nazarov, 1990; Haug and Jauho, 1996; Bergfield and Stafford, 2009; Härtle et al., 2009).

A necessary precursor of the method is a prediagonalization of the Hamiltonian (5.19) with respect to the electronic and vibrational subspaces. To this end, the small polaron transformation is employed (Lang and Firsov, 1963; Mahan, 1981; König et al., 1996; Galperin et al., 2006c), where the Hamiltonian is transformed by a canonical transformation,

$$\begin{aligned}\overline{H} &= \mathrm{e}^{S} H \mathrm{e}^{-S} \\ &= \sum_{m} \overline{\varepsilon}_m (c_m^\dagger c_m - \delta_m) + \sum_{m<n} \overline{U}_{mn} (c_m^\dagger c_m - \delta_m)(c_n^\dagger c_n - \delta_n) \\ &\quad + \sum_{k} \varepsilon_k c_k^\dagger c_k + \sum_{k \in \mathrm{L,R}, m \in \mathrm{M}} (V_{mk} c_k^\dagger c_m X_m + \mathrm{H.c.}) + \sum_{\alpha} \Omega_\alpha a_\alpha^\dagger a_\alpha, \end{aligned} \tag{5.20}$$

using the transformation operator

$$S = -i \sum_{m\alpha} \frac{\lambda_{m\alpha}}{\Omega_\alpha} P_\alpha (c_m^\dagger c_m - \delta_m) \tag{5.21}$$

and $P_\alpha = -i(a_\alpha - a_\alpha^\dagger)$. Since $S^\dagger = -S$, it represents a unitary transformation. The transformed Hamiltonian $\overline{H}$ includes polaron-shifted molecular orbital energies

$$\overline{\varepsilon}_m = \varepsilon_m + (2\delta_m - 1) \sum_{\alpha} \frac{\lambda_{m\alpha}^2}{\Omega_\alpha}, \tag{5.22}$$

vibrationally induced electron–electron interactions

$$\bar{U}_{mn} = U_{mn} - 2\sum_{\alpha} \frac{\lambda_{m\alpha}\lambda_{n\alpha}}{\Omega_{\alpha}} \tag{5.23}$$

and a molecule–lead coupling term, which is renormalized by the shift operators $X_m = \exp\left(i\sum_{\alpha}\lambda_{m\alpha}P_{\alpha}\right)$.

The central objects of NEGF theory are the single-particle Green's function, which are defined by

$$G_{mn}(t,t') = -i\langle \mathcal{T} c_m(t) c_n^{\dagger}(t')\rangle_H, \tag{5.24}$$

where $\mathcal{T}$ denotes time-ordering on the Keldysh contour and the subscript H denotes the Hamiltonian that is used to evaluate the respective expectation value. These functions transform under the small polaron transformation as follows:

$$\begin{aligned} G_{mn}(t,t') &= -i\langle \Psi_0 | \mathcal{T} e^{-S} e^{S} c_m(t) e^{-S} e^{S} c_n^{\dagger}(t') e^{-S} e^{S} | \Psi_0 \rangle_H \\ &= -i\langle \mathcal{T} c_m(t) X_m(t) c_n^{\dagger}(t') X_n(t')\rangle_{\bar{H}}. \end{aligned} \tag{5.25}$$

If the dynamics of the nuclear and the electronic degrees of freedom decouple from each other, one can approximate the single-particle Green's function by the product (Galperin et al., 2006c; Härtle et al., 2008, 2009, 2011)

$$\begin{aligned} G_{mn}(t,t') &\approx -i\langle \mathcal{T} c_m(t) c_n^{\dagger}(t')\rangle_{\bar{H}} \langle \mathcal{T} X_m(t) X_n^{\dagger}(t')\rangle_{\bar{H}} \\ &\equiv G_{\mathrm{el},mn}(t,t') \langle \mathcal{T} X_m(t) X_n^{\dagger}(t')\rangle_{\bar{H}}. \end{aligned} \tag{5.26}$$

Thereby, $G_{\mathrm{el},mn}(t, t')$ denotes the "electronic" part of the single-particle Green's function $G_{mn}(t, t')$,

$$G_{\mathrm{el},mn}(t,t') = -i\langle \mathcal{T} c_m(t) c_n^{\dagger}(t')\rangle_{\bar{H}}. \tag{5.27}$$

The "vibrational" part is given by the correlation function of the shift operators

$$\langle \mathcal{T} X_m(t) X_n^{\dagger}(t')\rangle_{\bar{H}} = \left\langle \mathcal{T} \exp\left(i\sum_{\alpha}\lambda_{m\alpha}P_{\alpha}(t)\right)\exp\left(-i\sum_{\alpha'}\lambda_{n\alpha'}P_{\alpha'}(t')\right)\right\rangle_{\bar{H}}. \tag{5.28}$$

Note that in the limit $V_{mk} \to 0$, this factorization of the single-particle Green's function into an electronic and a vibrational part is exact (Mahan, 1981). For weak coupling to the leads, when electron tunneling events occur on much longer time scales than the nuclei need to adjust to the respective charge fluctuations, this factorization is also justified. This weak-coupling regime defines the anti-adiabatic regime of a molecular junction. In the opposite limit of strong coupling to the leads, that is the adiabatic regime, electron tunneling events occur on much shorter time scales such that the nuclei have no time to respond to single charge fluctuations. They move rather in a background potential induced by the continuous flow of electrons. In this regime, the above factorization of the single-particle Green's function corresponds to the adiabatic approximation. For intermediate coupling strengths to the leads, where the residence time of the electrons on the molecular bridge is comparable to the oscillation period of a vibrational mode, Eq. (5.26) ceases to be valid. In the following, we consider the anti-adiabatic regime, assuming weak coupling between the molecule and the leads.

We are left with the task to calculate the electronic Green's functions $G_{\mathrm{el},mn}(t,\ t')$ and the correlation function of the shift operators $\langle \mathcal{T} X_m(t) X_n^\dagger(t') \rangle_{\bar{H}}$. We will outline in the following how we obtain these functions. Thereby, we start with the electronic Green's functions, which we calculate employing the equation of motion technique (Haug and Jauho, 1996). The evaluation of the correlation functions $\langle \mathcal{T} X_m(t) X_n^\dagger(t') \rangle_{\bar{H}}$ will be given afterwards.

5.2.3.1 Electronic Green's functions

The equation of motion of the Green's functions $G_{\mathrm{el},mn}(t,\ t')$ is determined by the time dependence of the time-ordering operator $\mathcal{T}$ and the Heisenberg equation of motion of the annihilation operator c_m (or, equivalently, $c_m^\dagger$)

$$i\partial_t c_m = [c_m, \bar{H}] = \bar{\varepsilon}_m c_m + \sum_{o \neq m} \bar{U}_{mo} c_m (c_o^\dagger c_o - \delta_o) + \sum_k V_{mk}^* c_k X_{.} \qquad (5.29)$$

Due to the small polaron transformation, this EOM involves the transformed Hamiltonian $\bar{H}$. The derivative of $G_{\mathrm{el},mn}(t,\ t')$ with respect to time t is therefore given by

$$\partial_t G_{\mathrm{el},mn}(t,t') = -i\delta(t,t') \langle [c_m, c_n^\dagger]_+ \rangle_{\bar{H}} - \langle \mathcal{T}[c_m(t), \bar{H}] c_n^\dagger(t') \rangle_{\bar{H}}, \quad (5.30)$$

where the $\delta(t, t')$-term originates from the derivative of the time-ordering operator. Employing the inverse operator $(G^0_{\text{el},m})^{-1} = i\partial_t - \overline{\varepsilon}_m + \sum_{o \neq m} \overline{U}_{mo}\delta_o$, and dropping the indices $\overline{H}$ in the following, we rewrite this equation of motion as

$$
\begin{aligned}
(G^0_{\text{el},m})^{-1} G_{\text{el},mn}(t,t') = {} & \delta(t,t')\delta_{mn} \\
& - i\sum_k V^*_{mk} \left\langle \mathcal{T} c_k(t) X^\dagger_m(t) c^\dagger_n(t') \right\rangle \\
& - i\sum_{o \neq m} \overline{U}_{mo} \left\langle \mathcal{T} c_m(t) c^\dagger_o(t) c_o(t) c^\dagger_n(t') \right\rangle.
\end{aligned} \tag{5.31}
$$

Similarly, we apply the operator $(G'^0_{\text{el},n})^{-1} = -i\partial_{t'} - \overline{\varepsilon}_n + \sum_{p \neq n} \overline{U}_{np}\delta_p$ from the right hand side of Eq. (5.31) and obtain

$$
\begin{aligned}
& (G^0_{\text{el},m})^{-1} G_{\text{el},mn}(t,t')(G'^0_{\text{el},n})^{-1} \\
& \quad = \delta(t,t')\delta_{mn}(G'^0_{\text{el},n})^{-1} \\
& \quad - i \sum_{o \neq m, p \neq m} \overline{U}_{mo}\overline{U}_{np} \left\langle \mathcal{T} c_m(t) c^\dagger_o(t) c_o(t) c^\dagger_n(t') c^\dagger_p(t') c_p(t') \right\rangle \\
& \quad + \delta(t,t') \sum_{o \neq m} \left(\delta_{no}\overline{U}_{mo} \left\langle c_m c^\dagger_n \right\rangle + \delta_{mn}\overline{U}_{mo} \left\langle c^\dagger_o c_o \right\rangle \right) \\
& \quad - i\sum_{kk'} V^*_{mk} V_{nk'} \left\langle \mathcal{T} c_k(t) X^\dagger_m(t) c^\dagger_{k'}(t') X_n(t') \right\rangle \\
& \quad - i\sum_{k, p \neq n} V^*_{mk} \overline{U}_{np} \left\langle \mathcal{T} c_k(t) X^\dagger_m(t) c^\dagger_n(t') c^\dagger_p(t') c_p(t') \right\rangle \\
& \quad - i\sum_{k, o \neq m} \overline{U}_{mo} V_{nk} \left\langle \mathcal{T} c_m(t) c^\dagger_o(t) c_o(t) c^\dagger_k(t') X_n(t') \right\rangle,
\end{aligned} \tag{5.32}
$$

where the derivative $\partial_{t'}$ acts to the left. Due to the electron–electron interaction term in $\overline{H}_{\text{M}}$, this equation of motion includes a hierarchy of different correlation terms. In the next two paragraphs, we outline how we treat these correlations in an approximate but non-perturbative way, that is, the elastic co-tunneling approximation (Groshev et al., 1991; Averin and Nazarov, 1990; Haug and Jauho, 1996; Bergfield and Stafford, 2009; Härtle et al., 2009).

To this end, we disregard electron–electron interactions for the time being and continue to evaluate the EOM (5.32) up to second order in the molecule–lead coupling ($\mathcal{O}(V^2_{mk})$):

$$(G^0_{\mathrm{el},m})^{-1}G_{\mathrm{el},mn}(t,t')(G'^0_{\mathrm{el},n})^{-1} = \delta(t,t')\delta_{mn}(G'^0_{\mathrm{el},n})^{-1}$$
$$+\sum_k V^*_{mk}V_{nk}g_k(t,t')\langle TX^\dagger_m(t)X_n(t')\rangle, \quad (5.33)$$

where $g_k(t, t')$ denotes the free Green's function of lead state k. This defines an approximate expression for the electronic self-energy $\Sigma_{mn}(t, t')$,

$$\Sigma_{mn}(t,t') = \Sigma_{\mathrm{L},mn}(t,t') + \Sigma_{\mathrm{R},mn}(t,t'), \quad (5.34)$$

$$\Sigma_{\mathrm{L/R},mn}(t,t') = \sum_{k\in\mathrm{L/R}} V^*_{mk}V_{nk}g_k(t,t')\langle TX^\dagger_m(t)X_n(t')\rangle \quad (5.35)$$

which describes the coupling of the electronic states located on the molecular bridge to the continuum of electronic states in the left and the right lead. Using this self-energy expression, we integrate the equation of motion (5.33) in time, and obtain

$$G_{\mathrm{el},mn}(t,t') = G^0_{\mathrm{el},mn}(t,t')$$
$$+\sum_{op}\int_{-\infty}^{\infty}\mathrm{d}t_1\int_{-\infty}^{\infty}\mathrm{d}t_2 G^0_{\mathrm{el},mo}(t,t_1)\Sigma_{op}(t_1,t_2)G^0_{\mathrm{el},pn}(t_2,t'), \quad (5.36)$$

where $G^0_{\mathrm{el},mo}(t,t_1)$ denotes the single-particle Green's function of the isolated molecular bridge. This integral equation for $G_{\mathrm{el},mn}(t, t')$ is valid to $\mathcal{O}(V^2_{mk})$. Higher-order terms are easily generated by replacing one of the $G^0_{\mathrm{el},mn}(t,t')$ in the kernel of the time integral by $G_{\mathrm{el},mn}(t, t')$, for example

$$G_{\mathrm{el},mn}(t,t') = G^0_{\mathrm{el},mn}(t,t')$$
$$+\sum_{op}\int_{-\infty}^{\infty}\mathrm{d}t_1\int_{-\infty}^{\infty}\mathrm{d}t_2 G^0_{\mathrm{el},mo}(t,t_1)\Sigma_{op}(t_1,t_2)G_{\mathrm{el},pn}(t_2,t'). \quad (5.37)$$

The lesser, greater and retarded projection of the corresponding contour-ordered self energy $\Sigma^C_{mn}(t, t')$ determine the lesser/greater Green's function via the Keldysh equation (Haug and Jauho, 1996),

$$G^{</>}_{\mathrm{el},mn}(\varepsilon) = \sum_{op} G^{\mathrm{r}}_{\mathrm{el},mo}(\varepsilon)\Sigma^{</>}_{op}(\varepsilon)G^{\mathrm{a}}_{\mathrm{el},pn}(\varepsilon), \quad (5.38)$$

and the retarded/advanced Green's function via the Dyson equation (Haug and Jauho, 1996),

$$G^{r/a}_{el,mn}(\varepsilon)=G^{0,r/a}_{el,mn}(\varepsilon)+\sum_{op}G^{0,r/a}_{el,mo}(\varepsilon)\Sigma^{r/a}_{op}(\varepsilon)G^{r/a}_{el,pn}(\varepsilon). \tag{5.39}$$

Since the tunnel couplings, H_{ML} and H_{MR}, are bilinear in $c_k^\dagger$ and c_m, this replacement yields the exact result in the non-interacting limit, where neither electron–electron interactions (U_{mn} = 0) nor electronic-vibrational coupling ($\lambda_{m\alpha}$ = 0) is present.

At this point, we include electron–electron interactions $\overline{U}_{mn}$ again. To this end, we employ the exact result for the retarded Green's function $G^{0,r}_{el,mn}$ of the isolated molecular bridge (V_{mk} = 0). In order to derive this expression, we consider first a unit-cube in N_{el}-dimensions, where N_{el} corresponds to the number of the electronic states located at the molecular bridge. One corner of this cube coincides with the origin, while all other corners are located in the positive coordinate space. There are $2^{N_{el}}$ vectors, $\mathbf{p}_\alpha^{N_{el}}$, pointing to the corners of this unit-cube. All coefficients of these vectors are either 0 or 1. The exact Green's function, $G^{0,r}_{el,mn}$, can be written in terms of these vectors as

$$\begin{aligned}G^{0,r}_{el,mn}(t,t')=-i\Theta(t-t')\delta_{mn}\sum_{\alpha=1\ldots2^{N_{el}}}\mathrm{e}^{-i(\overline{\varepsilon}_m-\sum_{o\neq m}\overline{U}_{mo}(\mathbf{p}_\alpha^{N_{el}})_o)(t-t')}\\ \cdot\prod_n(1-n_n)^{1-(\mathbf{p}_\alpha^{N_{el}})_n}n_n^{(\mathbf{p}_\alpha^{N_{el}})_n},\end{aligned} \tag{5.40}$$

$$G^{0,r}_{el,mn}(\varepsilon)=\delta_{mn}\sum_{\alpha=1\ldots2^{N_{el}}}\frac{\prod_n(1-n_n)^{1-(\mathbf{p}_\alpha^{N_{el}})_n}n_n^{(\mathbf{p}_\alpha^{N_{el}})_n}}{\varepsilon-\overline{\varepsilon}_m-\sum_{o\neq m}\overline{U}_{mo}(\mathbf{p}_\alpha^{N_{el}})_o+i0^+}, \tag{5.41}$$

where $i0^+$ represents a positive imaginary infinitesimal and the n_m denote the populations of the electronic states m. This expression can easily be verified evaluating Eq. (5.32) in the limit $V_{mk}\to 0$. For a single-molecule contact ($V_{mk}\neq 0$), the populations n_m are determined with respect to the stationary state of the junction. The retarded Green's function $G^{r}_{el,mn}$ is obtained according to the Dyson equation (5.39), using the exact expression for the retarded zeroth-order Green's function (5.41) and the retarded projection of the self-energy $\Sigma_{mn}(t,t')$ (cf. Eqs. (5.34) and (5.35)). Using that $G^{a}_{el,mn}=(G^{r}_{el,mn})^*$ and the lesser projection of $\Sigma_{mn}(t,t')$, the corresponding lesser and greater Green's functions $G^{</>}_{el,mn}$ are determined according to the Keldysh equation (5.38).

This approximate approach treats electron–electron interactions non-perturbatively but on a mean-field level. It is referred to as the elastic co-tunneling approximation (Groshev et al., 1991; Averin and Nazarov, 1990; Haug and Jauho, 1996; Bergfield and Stafford, 2009; Härtle et al., 2009). For non-degenerate electronic states, that is, if the off-diagonal elements of the self-energy Σ_{mn} can be disregarded, this approach corresponds effectively to the Hartree–Fock approximation (as outlined, for example, in Haug and Jauho (1996)). The mean-field character of the approach needs to be considered with care. For example, Kondo physics cannot be accounted for (Haug and Jauho, 1996). Moreover, the approach can fail to describe transport through quasidegenerate electronic states, for which the off-diagonal elements of the self-energy are needed. This is particularly important in the context of quantum interference effects (cf. appendix A of Härtle (2012)). Apart from these deficiencies, however, the elastic co-tunneling approximation allows a rather accurate description of electron–electron interactions. This becomes evident by comparison with the master equation approach (see Härtle et al., 2009; Volkovich et al., 2011; Härtle, 2012), where to the given order in the molecule–lead coupling, $\mathcal{O}(V_{mk}^2)$, electron–electron interactions are exactly accounted for.

5.2.3.2 Vibrational Green's functions

We now turn to the calculation of the correlation functions of the shift operators $\langle TX_m(t)X_n^\dagger(t')\rangle_{\bar{H}}$. To this end, we employ a cumulant expansion to second order in the electronic-vibrational coupling strengths $\lambda_{m\alpha}/\Omega_\alpha$ (Galperin et al., 2006c; Härtle et al., 2008, 2009, 2010), which allows us to express these functions in terms of the vibrational single-particle Green's function $D_{\alpha\alpha'}(t, t') = -i\langle TP_\alpha(t)P_{\alpha'}(t')\rangle$. Thereby, the cumulant $\Phi_{mn}(t, t')$ of the shift-operator correlation function, $\langle TX_m(t)X_n^\dagger(t')\rangle_{\bar{H}} \equiv e^{\Phi_{mn}(t,t')}$, is expanded in a Taylor series

$$
\begin{aligned}
\Phi_{mn}(t,t') \approx 1 &+ \sum_\alpha \left(\frac{\lambda_{m\alpha}}{\Omega_\alpha} \phi_{mn,\alpha}^{1,m}(t,t') + \frac{\lambda_{n\alpha}}{\Omega_\alpha} \phi_{mn,\alpha}^{1,n}(t,t') \right) \\
&+ \sum_{\alpha\alpha'} \frac{\lambda_{m\alpha}\lambda_{m\alpha'}}{2\Omega_\alpha\Omega_{\alpha'}} \phi_{mn,\alpha\alpha'}^{2,m}(t,t') + \sum_{\alpha\alpha'} \frac{\lambda_{n\alpha}\lambda_{n\alpha'}}{2\Omega_\alpha\Omega_{\alpha'}} \phi_{mn,\alpha\alpha'}^{2,n}(t,t') \\
&+ \sum_{\alpha\alpha'} \frac{\lambda_{m\alpha}\lambda_{n\alpha'}}{2\Omega_\alpha\Omega_{\alpha'}} \phi_{mn,\alpha\alpha'}^{2,mn}(t,t'),
\end{aligned}
\tag{5.42}
$$

which is terminated at $\mathcal{O}(\lambda_{m\alpha}^2)$ and involves the derivatives $\phi_{mn,\alpha}^{1,m}(t,t') = \partial_{\lambda_{m\alpha}} \Phi_{mn}(t,t')$, $\phi_{mn,\alpha\alpha'}^{2,m}(t,t') = \partial_{\lambda_{m\alpha}} \partial_{\lambda_{m\alpha'}} \Phi_{mn}(t,t')$ and $\phi_{mn,\alpha\alpha'}^{2,mn}(t,t') = \partial_{\lambda_{m\alpha}} \partial_{\lambda_{n\alpha'}} \Phi_{mn}(t,t')$. We determine these derivatives using an expansion of the shift-operator correlation function to second order in the electronic-vibrational coupling strengths

$$\langle \mathcal{T} X_m(t) X_n^\dagger(t') \rangle_{\bar{H}} = \langle \mathcal{T} \mathrm{e}^{i\sum_\alpha \frac{\lambda_{m\alpha}}{\Omega_\alpha} P_\alpha(t)} \mathrm{e}^{-i\sum_\alpha \frac{\lambda_{n\alpha}}{\Omega_\alpha} P_\alpha(t)} \rangle_{\bar{H}} \tag{5.43}$$

$$\begin{aligned} &\approx 1 + \sum_{\alpha\alpha'} \frac{\lambda_{m\alpha}\lambda_{n\alpha'}}{\Omega_\alpha \Omega_{\alpha'}} \langle \mathcal{T} P_\alpha(t) P_{\alpha'}(t') \rangle \\ &- \sum_{\alpha\alpha'} \frac{\lambda_{m\alpha}\lambda_{m\alpha'}}{2\Omega_\alpha \Omega_{\alpha'}} \langle \mathcal{T} P_\alpha(t) P_{\alpha'}(t') \rangle \\ &- \sum_{\alpha\alpha'} \frac{\lambda_{n\alpha}\lambda_{n\alpha'}}{2\Omega_\alpha \Omega_{\alpha'}} \langle \mathcal{T} P_\alpha(t) P_{\alpha'}(t') \rangle. \end{aligned} \tag{5.44}$$

Thereby, we have used that $\langle P_\alpha(t) \rangle_{\bar{H}} = \langle P_\alpha(t') \rangle_{\bar{H}}$ and $\langle P_\alpha(t) P_{\alpha'}(t) \rangle_{\bar{H}} = \langle P_\alpha(t') P_{\alpha'}(t') \rangle_{\bar{H}}$ in the steady-state transport regime. We require that the expression $e^{\Phi_{mn}(t,t')}$ has the same second order expansion with respect to the dimensionless electronic-vibrational coupling strengths $\lambda_{m\alpha}/\Omega_\alpha$ and, thus, obtain

$$\langle \mathcal{T} X_m(t) X_n^\dagger(t') \rangle_{\bar{H}} \approx \mathrm{e}^{i\sum_{\alpha\alpha'} \frac{\lambda_{m\alpha}\lambda_{n\alpha'}}{\Omega_\alpha \Omega_{\alpha'}} D_{\alpha\alpha'}(t,t') - i\sum_{\alpha\alpha'} \frac{\lambda_{m\alpha}\lambda_{m\alpha'} + \lambda_{n\alpha}\lambda_{n\alpha'}}{2\Omega_\alpha \Omega_{\alpha'}} D_{\alpha\alpha'}(t,t')}. \tag{5.45}$$

Note that this expansion yields the exact result in the limit of an isolated molecule, $V_{mk} \to 0$ (Mahan, 1981).

To obtain the Green's functions $D_{\alpha\alpha'}$, we employ the Heisenberg equation of motion of the vibrational momentum operator

$$i\partial_t P_\alpha = -i\Omega_\alpha Q_\alpha. \tag{5.46}$$

In contrast to the electronic Green's function, however, the inverse of the non-interacting vibrational Green's function,

$$(D_\alpha^0)^{-1} = -\frac{1}{2\Omega_\alpha}(\partial_t^2 + \Omega_\alpha^2), \tag{5.47}$$

involves two derivatives with respect to the time t. Therefore, in order to derive a Dyson-like equation of motion for $D_{\alpha\alpha'}$, as for $G_{\mathrm{el},mn}$ (cf. Eq. (5.32)), we use the differential operator $(D_\alpha^0)^{-1}$ in the following and obtain

$$
\begin{aligned}
(D_\alpha^0)^{-1} D_{\alpha\alpha'}(t,t') = \delta_{\alpha\alpha'}\delta(t,t') \\
&+ \sum_{mk} \frac{\lambda_{m\alpha}}{\Omega_\alpha} V_{mk} \left\langle \mathcal{T} c_k^\dagger(t) c_m(t) X_m(t) P_{\alpha'}(t') \right\rangle_{\bar{H}} \\
&- \sum_{mk} \frac{\lambda_{m\alpha}}{\Omega_\alpha} V_{mk}^* \left\langle \mathcal{T} c_m^\dagger(t) c_k(t) X_m^\dagger(t) P_{\alpha'}(t') \right\rangle_{\bar{H}},
\end{aligned}
\tag{5.48}
$$

using the Heisenberg equation of motion

$$
i\partial_t Q_\alpha = i\Omega_\alpha P_\alpha - 2\sum_{mk} \frac{\lambda_{m\alpha}}{\Omega_\alpha} V_{mk} c_k^\dagger c_m X_m + 2\sum_{mk} \frac{\lambda_{m\alpha}}{\Omega_\alpha} V_{mk}^* c_m^\dagger c_k X_m^\dagger \tag{5.49}
$$

for the vibrational displacement operators Q_α. Applying $(D'^0_{\alpha'})^{-1} = -(2\Omega_{\alpha'})^{-1}(\partial_{t'}^2 - \Omega_{\alpha'}^2)$ from the right, where the derivative $\partial_{t'}^2$, acts to the left, we obtain the following equation of motion for the vibrational Green's function

$$
\begin{aligned}
(D_\alpha^0)^{-1} D_{\alpha\alpha'}(t,t')(D'^0_{\alpha'})^{-1} = \delta_{\alpha\alpha'}\delta(t,t')(D'^0_{\alpha'})^{-1} \\
&+ i\sum_{mnkk'} \frac{\lambda_{m\alpha}\lambda_{n\alpha'}}{\Omega_\alpha\Omega_{\alpha'}} V_{mk} V_{nk'} \left\langle \mathcal{T} c_k^\dagger(t) c_m(t) X_m(t) c_{k'}^\dagger(t') c_n(t') X_n(t') \right\rangle \\
&- i\sum_{mnkk'} \frac{\lambda_{m\alpha}\lambda_{n\alpha'}}{\Omega_\alpha\Omega_{\alpha'}} V_{mk} V_{nk'}^* \left\langle \mathcal{T} c_k^\dagger(t) c_m(t) X_m(t) c_n^\dagger(t') c_{k'}(t') X_n^\dagger(t') \right\rangle \\
&- i\sum_{mnkk'} \frac{\lambda_{m\alpha}\lambda_{n\alpha'}}{\Omega_\alpha\Omega_{\alpha'}} V_{mk}^* V_{nk'} \left\langle \mathcal{T} c_m^\dagger(t) c_k(t) X_m^\dagger(t) c_{k'}^\dagger(t') c_n(t') X_n(t') \right\rangle \\
&+ i\sum_{mnkk'} \frac{\lambda_{m\alpha}\lambda_{n\alpha'}}{\Omega_\alpha\Omega_{\alpha'}} V_{mk}^* V_{nk'}^* \left\langle \mathcal{T} c_m^\dagger(t) c_k(t) X_m^\dagger(t) c_n^\dagger(t') c_{k'}(t') X_n^\dagger(t') \right\rangle.
\end{aligned}
\tag{5.50}
$$

We further evaluate the correlation functions on the right hand side of this equation to second order in the molecule–lead coupling ($\mathcal{O}(V_{mk}^2)$). We thus arrive at the following approximate expression for the vibrational self-energy $\Pi_{\alpha\alpha'}(t, t')$ (similar to Eqs. (5.32) and (5.33)):

$$
(D_\alpha^0)^{-1} D_{\alpha\alpha'}(t,t')(D'^0_{\alpha'})^{-1} \approx \delta_{\alpha\alpha'}\delta(t,t')(D'^0_{\alpha'})^{-1} + \Pi_{\alpha\alpha'}(t,t'), \tag{5.51}
$$

$$
\begin{aligned}
\Pi_{\alpha\alpha'}(t, t') = -i\sum_{mn} \frac{\lambda_{m\alpha}\lambda_{n\alpha'}}{\Omega_\alpha\Omega_{\alpha'}} \big(\Sigma_{nm}(t', t) G_{\text{el},mn}(t, t') \\
+ \Sigma_{mn}(t, t') G_{\text{el},nm}(t', t) \big).
\end{aligned}
\tag{5.52}
$$

As for the electronic Green's function, we use the retarded projection of the self-energies, $\Pi^{\mathrm{r}}_{\alpha\alpha'}(\varepsilon)$, to compute the retarded projection of the vibrational Green's function, $D^{\mathrm{r}}_{\alpha\alpha'}(\varepsilon)$ according to the Dyson equation:

$$D^{\mathrm{r}}_{\alpha\alpha'}(\varepsilon) = D^{0,\mathrm{r}}_{\alpha\alpha'}(\varepsilon) + \sum_{\alpha_1\alpha_2} D^{0,\mathrm{r}}_{\alpha\alpha_1}(\varepsilon)\Pi^{\mathrm{r}}_{\alpha_1\alpha_2}(\varepsilon)D^{\mathrm{r}}_{\alpha_2\alpha'}(\varepsilon). \tag{5.53}$$

Accordingly, the Keldysh equations

$$D^{</>}_{\alpha\alpha'}(\varepsilon) = \sum_{\alpha_1\alpha_2} D^{\mathrm{r}}_{\alpha\alpha_1}(\varepsilon)\Pi^{</>}_{\alpha_1\alpha_2}(\varepsilon)D^{\mathrm{a}}_{\alpha_2\alpha'}(\varepsilon) \tag{5.54}$$

give the lesser and the greater projections of the vibrational Green's function. Similar equations apply in the time domain. Having determined the vibrational Green's functions, we can readily compute the correlation function of the shift operators according to Eq. (5.45). To elucidate the role of (current-induced) non-equilibrium effects, it is sometimes useful to describe the vibrational degrees of freedom as being in thermal equilibrium. This can be achieved in the framework of this NEGF method by using $\Pi_{\mathrm{el},\alpha\alpha'}(t, t') = 0$.

5.2.3.3 Self-consistent solution scheme

The vibrational self-energy $\Pi_{\alpha\alpha'}(t, t')$ (Eq. (5.52)) depends on both the electronic self-energy $\Sigma_{mn}(t, t')$ (Eqs. (5.34) and (5.35)) and the electronic Green's function $G_{\mathrm{el},mn}(t, t')$ (Eq. (5.38)). The corresponding vibrational Green's function $D_{\alpha\alpha'}(t, t')$ enters the correlation function of the shift operators X_m and $X^{\dagger}_n$ (Eq. (5.45)). This correlation function, in turn, is required to determine the electronic self-energy $\Sigma_{mn}(t, t')$ and the electronic Green's function $G_{\mathrm{el},mn}(t, t')$ (cf. Eqs. (5.38) and (5.39)). It is therefore not possible to determine these Green's functions and self-energies straightforwardly. The solution of this closed set of equations requires a self-consistent solution scheme, which is outlined in the following.

The starting point is the electronic self-energy $\Sigma^{(0)}_{mn}$, which represents the self-energy of a non-interacting molecular bridge. In the stationary state of a molecular junction, the non-interacting electronic self-energy is given by

$$\Sigma^{(0)}_{\mathrm{L/R},mn}(t) = \sum_{k\in \mathrm{L/R}} V^{*}_{mk} V_{nk} g_k(t), \tag{5.55}$$

where g_k is the free Green's function associated with lead state k. It has the Fourier transformed real-time projections

$$\begin{aligned}\Sigma^{(0),\mathrm{r}}_{\mathrm{L/R},mn}(\varepsilon) &= \sum_{k\in \mathrm{L/R}} V^{*}_{mk} V_{nk} g^{\mathrm{r}}_k(\varepsilon) \\ &\equiv \Delta_{\mathrm{L/R},mn}(\varepsilon) - \frac{i}{2}\Gamma_{\mathrm{L/R},mn}(\varepsilon),\end{aligned} \tag{5.56}$$

$$\Sigma^{(0),<}_{\mathrm{L/R},mn}(\varepsilon) = i f_{\mathrm{L/R}}(\varepsilon)\Gamma_{\mathrm{L/R},mn}(\varepsilon), \tag{5.57}$$

$$\Sigma^{(0),>}_{\mathrm{L/R},mn}(\varepsilon) = -i(1 - f_{\mathrm{L/R}}(\varepsilon))\Gamma_{\mathrm{L/R},mn}(\varepsilon). \tag{5.58}$$

Accordingly, a first estimate for the population of the electronic states can be given by

$$n_m = \frac{\Gamma_{\mathrm{L},mm}(\varepsilon_m) f_{\mathrm{L}}(\varepsilon_m) + \Gamma_{\mathrm{R},mm}(\varepsilon_m) f_{\mathrm{R}}(\varepsilon_m)}{\Gamma_{\mathrm{L},mm}(\varepsilon_m) + \Gamma_{\mathrm{R},mm}(\varepsilon_m)}. \tag{5.59}$$

Next, we compute the Fourier transforms of the real-time projections of the non-interacting vibrational Green's function $D^{(0),\mathrm{r}}(\varepsilon)$ and $D^{(0),</>}_{\alpha\alpha'}(\varepsilon)$.

At this point, we employ a self-consistent cycle, including the following steps:

(1) We compute the lesser and greater projection of the electronic self-energy in the time domain

$$\Sigma^{</>}_{mn}(t) = \Sigma^{(0),</>}_{mn}(t)\, \mathrm{e}^{i\sum_{\alpha\alpha'}\frac{\lambda_{n\alpha}\lambda_{m\alpha'}}{\Omega_\alpha\Omega_{\alpha'}} D^{>/<}_{\alpha\alpha'}(-t) - i\sum_{\alpha\alpha'}\frac{\lambda_{m\alpha}\lambda_{m\alpha'} + \lambda_{n\alpha}\lambda_{n\alpha'}}{2\Omega_\alpha\Omega_{\alpha'}} D^{>/<}_{\alpha\alpha'}(0)}, \tag{5.60}$$

using Eq. (5.45) and the Langreth rules (Langreth, 1976; Haug and Jauho, 1996; Bonitz, 1998). If we perform this step for the first time, we use $D^{(0),</>}_{\alpha\alpha'}(t)$ instead of $D^{</>}_{\alpha\alpha'}(t)$. The Fourier transformed lesser and greater projections of the electronic self-energy are then used to calculate the corresponding retarded and advanced projections.

(2) The electronic populations n_m are used to determine the retarded Green's function $G^{0,\mathrm{r}}_{\mathrm{el},mn}(E)$ according to Eq. (5.41). This result, in conjunction with the self-energies determined

in the preceding step, are used to compute the retarded projection of the electronic Green's function, $G^{\mathrm{r}}_{\mathrm{el},mn}(E)$, according to the Dyson equation (5.39). The respective Keldysh equation (5.38) gives $G^{</>}_{\mathrm{el},mn}(E)$.

(3) With the Fourier transforms of the lesser and the greater projections of the electronic Green's, $G^{</>}_{\mathrm{el},mn}(E) \xrightarrow{\mathrm{FT}} G^{</>}_{\mathrm{el},mn}(t)$, and the self-energy functions $\Sigma^{</>}_{mn}(E) \xrightarrow{\mathrm{FT}} \Sigma^{</>}_{mn}(t)$, we determine the vibrational self-energy $\Pi_{\alpha\alpha'}(t)$ from Eqs. (5.52).

(4) We update the real-time projections of the vibrational Green's function, $D_{\alpha\alpha'}(E)$, with the Fourier transforms of the real time projections of the self-energies $\Pi_{\alpha\alpha'}(t)$ that we obtained in step (3), employing the Dyson- and the Keldysh equations (5.53) and (5.54), respectively. The Fourier transforms of the thus updated vibrational Green's functions $D^{</>}_{\alpha\alpha'}(t)$ is used in step (1) of the next cycle.

(5) In the last step, we recalculate the electronic populations $n_m = \mathrm{Im}[G^{<}_{\mathrm{el},mm}(t=0)]$ from the lesser Green's functions $G^{<}_{\mathrm{el},mn}(t)$ that are obtained in step (3). We terminate the self-consistent cycle, if the difference between the electronic populations obtained in two subsequent cycles is smaller than 10^{-7}.

Thus, we obtain a self-consistent solution for the single-particle Green's functions, $G_{\mathrm{el},mn}$ and $D_{\alpha\alpha'}$, and the corresponding self-energies, Σ_{mn} and $\Pi_{\alpha\alpha'}$.

5.2.3.4 Observables of interest

We characterize electron transport through a single-molecule junction by the average level of vibrational excitation and the electrical current that is flowing through the junction as functions of the applied bias voltage Φ. In this section, we outline how these observables are computed, once the single-particle Green's functions, $G_{\mathrm{el},mn}$ and $D_{\alpha\alpha'}$, and the respective self-energies, Σ_{mn} and $\Pi_{\alpha\alpha'}$, have been determined according to the non-equilibrium Green's function approach outlined in the sections before.

The lesser and greater real-time projections of the single-particle Green's function are directly related to physical observables. For example, the population n_m of the electronic levels of a molecular junction are given by the lesser Green's function $G^{<}_{\mathrm{el},mn}$:

$$n_m = \langle c_m^\dagger c_m \rangle_H = \langle c_m^\dagger c_m \rangle_{\bar{H}}$$
$$= \mathrm{Im}[G^<_{\mathrm{el},mm}(t=0)] = \int_{-\infty}^{\infty} \frac{d\varepsilon}{2\pi} \mathrm{Im}[G^<_{\mathrm{el},mm}(\varepsilon)]. \tag{5.61}$$

A similar expression can be derived for the average level of vibrational excitation (Härtle et al., 2008; Volkovich et al., 2011; Härtle, 2012)

$$\langle a_\alpha^\dagger a_\alpha \rangle_H = \langle a_\alpha^\dagger a_\alpha \rangle_{\bar{H}} + \sum_{mn} \frac{\lambda_{m\alpha}\lambda_{n\alpha}}{\Omega_\alpha^2} \langle (c_m^\dagger c_m - \delta_m)(c_n^\dagger c_n - \delta_n) \rangle_{\bar{H}} \tag{5.62}$$

$$= -\frac{1}{2}\mathrm{Im}[D^<_{\alpha\alpha}(t=0)] - \frac{1}{2} + \sum_{mn} \frac{\lambda_{m\alpha}\lambda_{n\alpha}}{\Omega_\alpha^2} \langle (c_m^\dagger c_m - \delta_m)(c_n^\dagger c_n - \delta_n) \rangle_{\bar{H}}, \tag{5.63}$$

which involves the lesser projection of the vibrational Green's functions $D_{\alpha\alpha}$ and a contribution $\sim \langle (c_m^\dagger c_m - \delta_m)(c_n^\dagger c_n - \delta_n) \rangle_{\bar{H}}$ that describes the displacement of the vibrational degrees of freedom due to charging of the molecule (i.e., polaron formation) (Härtle et al., 2008, 2010; Härtle, 2012). The latter is computed, using the Hartree–Fock-like approximation $\langle c_m^\dagger c_m c_{m'}^\dagger c_{m'} \rangle \approx \langle c_m^\dagger c_m \rangle \langle c_{m'}^\dagger c_{m'} \rangle$ – $\langle c_m^\dagger c_{m'} \rangle \langle c_{m'}^\dagger c_m \rangle$, that is

$$\sum_{mn} \frac{\lambda_{m\alpha}\lambda_{n\alpha}}{\Omega_\alpha^2} \langle (c_m^\dagger c_m - \delta_m)(c_n^\dagger c_n - \delta_n) \rangle_{\bar{H}}$$
$$= \sum_m \frac{\lambda_{m\alpha}^2}{\Omega_\alpha^2} \mathrm{Im}[G^<_{\mathrm{el},mm}(t=0)]$$
$$+2 \sum_{m<m'} \frac{\lambda_{m\alpha}\lambda_{m'\alpha}}{\Omega_\nu^2} \mathrm{Im}[G^<_{\mathrm{el},mm}(t=0)]\mathrm{Im}[G^<_{\mathrm{el},m'm'}(t=0)]$$
$$-2 \sum_{m<m'} \frac{\lambda_{m\alpha}\lambda_{m'\alpha}}{\Omega_\nu^2} \mathrm{Im}[G^<_{\mathrm{el},mm'}(t=0)]\mathrm{Im}[G^<_{\mathrm{el},m'm}(t=0)]. \tag{5.64}$$

The electrical current flowing through the junction can be determined in the steady-state transport regime, according to the Meir–Wingreen-like formula:

$$I_K = 2e\sum_{mn}\int_{-\infty}^{\infty}\frac{d\varepsilon}{2\pi}\left(\Sigma^{<}_{\mathrm{L},mn}(\varepsilon)G^{>}_{\mathrm{el},nm}(\varepsilon)-\Sigma^{>}_{\mathrm{L},mn}(\varepsilon)G^{<}_{\mathrm{el},nm}(\varepsilon)\right), \tag{5.65}$$

where we employ the same separation of electronic and vibrational time scales as in Eq. (5.26).

5.3 Processes and Mechanisms in Vibrationally Coupled Electron Transport

In this section, we give a comprehensive overview of the processes and mechanisms that are relevant in vibrationally coupled electron transport. Implications for possible device applications of single-molecule junctions are reviewed in Section 5.4.

Vibrationally coupled transport can be understood in terms of transport and electron–hole pair creation processes. The current–voltage characteristic of a molecular junction and the corresponding population of the electronic states can be readily understood in terms of transport processes, at least on a qualitative level (vide infra). We therefore study these processes and transport characteristics in the first part of this section, Sections 5.3.1 and 5.3.2. Electron–hole pair creation processes do not directly contribute to the current flowing through a molecular junction but to the respective level of vibrational excitation. As the efficiency of transport processes, however, is also determined by the level of vibrational excitation, pair creation processes have an indirect influence on the current–voltage characteristics and the corresponding population of the electronic states. Electron–hole pair creation processes and vibrational excitation characteristics are the subject of Section 5.3.3.

In the discussion below, we distinguish processes in the non-resonant (Section 5.3.1) and the resonant transport regime of a molecular junction (Section 5.3.2). Non-resonant electron transport processes are typically important at low bias voltages. There, electrons can tunnel through the molecular bridge only by virtually occupying states of the molecular bridge. As the corresponding tunnel current is rather low (pA – nA), this type of tunneling is also referred to as co-tunneling. In contrast, resonant electron transport is characterized by larger currents (nA – μA). In this regime, an electron tunnels through the junction in two sequential tunneling events, populating an eigenstate of the molecular bridge as an intermediate step.

5.3.1 Non-Resonant Transport Processes/Co-Tunneling

We begin our discussion with the current–voltage and conductance–voltage characteristics[2] of a molecular junction in the non-resonant transport regime. This regime is defined as the range of bias voltages, where the electronic levels of the molecular bridge are located outside the bias window, that is, for bias voltages, where the energies of the electronic states are either above the chemical potentials in the leads, $\varepsilon_m > \mu_{L/R}$, or below, $\varepsilon_m < \mu_{L/R}$. In this regime, only "virtual" tunneling processes from the left to the right lead occur. They are commonly referred to as co-tunneling processes. Examples of co-tunneling processes are depicted in Fig. 5.1.

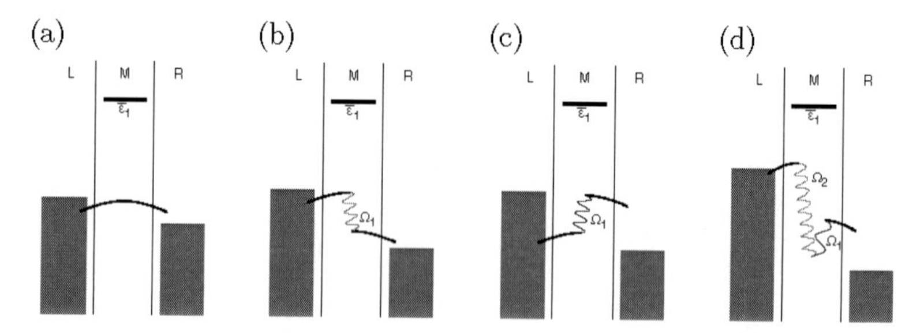

Figure 5.1 Examples of co-tunneling processes that occur in the non-resonant transport regime of a molecular junction. Panel (a) depicts a direct "virtual" tunneling process of an electron from the left to the right lead. Panel (b) and (c) show inelastic co-tunneling processes, where the tunneling electron excites and deexcites one of the vibrational degrees of freedom, respectively. Panel (d) depicts a co-tunneling process, which involves the simultaneous excitation of a vibrational mode and the de-excitation of another.

To discuss these processes, we employ a rather simple model of a molecular junction. This model involves a single electronic state and two vibrational modes (model E1V2). It allows us to discuss the basic processes but also multimode vibrational effects.

[2]Note that we consider the differential conductance $g = dI/d\Phi$.

The latter is crucial for an understanding of real systems, which typically involve tens or hundreds of vibrational modes. The electronic state is located $\varepsilon_1 - \varepsilon_F = 1$ eV above the Fermi level of the junction and coupled to a left and a right lead with coupling strength $\nu_{L/R,1} = 0.1$ eV. Each lead is modeled by a semi-elliptic conduction band with a bandwidth that is determined by the internal hopping parameter $\gamma = 2$ eV (Cizek et al., 2004; Härtle et al., 2008). Accordingly, the level-width functions of the left and the right lead are given by

$$\Gamma_{L/R,mn}(\varepsilon) = \frac{\nu_{L/R,m}\nu_{L/R,n}}{\gamma^2}\sqrt{4\gamma^2 - (\varepsilon - \mu_{L/R})^2}, \tag{5.66}$$

where $m = n = 1$ corresponding to a single electronic state (in general $m, n \in \{1, \ldots, N_{el}\}$ for N_{el} electronic states). The electronic state is also coupled to the vibrational modes 1 and 2 with coupling strength $\lambda_{11} = 0.06$ eV and $\lambda_{12} = 0.15$ eV. The frequency of the modes is given by $\Omega_1 = 0.1$ eV and $\Omega_2 = 0.25$ eV, respectively. These parameters constitute typical values for a molecule (cf., for example, the list of frequencies and electronic-vibrational coupling constants given in the supplemental material of Härtle et al. (2011)). Thus, the antiadiabatic condition $\Gamma_{L/R,11} < \Omega_{1/2}$ is fulfilled. Moreover, the electronic state is located well above the Fermi-level of the junction that is by several units of $\Omega_{1/2}$, which enables us to give a detailed study of vibrational effects in the non-resonant transport regime. The parameters of this model molecular junction are also summarized in Table 5.1.

Figure 5.2 represents current–voltage and conductance–voltage characteristics of this junction. The three lines shown in this figure correspond to three different transport scenarios: electronic transport, where we do not consider electronic-vibrational coupling, vibronic transport, where we take into account electronic-vibrational coupling, and thermally equilibrated transport, where the vibrational mode is enforced to relax to its thermal equilibrium state after every electron transmission event (i.e., we use $\Pi_{\alpha\alpha'} = 0$). This thermal equilibrium state is characterized by the same temperature as used in the leads, that is 10 K. Thus, the vibrational mode returns effectively to its ground state, even if it has been excited in the course of an electron tunneling event. We use the comparison between the vibronic and the thermally equilibrated transport scenario in the following to highlight vibrational non-equilibrium effects.

Table 5.1 Model parameters for the molecular junctions considered in the text

Electronic parameters for model	ε_1	ε_2	U_{12}	$\nu_{L,1}$	$\nu_{R,1}$	$\nu_{L,2}$	$\nu_{L,2}$	γ
E1V2	1.0	—	—	0.1	0.1	—	—	2
E2V1	0.15	0.8	0	0.1	0.1	0.1	0.1	3
Vibrational parameters for model	Ω_1	Ω_2	λ_{11}	λ_{12}	λ_{21}			
E1V2	0.1	0.25	0.06	0.15	—			
E2V1	0.1	—	0.06	—	-0.06			

Note: Energy values are given in eV. The temperature in the leads, $T = 10$ K, is the same for all these model molecular junctions. These parameters represent typical values for molecular junctions as they are found for example in experiments (Böhler et al., 2004; Elbing et al., 2005; Sapmaz et al., 2005; Tao, 2006; Keane et al., 2006; Osorio et al., 2010; Martin et al., 2010; Ballmann et al., 2010; Choi et al., 2008; Huisman et al., 2009; Trouwborst et al., 2011; Secker et al., 2011; Pump et al., 2008; Lafferentz et al., 2009; Brumme et al., 2011) or employing ab initio calculations (Xue and Ratner, 2003a,b; Evers et al., 2004; Benesch et al., 2006; Frederiksen et al., 2007a,b; Troisi and Ratner, 2006; Gagliardi et al., 2007; Wheeler and Dahnovsky, 2008; Benesch et al., 2008, 2009; Chen et al., 2009; Monturet et al. (2010).

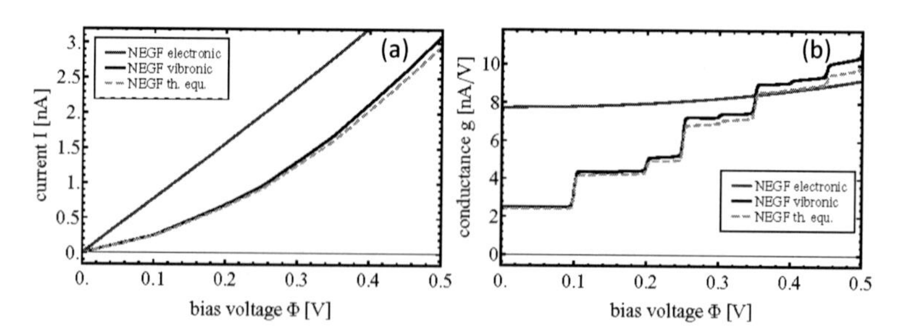

Figure 5.2 Current–voltage and conductance–voltage characteristics of the model molecular junction E1V2 in the non-resonant transport regime ($e\Phi << 2\bar{\varepsilon}_1$).

In the non-resonant transport regime, electronic transport (solid purple lines) is characterized by an almost linear increase of the current with increasing bias voltage.[3] This can be understood by the number of electrons that can directly tunnel from the left to the right lead (cf. Fig. 5.1a for an example of such a co-tunneling process). This number increases almost linearly with the applied bias voltage Φ. The respective conductance is therefore almost constant. In this regime, tunneling electrons populate the electronic state only virtually such that it remains essentially unoccupied. Only for larger bias voltages, where the chemical potential in the left lead approaches the electronic level of the molecular bridge, the conductance through the junction increases slightly. This is because the electronic state exhibits broadening due to the coupling to the leads and, thus, facilitates (resonant) tunneling of electrons even before the electronic state has entered the bias window (cf. Section 5.3.2.1).

Including electronic-vibrational coupling (solid black and dashed gray lines), the electronic state is polaron-shifted to lower energies, $\varepsilon_1 = 1$ eV $\rightarrow \bar{\varepsilon}_1 = 0.874$ eV, according to Eq. (5.22). Thus, electrons can tunnel through the junction more easily and one expects a larger current/conductance. On the other hand, however, the equilibrium position of the vibrational modes are also shifted, $\Delta Q_1 = -2\lambda_{11}/\Omega_1$ and $\Delta Q_2 = -2\lambda_{12}/\Omega_2$, when the electronic state becomes populated. This shift results in a suppression of electronic co-tunneling events (cf. Fig. 5.1a), because the different charge states of the molecular bridge that are involved in the virtual tunneling processes exhibit effectively less overlap. This can be quantified by the respective Franck–Condon (FC) matrix element $|X_1^{00,00}|^2 = \mathrm{e}^{-\lambda_{11}^2/\Omega_1^2 - \lambda_{12}^2/\Omega_2^2} \approx 0.5$ which is significantly smaller than one. Therefore, at least for low bias voltages, $e\Phi < \Omega_{1/2}$, where electronic co-tunneling processes are dominant, the vibronic and thermally equilibrated transport scenario give a lower current than the electronic transport scenario. It is noted, that recent results of other NEGF approaches and the ML-MCTDH method seem to indicate that the reduction of the current due to vibronic coupling is smaller than that obtained with the NEGF theory employed here.

[3]In general, there is, besides a linear dependence, also an exponential dependence of the current I on the applied bias voltage Φ (see for example, the Simmons formula for tunneling through a potential barrier (Simmons, 1963; Chapline and Wang, 2007). For low bias voltages and $\varepsilon_1 >> \varepsilon_F$; however, this exponential dependence is typically not significant.

At higher bias voltages, electronic-vibrational coupling facilitates inelastic co-tunneling processes (cf. Fig. 5.1b), where an electron emits one, two, or more vibrational quanta upon tunneling through the junction. Since these processes require electrons with sufficiently high energies, these processes become active one by one at bias voltages $e\Phi = m\Omega_1 + n\Omega_2$ ($m, n \in \mathbb{N}$), that is superpositions of the vibrational frequencies Ω_1 and Ω_2. These processes result in an increased transmission probability for electrons with an energy $\varepsilon > \min(\mu_L, \mu_R) + m\Omega_1 + n\Omega_2$. Accordingly, as the conductance for these electrons is also larger, the respective conductance–voltage characteristic exhibits a step-like increase at $e\Phi = m\Omega_1 + n\Omega_2$.

In fact, inelastic co-tunneling processes become active at slightly different bias voltages or energies, because the vibrational frequencies Ω_α are renormalized due to the coupling of the molecule to the leads. The corresponding frequency shifts are described, for example, by the real parts of the vibrational self-energy matrix $\Pi_{\alpha\alpha'}$. In the antiadiabatic regime ($\Gamma << \Omega_\alpha$) however, this renormalization of the vibrational frequencies is not very pronounced, and therefore, it is neglected whenever we study vibrational effects in the non-resonant transport regime. That way, we avoid numerical errors that spoil the analysis of the corresponding transport characteristics (cf. appendix B of Härtle (2012)).

The height of the steps due to the onset of inelastic co-tunneling processes correlates with the respective FC factors for a transition from the vibrational ground to its (m, n) excited state, $|X_1^{00,mn}|^2 = e^{-\lambda_{11}^2/\Omega_1^2 - \lambda_{12}^2/\Omega_2^2}(\lambda_{11}^{2m}/\Omega_1^{2m})/m!(\lambda_{12}^{2n}/\Omega_2^{2n})/n!$. For $\lambda_{11}/\Omega_1 < 1$ and $\lambda_{12}/\Omega_2 < 1$, these transition matrix elements become gradually smaller with an increasing number of vibrational quanta m and n involved in the respective co-tunneling processes. While for $e\Phi \lesssim \Omega_2$, the vibronic and the thermal conductance is smaller than the electronic one, they slightly exceed the electronic conductance for higher bias voltages. This behavior is a result of both the polaron-shift of the broadened electronic level and the fact that electrons entering the bias window at these bias voltages do not suffer from the reduced FC factors for electronic co-tunneling processes (cf. Fig. 5.1a), as in parallel a number of excitation and de-excitation channels (cf. Figs. 5.1b–d) are active.

The comparison of the solid black and the dashed gray line, which correspond to vibronic and thermally equilibrated transport, respectively, reveals that heating of the vibrational mode due to vibrational excitation processes (cf. Fig. 5.5b) is less significant in the non-resonant transport regime. In non-equilibrium, however, the current (solid black line) is slightly larger due to additional vibrational de-excitation processes (cf. Fig. 5.1c), which are suppressed in the thermally equilibrated transport scenario.

Note that in the adiabatic regime, $\Gamma > \Omega_{1/2}$, the opening of inelastic channels does not result in a step-wise increase of the conductance but rather in a step-wise decrease, if the transmission probability for an electron is larger than ≈ 0.5 (Avriller and Levy Yeyati, 2009; Haupt et al., 2009; Schmidt and Komnik, 2009; Entin-Wohlman et al., 2009, 2010). While in this regime the contribution of inelastic co-tunneling channels is still present, they also result in an effective Pauli-blocking of elastic co-tunneling processes (Lorente and Persson, 2000). Thus, instead of increasing the conductivity of the junction, these processes result in a decrease of the conductivity. This intriguing phenomenon has been found and verified in a number of experiments (Agrait et al., 2002a; Smit et al., 2002; Tal et al., 2008), and theoretical studies (Lorente and Persson, 2000; Frederiksen et al., 2004, 2007b; Kiguchi et al., 2008; Avriller and Levy Yeyati, 2009; Haupt et al., 2009; Schmidt and Komnik, 2009; Entin-Wohlman et al., 2009, 2010).

As we have seen, inelastic co-tunneling processes result in a step-wise increase of the conductance–voltage characteristic of a single-molecule junction at multiples of the vibrational frequencies Ω_α. Therefore, one can use these steps to determine the energy of the vibrational levels in a molecular junction. This is commonly referred to as Inelastic Electron Tunneling Spectroscopy (IETS) (Jaklevic and Lambe, 1966; Ho, 2002; Agrait et al., 2002a,b; Kushmerick et al., 2004; Yu et al., 2004; Secker, 2008; Hihath et al., 2008, 2010; Kim et al., 2011). Thereby, propensity rules (Troisi and Ratner, 2006; Solomon et al., 2006a; Gagliardi et al., 2007) determine the active part of the vibrational spectrum, where the multitude of vibrational modes in a molecular junction gives a characteristic fingerprint of the molecule bridging the gap between the left and the right lead. With the fine-tunable tip of a STM, vibrational signatures of a single molecule can be resolved even spatially (Ogawa et al., 2007; Monturet et al., 2010), as the tunnel current may flow through a specific part of the molecule only.

5.3.2 Resonant Transport Processes/Sequential Tunneling

Having analyzed the non-resonant transport regime in Section 5.3.1, we now address the current–voltage and conductance–voltage characteristics of molecular junctions in the resonant transport regime. Thereby, we focus in Section 5.3.2.1 on effects due to electronic-vibrational coupling to a single electronic state. In Section 5.3.2.2, we extend our considerations to vibronic effects with respect to multiple electronic states.

5.3.2.1 With respect to a single electronic state

To discuss resonant transport processes with respect to a single electronic state, we use the same model system as in Section 5.3.1, i.e., model E1V2 (see Table 5.1). The current–voltage and conductance–voltage characteristics of this system are shown in Fig. 5.4. In contrast to Fig. 5.2, however, the range of bias voltages is changed from $\Phi = 0–0.5$ V (corresponding to $e|\Phi| << 2\varepsilon_1$) to $\Phi = 1–3$ V (which includes voltages $> 2\varepsilon_1$) in order to capture resonant effects. Thereby, as in Section 5.3.1, the solid purple and the solid black line represent results for purely electronic and vibronic transport, respectively. In addition, the dashed gray line depicts results where the vibrational mode is confined (effectively) to its ground state.

The electronic current–voltage characteristics in Fig. 5.4a (solid purple lines) exhibit a single step at $e\Phi \simeq 2\varepsilon_1$. This step indicates the onset of resonant transport, where electrons in the left lead have enough energy to pass through the molecular junction by two sequential tunneling processes (as depicted by Fig. 5.3a). At higher bias voltages, the electrical current remains more or less at the same level, as the number of electrons in the left lead, which can resonantly tunnel through the molecular resonance, is also more or less constant.[4] The coupling of the

[4]This statement is based on the fact that in the given range of bias voltages, the density of states in both leads is not significantly influenced by the applied bias voltage.

molecule to the leads results in a broadening of the resonance. The corresponding width is given by $\Gamma_{L,11} + \Gamma_{R,11} + k_B T \approx 7$ meV.

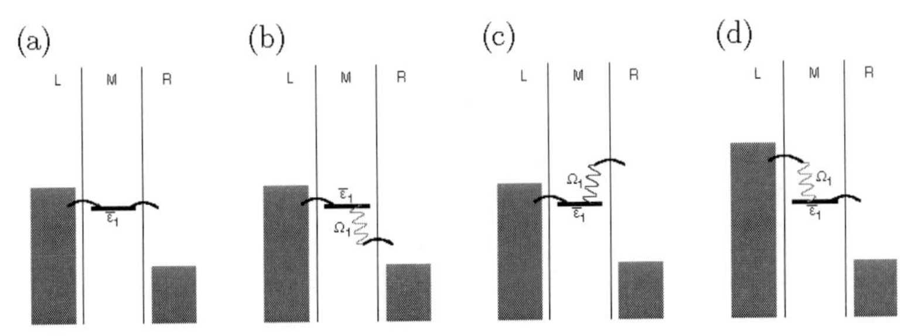

Figure 5.3 Schematic representation of example processes for sequential tunneling in a molecular junction. Panel (a) shows sequential tunneling of an electron from the left lead to the right lead that involves two consecutive tunneling processes: from the left lead onto the molecular bridge, and from the molecular bridge to the right lead. In panel (b)/(c) the latter of the two tunneling processes is accompanied by an excitation/deexcitation process, where due to electronic-vibrational coupling the vibrational mode is excited/deexcited by a single vibrational quantum. While the processes depicted by panels (a–c) become active at the same bias voltage, that is for $e\Phi \approx 2\bar{\varepsilon}_1$, resonant excitation processes like the one depicted by panel (d) require higher bias voltages, $e\Phi \gtrsim 2(\bar{\varepsilon}_1 + \Omega_1)$.

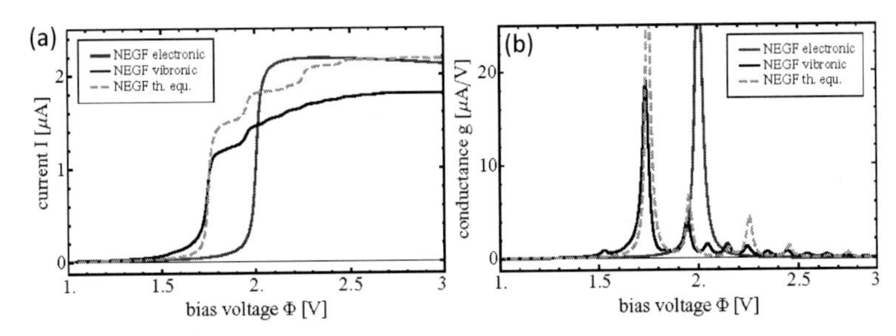

Figure 5.4 Current–voltage and conductance–voltage characteristics of a model molecular junction in the resonant transport regime (model E1V2, cf. Table 5.1).

Electronic-vibrational coupling triggers a variety of steps in the current–voltage characteristic of this junction (see the solid black line in Fig. 5.4a). The step at $e\Phi \simeq 2\bar{\varepsilon}_1$ indicates, as for electronic transport, the onset of resonant transport processes. Due to the polaron-shift of the electronic level (cf. Eq. (5.22)), however, this step appears at lower bias voltages. In addition, the height of this step is reduced compared to the one in the electronic current–voltage characteristic. This is related to the suppression of electronic transport processes (cf. Fig. 5.3a) due to the different equilibrium positions of the vibrational modes in different charge states of the molecular bridge, $\Delta Q_{1/2} = -2\lambda_{11/12}/\Omega_{1/2}$ (cf. Section 5.3.1). This shift leads to a reduced overlap between the charge states involved in these processes, and can be quantified by the FC-factor $|X_1^{00,00}|^2 = \mathrm{e}^{-\lambda_{11}^2/\Omega_1^2 - \lambda_{12}^2/\Omega_2^2} \approx 0.5$ (cf. Section 5.3.1). This suppression of the current at the onset of the resonant transport regime is also referred to as Franck–Condon blockade, and is particularly pronounced for large coupling strengths $\lambda_{11/12}$ (Koch and von Oppen, 2005; Koch et al., 2006b; May and Kühn, 2008b). Steps that appear at higher bias voltages, $e\Phi \simeq 2(\bar{\varepsilon}_1 + m\Omega_1 + n\Omega_2)$ $(m, n \in \mathbb{N})$, indicate the onset of resonant excitation processes (cf. Fig. 5.3d), where an electron with energy $\bar{\varepsilon}_1 + m\Omega_1 + n\Omega_2$ tunnels resonantly onto the molecular bridge exciting vibrational modes 1 and 2 by m and n vibrational quanta, respectively. The respective step heights can be qualitatively understood by the Franck–Condon matrix elements, $|X_1^{00,mn}|^2 = (n!m!)^{-1}(\lambda_{11}/\Omega_1)^{2m}(\lambda_{12}/\Omega_2)^{2n}\exp(-\lambda_{11}^2/\Omega_1^2 - \lambda_{12}^2/\Omega_2^2)$, which correspond to a transition from the vibrational ground to the (m, n) excited state. For a quantitative analysis, however, the exact number of sequential tunneling processes and the associated level of vibrational excitation needs to be taken into account (Härtle and Thoss, 2011a). Note that, at the onset of the resonant tunneling regime ($e\Phi = 2\bar{\varepsilon}_1$), not only electronic transport processes (Fig. 5.3a) become active, but also a number of vibrational excitation and deexcitation processes (such as the example processes shown in Figs. 5.3b,c). Deexcitation processes where n/m acquires negative values can also occur. Thereby, the resonantly tunneling electrons use vibrational energy that is generated in preceding inelastic tunneling processes (cf. Figs. 5.1b, 5.3b or 5.3d). Such processes can be inelastic sequential or co-tunneling processes, where the mechanism associated with the latter is referred to as co-tunneling assisted sequential electron tunneling (CoSET) (Koch et al., 2006b; Lüffe et al., 2008).

Considering thermally equilibrated transport, we obtain the current–voltage characteristic shown by the dashed gray line in Fig. 5.4. Recall that in thermally equilibrated transport the vibrational mode is effectively reset to its thermal equilibrium state after each electron transmission process. At the effective temperature of 10 K used in these calculations, this state corresponds essentially to the ground state of the vibrational mode. Electrons tunneling through the junction are thus not affected by inelastic processes that occurred in previous tunneling events. This is in contrast to vibronic transport, where current-induced vibrational excitation is accounted for. There, the efficiency of electron tunneling processes depends on inelastic processes of previous tunneling events. The thermal current–voltage characteristic, as the vibronic current–voltage characteristic, show a multitude of steps at $e\Phi = 2(\overline{\varepsilon}_1 + m\Omega_1 + n\Omega_1)$ $(m, n \in \mathbb{N}_0)$, which are associated with the onset of resonant transport processes (cf. Fig. 5.3a–d). In thermally equilibrated transport, however, the respective currents reach the electronic current level (≈1.6 μA) in a few steps, while the vibronic currents show significantly lower current levels in the resonant transport regime. This indicates a suppression of the current through a molecular junction due to (current-induced) vibrational excitation.

This is a rather general phenomenon in molecular junctions, and can be qualitatively understood by the following analysis. If, for example, a positive bias voltage is applied to a molecular junction and allows for $l_{1/2} = \mathrm{mod}(e\Phi/2 - \overline{\varepsilon}_1, \Omega_{1/2})$ resonant excitation processes (cf. Fig. 5.3d) and if the molecular bridge is in its ground state, the transition probability for an electron tunneling from the left lead onto the bridge can be written as $\sum_{m=0}^{l_1}\sum_{n=0}^{l_2} |X_1^{00,mn}|^2 \Theta(e\Phi/2 - \overline{\varepsilon}_1 - m\Omega_1 - n\Omega_2)$. This transition probability converges typically in a few steps to unity with increasing $l_{1/2}$, or bias voltage Φ, and explains why the thermally equilibrated current–voltage characteristic reaches the electronic current level in a few steps. For this transition probability, we do not consider the second tunneling process of the corresponding sequential tunneling events because the number of the relevant tunneling processes is not restricted by the bias voltage, at least as long as the electronic state is located above the Fermi level by several units of $\Omega_{1/2}$.

If the vibrational modes are in a non-equilibrium state, where the population of the (m, n)th vibrational level is given by $\alpha_{m,n} \neq \delta_{0m}\delta_{0n}(m, n \in \mathbb{N}_0)$, the transition probability associated with the tunneling of an electron from the left lead onto the molecular bridge is determined by $\Sigma_{m,n=0...\infty}\alpha_{mn}\Sigma_{m'}^{l_1+m}\Sigma_{n'}^{l_2+n}|X_1^{mn,m'n'}|^2$. This transition probability includes all excitation and deexcitation processes that occur, if the vibrational modes are in their ground state ($l_{1/2}$ = 0), first excited states ($l_{1/2}$ = 1), or in higher excited states ($l_{1/2}$ > 1). It is, in general, smaller than the transition probability $\Sigma_{m=0}^{l_1}\Sigma_{n=0}^{l_2}|X_1^{00,mn}|^2$ for thermally equilibrated vibrational modes (at 10 K). Accordingly, the vibronic transport scenario exhibits lower current levels than the thermally equilibrated one in this molecular junction.

5.3.2.2 With respect to multiple electronic states

At this point, we extend our considerations to transport through multiple electronic states. To this end, we employ a model for a molecular junction (model E2V1) with two electronic states: a lower- and a higher-lying electronic state located ε_1 = 0.15 eV and ε_2 = 0.8 eV above the Fermi level of the junction, respectively. Both states are coupled to a single vibrational mode with coupling strengths λ_{11} = 0.06 eV and λ_{21} = −0.06 eV.[5] Due to the different sign in the electronic-vibrational coupling strengths, electron–electron interactions, which are effectively induced in this model molecular junction by the coupling of the two electronic states to the same vibrational mode (cf. Eq. (5.23)), are repulsive: $\overline{U}_{12} = -2\lambda_{11}\lambda_{21}/\Omega_1 > 0$. All parameters of this model molecular junction are summarized in Table 5.1. Respective current–voltage/conductance–voltage characteristic are shown in Fig. 5.6.

The electronic current–voltage/conductance–voltage characteristic of this model molecular junction is depicted by the solid purple line. According to the analysis given for a single electronic state, the two steps in this current–voltage characteristic, at $e\Phi = 2\varepsilon_1$, and $e\Phi = 2\varepsilon_2$, indicate the onset of resonant transport processes through states 1 and 2, respectively. Since both states are coupled to the leads with the same coupling strengths, and because the two states are non-degenerate, $|\varepsilon_1 - \varepsilon_2| >> \Gamma_{\mathrm{L/R},mn}$ these steps, as well as the peaks in the corresponding conductance–voltage characteristic, are very similar in height and shape.

[5]This is a quite common electronic-vibrational coupling scenario, as can be inferred, for example, from the ab initio data given in the supplementary material of Ref. Härtle et al. (2011).

The solid black lines of Fig. 5.6 depict the current–voltage/conductance–voltage characteristic associated with vibronic transport through this model molecular junction. In contrast to electronic transport, where the analysis given for a single electronic state (one state $\rightarrow$ one step) readily applies to two electronic states (two states $\rightarrow$ two steps), these characteristics cannot be understood solely in terms of vibrationally coupled electron transport through a single electronic state (cf. Section 5.3.2.1).

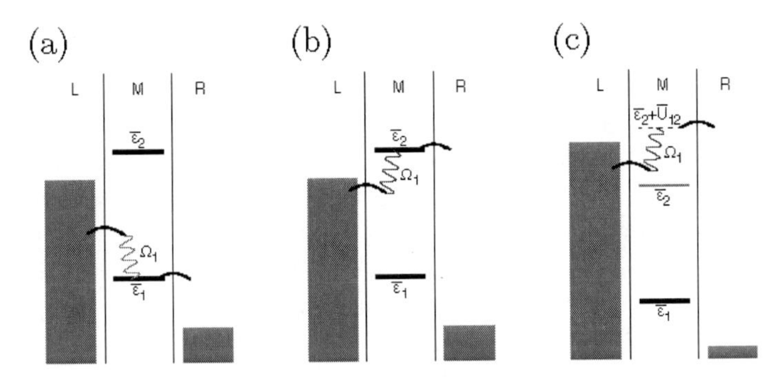

Figure 5.5 Examples of sequential tunneling processes involving two electronic states. Panel (a) depicts a sequential tunneling process, which involves a resonant excitation process with respect to state 1. Such processes occur also in the presence of a single electronic state. Examples of sequential tunneling processes that involve deexcitation processes with respect to state 2 are depicted in Panels (b) and (c), where the lower-lying electronic state is unoccupied and occupied, respectively.

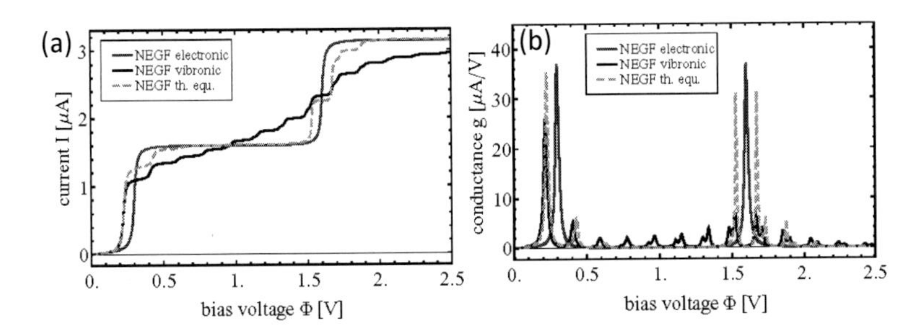

Figure 5.6 Current–voltage and conductance–voltage characteristics for a model molecular junction with two electronic states (model E2V1, cf. Table 5.1).

In accordance with the analysis given for a single electronic state in Section 5.3.2.1, we find steps/peaks in the current–voltage/conductance–voltage characteristic of this model molecular junction at $e\Phi = 2(\bar{\varepsilon}_1 + n\Omega_1)$ and $e\Phi = 2(\bar{\varepsilon}_2 + n\Omega_1)$ $(n \in \mathbb{N}_0)$. These are associated with the onset of resonant transport processes through states 1 and 2, respectively (cf., for example, Fig. 5.5a, which represents a vibrational excitation process with respect to state 1). However, the steps/peaks associated with state 2 are much less pronounced than the ones associated with state 1, although both states are coupled to the leads and to the vibrational mode with coupling strengths that have the same absolute value. Moreover, additional steps at $e\Phi = 2(\bar{\varepsilon}_2 + \bar{U}_{12} + n\Omega_1)$ indicate a third electronic resonance at $\bar{\varepsilon}_2 + \bar{U}_{12}$. These findings can be explained by (vibrationally induced) electron–electron interactions, $\bar{U}_{12}$, which become effective once the applied bias voltage allows for sequential tunneling through both electronic states, $e\Phi > 2\bar{\varepsilon}_2$. Due to these interactions, sequential tunneling depends on the population of the electronic states. For example, electrons with an energy of $\bar{\varepsilon}_2$ can resonantly tunnel through state 2 only if state 1 is unoccupied, while electrons with an energy of $\bar{\varepsilon}_2 + \bar{U}$ an resonantly tunnel through state 2 only if state 1 is occupied. This results effectively in a splitting of resonances associated with state 2, since the corresponding tunneling processes become active at different bias voltages and because state 1 is neither empty nor fully occupied but $n_1 \approx 1/2$ for $e\Phi \simeq 2\bar{\varepsilon}_2$. While steps/peaks in the current/conductance at $e\Phi = 2(\bar{\varepsilon}_2 + n\Omega_1)$ are thus suppressed by a factor $1 - n_1$, the steps/peaks at $e\Phi = 2(\bar{\varepsilon}_2 + n\Omega_1 + \bar{U}_{12})$ are linearly correlated with the population of state 1, n_1. That way, electron transport processes through the two electronic states do not occur independently from each other, but are coupled by electron–electron interactions. This applies for electron transport processes through molecular junctions with multiple electronic states in general.

Another source for such coupling is current-induced vibrational excitation. This can be demonstrated comparing results for vibronic and thermally equilibrated transport, that is a comparison of the black and dashed gray line in Fig. 5.6. As shown in Section 5.3.2.1, vibronic transport is characterized by a suppression of current due to vibrational excitation. This trend is observed at the onset of the resonant transport regime at $e\Phi \simeq 2\bar{\varepsilon}_1$

as well as for higher bias voltages $e\Phi \gtrsim 2\bar{\varepsilon}_1$. In the intermediate voltage regime, however, where state 2 is close to, but still outside the bias window, vibronic transport yields a larger current level than thermally equilibrated transport. This is a result of current-induced vibrational excitation, which is induced by resonant excitation processes with respect to state 1 (cf. Fig. 5.5a). The vibrational energy thus generated facilitates resonant tunneling of electrons through state 2 by vibrational deexcitation processes, even though this state is still outside the bias window (cf. Figs. 5.5b,c). These sequential tunneling processes result in pronounced steps at $e\Phi = 2(\bar{\varepsilon}_2 - n\Omega_1)$ and $e\Phi = 2(\bar{\varepsilon}_2 + \bar{U}_{12} - n\Omega_1)$ ($n \in \mathbb{N}$). While transport processes through the two electronic states are thus coupled in thermally equilibrated transport by electron–electron interactions only, they are coupled in vibronic transport by both electron–electron interactions as well as vibrational excitation.

5.3.3 Local heating and cooling in a molecular junction

In an inelastic transport process (cf. Figs. 5.1, 5.3, and 5.5), the molecular bridge exchanges simultaneously an electron and quanta of vibrational energy with the leads. Vibrationally coupled electron transport thus involves both charge and energy exchange processes with the leads. While in Sections 5.3.1 and 5.3.2 we have discussed the role of charge exchange processes for the current–voltage characteristics of a molecular junction, we focus in this section on the role of energy exchange processes (local heating and cooling [Mitra et al., 2004; Chen et al., 2005; Koch et al., 2006a; D'Agosta et al., 2006; Siddiqui et al., 2007; Galperin et al., 2007b; Härtle et al., 2008; Schulze et al., 2008; McEniry et al., 2009; Galperin et al., 2009; Härtle et al., 2009; Schiff and Nitzan, 2010; Romano et al., 2010; Härtle and Thoss, 2011a,b; Volkovich et al., 2011; Kast et al., 2011]) and the corresponding vibrational excitation characteristic.

To this end, we consider another class of processes, that is electron–hole pair creation processes (see Fig. 5.7). In such a process, an electron tunnels from one lead onto the molecular bridge and back again to the same lead. Thereby, the electron absorbs vibrational energy from the molecular bridge. Thus, effectively, an electron–hole pair is created in the respective lead.

Note that in symmetric systems such pair creation processes occur with the same probability as corresponding transport process (for example the deexcitation process shown in Fig. 5.3c is as probable as the electron–hole pair creation process depicted in Fig. 5.7a). In asymmetrically coupled junctions, however, where for example $\Gamma_{\mathrm{L},mn} > \Gamma_{\mathrm{R},mn}$, pair creation processes with respect to the left/right lead are more/less probable than transport processes (because they depend on both $\Gamma_{\mathrm{L},mn}$ and $\Gamma_{\mathrm{R},mn}$ while pair creation processes involve the coupling to one of the leads only). These processes, in combination with inelastic transport processes, constitute all possible energy exchange processes between the molecular bridge and the leads, which involve electron tunneling.[6]

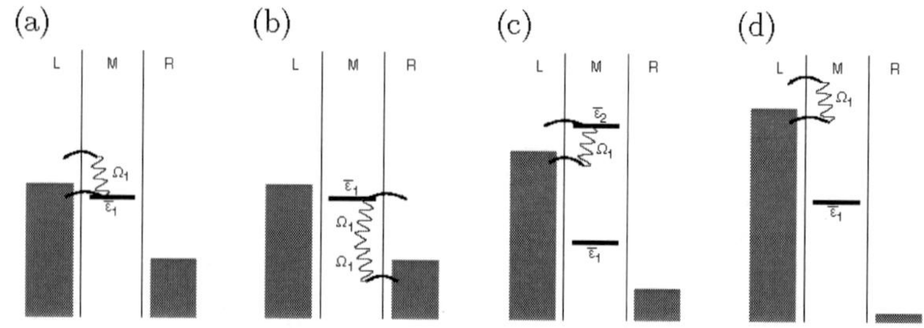

Figure 5.7 Examples of electron–hole pair creation processes. Panel (a) shows a resonant electron–hole pair creation process, where an electron tunnels from the left lead onto the molecular bridge and back again to the left lead in two sequential tunneling events. Thereby, the electron takes up a quantum of vibrational energy. Panel (b) depicts a similar process with respect to the right lead. As in this configuration the chemical potential in the right lead is located further away from the electronic level, electron–hole pair creation requires the absorption of two vibrational quanta. This, however, is typically unfavorable and results in less efficient cooling. A pair creation process with respect to a higher-lying electronic state is shown in Panel (c). Panel (d) represents an off-resonant pair creation processes. These processes constitute the pair creation analog of deexcitation via inelastic co-tunneling (cf. Fig. 5.1c).

[6]Note that a direct coupling between the vibrational modes of the molecular bridge and the phonon modes in the leads (Jorn and Seidemann, 2009) constitutes another mechanism for energy exchange with the leads.

Electron–hole pair creation processes are crucial to understand the vibrational excitation characteristic of a molecular junction. This is demonstrated, for example, in Section 5.3.3.1, where we show that electron–hole pair creation processes facilitate an understanding why a higher level of vibrational excitation is obtained for weaker electronic-vibrational coupling. The role of off-resonant pair-creation processes is elucidated for example in Volkovich et al., 2011; Härtle (2012). Extending our considerations to transport through multiple electronic states (Section 5.3.3.2), we study the bias dependence of local heating and cooling, which turns out to be highly correlated with the location of the electronic states with respect to the Fermi level of the junction.

Although electron–hole pair creation processes do not directly contribute to the current that is flowing through a molecular junction, they have a significant influence on the respective current–voltage/conductance–voltage and the population characteristic of the electronic states (Härtle et al., 2010; Härtle and Thoss, 2011a; Volkovich et al., 2011; Härtle, 2012; Härtle et al., 2013c). Electron–hole pair creation processes are well known from spectroscopic (Eigler and Schweizer, 1991; Eigler et al., 1991; Stipe et al., 1998; Pascual et al., 2003) and theoretical studies (Gao et al., 1992; Persson and Ueba, 2002; Mii et al., 2002) of molecular adsorbates at metal surfaces. Different mechanisms for electron–hole pair creation, which involve an electronically excited instead of a vibrationally excited state of the junction, are discussed, for example, in Refs. Galperin et al., 2006b; Fainberg et al. (2007).

5.3.3.1 Role of resonant electron–hole pair creation processes

The most fundamental mechanisms of local heating and cooling in a molecular junction can be studied by the model molecular junction E1V2. The corresponding current–voltage/conductance–voltage and the electronic populations characteristics have been analyzed in the non-resonant (Section 5.3.1) and in the resonant transport regime (Section 5.3.2.1). In this and in the following section, we discuss the respective vibrational excitation characteristics, which are depicted in Fig. 5.8. Thereby, the solid and dashed black lines represent the average level of excitation of modes 1 and 2, respectively. Distinct steps appear at $e\Phi = 2(\bar{\varepsilon}_1 + m\Omega_1 + n\Omega_2)$ $(m, n \in \mathbb{N}_0)$. Similar steps appear in the respective current–voltage characteristic at the same bias voltages (cf. Fig. 5.4). They are associated with the onset of resonant excitation processes, where electrons tunneling from the left lead onto the molecular bridge, exciting the vibrational modes by m and n vibrational quanta, respectively. Such a resonant excitation process with $m = 1$ and $n = 0$ (or $m = 0$ and $n = 1$) is shown in Fig. 5.3d. Whenever such an excitation channel becomes active, the level of vibrational excitation increases. As these processes involve an increasing number of vibrational quanta, the relative step heights in the vibrational excitation characteristic can be expected to be larger than the relative step heights in the current–voltage characteristic.

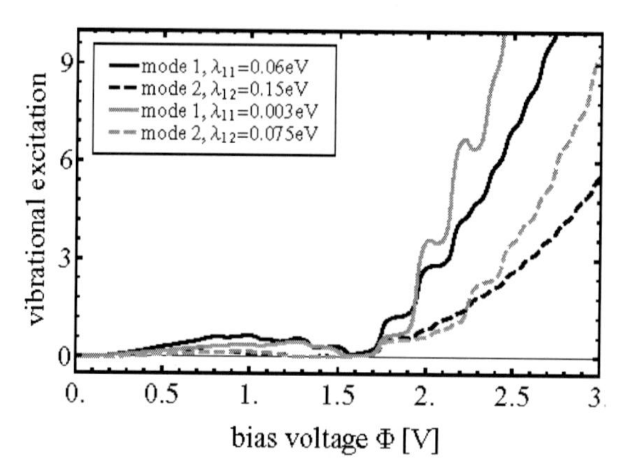

Figure 5.8 Vibrational excitation characteristics of a model molecular junction with a single electronic state coupled to two vibrational modes (model E1V2, cf. Table 5.1). For comparison, we also show the average level of vibrational excitation for a system where the electronic-vibrational coupling is reduced by a factor of 1/2 (orange lines). Thereby, the position of the polaron-shifted electronic level $\bar{\varepsilon}_0$ is adjusted such that it is the same in both systems.

So far, the vibrational excitation characteristics can be understood on the same footing as the corresponding current–voltage characteristic. There is, however, a qualitative difference between the relative step heights in both characteristics. While the step heights in the current–voltage characteristic become smaller with increasing bias voltage, the step heights in the vibrational excitation characteristics become larger. This can only be understood, if we take into account that the level of vibrational excitation is determined by the number and the efficiency of both excitation and deexcitation processes, and not only by the number and the efficiency of transport processes, such as the current. We therefore need to consider both inelastic transport and electron–hole pair creation processes (cf. Fig. 5.7). Thereby, low levels of vibrational excitation indicate that deexcitation processes are dominant, while high levels of vibrational excitation are obtained, if the probability for exciting and deexciting the vibrational modes become similar. In that case, the system probes the ladder of vibrational states like in a "random-walk" scenario (Koch et al., 2006a). As a result, the vibrational states are populated with very similar probabilities, leading to high levels of vibrational excitation (Härtle and Thoss, 2011b). Note that, due to the harmonic approximation, deexcitation of the vibrational modes is always more favorable than exciting it, at least in the steady state and for finite bias voltages. Excitation processes can only become dominant in situations where the junction/molecule dissociates (Pascual et al., 2003; Dzhioev and Kosov, 2011)).

Similar to inelastic transport processes, electron–hole pair creation processes are more likely to occur in this molecular junction the less vibrational quanta they involve. Thus, at the onset of the resonant transport regime, cooling by electron–hole pair creation processes is very efficient, as electron–hole pairs can be created in the left lead by the absorption of just a single vibrational quantum (cf. Fig. 5.7a). This leads to a rather low level of vibrational excitation at $e\Phi \approx 2\overline{\varepsilon}_1$, although the number and the efficiency of inelastic transport processes, where the vibrational mode is excited (Fig. 5.3b) and deexcited (Fig. 5.3c), is very similar.

In contrast to inelastic transport processes, the number of resonant electron–hole pair creation processes does not increase with increasing bias voltage but decreases. At $\Phi > 2(\overline{\varepsilon}_1 + m\Omega_1 + n\Omega_2)$, for example, resonant electron–hole pair creation processes involving m/n vibrational quanta $\Omega_{1/2}$ are suppressed. As a result, the probability for deexciting the vibrational modes decreases at these bias voltages. In addition, resonant excitation channels

(Fig. 5.3d) become active, which increase the probability for exciting the vibrational mode. Accordingly, the level of vibrational excitation increases at these bias voltages due to both the onset of resonant excitation channels and the suppression of electron–hole pair creation processes. Thereby, the level of vibrational excitation increases by increasingly larger steps, which is related to the fact that processes involving less vibrational quanta are more likely to occur than processes involving more vibrational quanta.

This can be demonstrated considering a very similar model system, which differs from junction E1V2 only by a weaker electronic-vibrational coupling: λ_{11} = 0.03 eV and λ_{12} = 0.075 eV. The respective vibrational excitation characteristics are shown in Fig. 5.8 by solid and dashed orange lines. Comparing the levels of vibrational excitation in both systems reveals that higher levels of vibrational excitation are obtained for weaker electronic-vibrational coupling. This counterintuitive phenomenon can be understood in terms of the analysis we have given before. For smaller electronic-vibrational coupling, the electron–hole pair creation processes that involve a smaller number of vibrational quanta are even more important than for stronger electronic-vibrational coupling. The same holds true for the corresponding resonant excitation channels. Thus, at $\Phi > 2(\overline{\varepsilon}_1 + m\Omega_1 + n\Omega_2)$, the level of vibrational excitation increases faster for a weaker electronic-vibrational coupling, because the probabilities for exciting and deexciting the vibrational mode faster increase and decrease, respectively.

We thus arrive at the conclusion that the vibrational excitation characteristic of a single-molecule junction can only be understood, if we account for all excitation and deexcitation processes, including electron–hole pair creation processes. This is in contrast to the corresponding current–voltage/conductance–voltage, which can be understood, at least qualitatively, in terms of transport processes. A quantitative analysis of these transport characteristics, however, can only be achieved, if local cooling of the vibrational degrees of freedom by electron–hole pair creation processes is taken into account (vide infra).

Note that the arguments given above strictly apply only for weak to intermediate electronic-vibrational coupling strengths,

$\lambda_{11}/\Omega_1 \lesssim 1$. For larger coupling strengths, the relative step heights in the vibrational excitation characteristic show deviations from the behavior described above. Nevertheless, electron–hole pair creation processes are also of importance to understand the vibrational excitation characteristic of these systems.

Electron–hole pair creation processes do also result in low levels of vibrational excitation in the non-resonant transport regime ($e\Phi < 2\bar{\varepsilon}_1$). In this regime (cf. the discussion in Section 5.3.1), inelastic co-tunneling processes lead to a level of vibrational excitation that is not as large as in the resonant transport regime but nevertheless significant. At the onset of inelastic co-tunneling, that is, at $e\Phi = \Omega_{1/2}$, vibrational excitation increases almost linearly with the applied bias voltage Φ. This behavior is very similar to the behavior of the respective current–voltage characteristic. Resonant electron–hole pair creation processes and resonant absorption processes, which are already active in this regime, control these levels of excitation. They lead to a stepwise decrease of vibrational excitation at $e\Phi = 2(\bar{\varepsilon}_1 - m\Omega_1 - n\Omega_2)$ (see Fig. 5.8).

If the electronic state is located far away from the chemical potential in the leads, resonant electron–hole pair creation processes (cf. Section 5.3.3.1) or CoSET (cf. Section 5.3.2.1) are not very efficient. Nevertheless, we observe rather low levels of vibrational excitation in the non-resonant transport regime of a molecular junction. This is, inter alia, due to the effect of off-resonant electron–hole pair creation processes (Fig. 5.7d), which become important, whenever resonant electron–hole pair creation processes are inefficient. This is also the case at high bias voltages, where the resonant creation of an electron–hole pair requires the absorption of many vibrational quanta. In contrast, off-resonant electron–hole pair creation processes can always occur upon the absorption of a single vibrational quantum, (almost) irrespective of the applied bias voltage.[7] The effect of off-resonant electron–hole pair creation processes is beyond the scope of our presentation here, but has been elucidated in some detail by a comparison of NEGF results and results that have been obtained using Born–Markov theory (Volkovich et al., 2011; Härtle, 2012).

[7]At $\Phi \to \infty$, however, off-resonant pair creation processes are also suppressed.

5.3.3.2 Cooling in the presence of multiple electronic states

So far, we have discussed local cooling with respect to a single electronic state. While, in principle, processes that occur in transport through a single electronic state do also occur with respect to multiple electronic states, the interplay of these processes is non-trivial. For example, the current–voltage characteristic of a junction with multiple electronic states is not simply given by the sum of currents that is obtained for each of these states separately. As discussed in Section 5.3.2.2, local heating of the vibrational degrees of freedom with respect to lower-lying electronic states of a molecular junction facilitates transport through the respective higher-lying electronic states, even before these states have entered the bias window. Steps in the current–voltage characteristic, which are associated with higher-lying electronic states, are therefore much less pronounced than the steps associated with lower-lying electronic states. This behavior has already been analyzed for the current–voltage characteristics of the two-state model molecular junction E2V1 (cf. Fig. 5.6, Section 5.3.2.2).

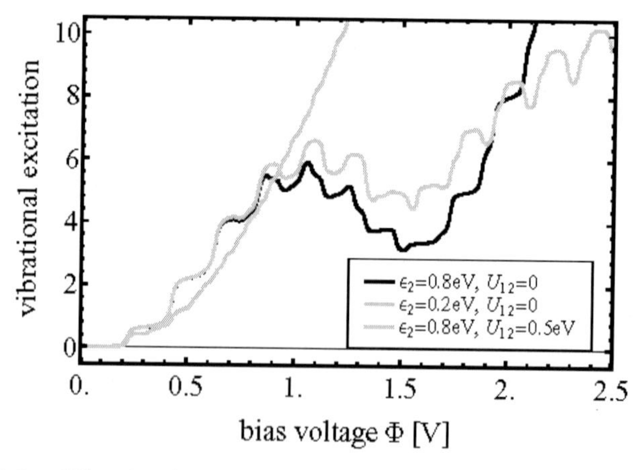

Figure 5.9 Vibrational excitation characteristics for a model molecular junction with two electronic states coupled to a single vibrational mode (model E2V1, cf. Table 5.1).

The corresponding excitation characteristic is shown in Fig. 5.9 by the solid black line. In addition, the solid green and blue lines show the vibrational excitation characteristic of a model molecular junction that deviates from model E2V1 just by the position of the higher-lying electronic state, that is $\varepsilon_2 = 0.2$ eV instead of $\varepsilon_2 = 0.8$ eV, and additional electron–electron interactions $U_{12} = 0.5$ eV instead of $U_{12} = 0$, respectively. For small bias voltages, $\Phi < 1$ V, one observes a continuous step-wise increase of vibrational excitation in these junctions, which is a result of local heating by inelastic transport processes with respect to the lower-lying electronic state ($\bar{\varepsilon}_1 = 0.114$ eV). Thereby, the solid green line exhibits additional steps, which are associated with local heating by the higher-lying electronic state at $\bar{\varepsilon}_2 = 0.164$ eV. Apart from these additional steps, the respective levels of vibrational excitation are very similar to the one of model E2V1 for $\Phi < 1$ V, and even to the one including just a single electronic state (cf. for example, the excitation characteristic of model E1V1 depicted in Fig. 5.8). This demonstrates that the level of vibrational excitation in a molecular junction is determined by the ratio between the number of excitation and deexcitation processes, which is almost the same for these three cases, rather than by the total number of processes.

For larger bias voltages, $\Phi > 1$ V, the levels of vibrational excitation that are obtained for these three systems strongly deviate from each other. This is due to the onset of resonant deexcitation processes with respect to sequential tunneling through (Fig. 5.5b) and electron–hole pair creation at the higher-lying electronic state (Fig. 5.7c). These processes result in a significantly lower level of vibrational excitation in junction E2V1, where the slope of the respective excitation characteristic can be even negative in the intermediate bias regime $2\bar{\varepsilon}_1 < e\Phi < 2\bar{\varepsilon}_2$. Although similar processes are also active for the junction with $\varepsilon_2 = 0.2$ eV, the efficiency of these processes is much larger in junction E2V1. This is because these processes require much less vibrational quanta in junction E2V1 at bias voltages $\Phi > 1$ V. Thus, the higher-lying electronic state in this molecular junction facilitates efficient cooling mechanisms for the vibrational degrees of freedom in this bias regime. These cooling mechanisms extend over a broad range of bias voltages, or in other words, stabilize the molecular junction. As the comparison of the black and the blue lines shows, electron–electron interactions greatly extend this range of bias voltages (Coulomb Cooling) (Härtle and Thoss, 2011a). These cooling or stabilization mechanisms are quite general phenomena in electron transport through molecular junctions (Härtle et al., 2009). They do also occur, for example, if the two electronic states are located below the Fermi level of the junction (see, for example, the benzenedibutanedithiolate molecular junction studied in Härtle et al. (2009)). In general, these mechanisms are facilitated by resonant deexcitation processes with respect to electronic states that are located farther away from the Fermi level of the junction.

5.4 The Role of Electronic-Vibrational Coupling for Possible Device Applications

In this section, we briefly discuss the relevance of vibronic effects for future technological applications. This includes first and foremost the possible use of single-molecule junctions in logic circuits (molecular electronics [Cuniberti et al., 2005; Cuevas and Scheer, 2010]). Diode-(Dhirani et al., 1997; Wu et al., 2004; Elbing et al., 2005; Yeganeh et al., 2007) and transistor-like behavior of single-molecule junctions (Piva et al., 2005; Song et al., 2009) have already been demonstrated. Even switches based on single molecules have been realized (Ho, 2002; Blum et al., 2005; Tao, 2006; Keane et al., 2006; Lörtscher et al., 2006; Choi et al., 2006; Quek et al., 2009; Wang et al., 2009; Meded et al., 2009; van der Molen et al., 2009; Mohn et al., 2010; van der Molen and Liljeroth, 2010). Thereby, one exploits the non-linear response of molecules to external bias voltages and/or electromagnetic fields. Vibrations can have both a constructive and a detrimental influence on this response.

A negative influence of vibrations can be observed when the device property is defined by sharp electronic resonances such as, for example, the step-like increase of the current at the onset of the resonant transport regime (cf. Section 5.3.2), rectification or negative differential resistance due to narrow conduction bands Davies et al., 1993, 1994) or blocking states (Hettler et al., 2002, 2003; Muralidharan and Datta, 2007; Härtle and Thoss, 2011a). Electronic-vibrational coupling can lead to a significant broadening of electronic resonances (Secker et al., 2011) and, therefore, quench the associated transport properties (cf., for example, the discussion of the second electronic resonance in Fig. 5.6 that is seen to be broadened by electronic-vibrational coupling). In contrast, we have shown that electronic-vibrational coupling can also give rise to negative differential resistance and rectification (Härtle and Thoss, 2011a; Härtle, 2012; Härtle et al., 2013c). This occurs in situations where cooling via electron–hole pair creation processes becomes blocked at specific bias voltages (leading to NDR) or inefficient, for example, due to an asymmetric coupling of the molecule to the electrodes (leading to rectification).

The asymmetric influence of electron–hole pair creation processes is not only relevant for logic circuits (or the spectroscopy of molecular junctions (Härtle et al., 2013c)). We also demonstrated that it can be used to control the level of vibrational excitation via the externally applied bias voltage (Härtle et al., 2010; Volkovich et al., 2011). Given suitable couplings of the electronic states to both the electrodes and the vibrational degrees of freedom, this control can be even mode-selective (Härtle et al., 2010; Volkovich et al., 2011), pointing towards chemical applications of molecular junctions (Jortner et al., 1991; Brumer and Shapiro, 2003; Pascual et al., 2003; Dzhioev and Kosov, 2011).

Many single-molecule junctions exhibit quasi-degenerate electronic states. They are often related to the symmetry of a junction, which can be a simple L $\leftrightarrow$ R symmetry or a more complex symmetry such as, for example, the D_{6h}-symmetry of a benzene molecule (Begemann et al., 2008; Donarini et al., 2009; Darau et al., 2009; Schultz, 2010). The coupling to the electrodes breaks these symmetries, which are required for exact degeneracies of electronic states. Thus, exact degeneracies are unlikely but quasidegeneracies are rather common. In these situations, the transport properties are, in principle, governed by quantum interference effects (Solomon et al., 2006b; Brisker et al., 2008; Begemann et al., 2008; Donarini et al., 2009; Darau et al., 2009; Schultz, 2010; Markussen et al., 2010; Härtle et al., 2013b). These effects may facilitate efficient means to control electron transport through molecules, or even quantum interference based molecular transistors (Stafford et al., 2007, issued August 31, 2010). Electronic-vibrational coupling, however, can lead to significant decoherence effects (Härtle et al., 2011, 2013a). While this suggests a detrimental influence, vibrationally induced decoherence, in conjunction with the effect of electron–hole pair creation processes, can also have a constructive effect. It gives rise, for example, to a pronounced temperature dependence of the current (Ballmann et al., 2012; Härtle et al., (2013a,c) or, in other words, a mean to control quantum interference effects in single-molecule junctions via the temperature in the electrodes.

5.5 Concluding Remarks

In this chapter, we have reviewed fundamental mechanisms of vibrationally coupled electron transport in molecular junctions. Employing generic models and a NEGF transport approach, we have discussed the signatures of electron-vibrational interaction in the current–voltage characteristics of a molecular junction and shown that the coupling to the vibrations results in a multitude of interesting non-equilibrium effects.

Due to great progress in the experimental techniques and theoretical methodology in recent years, a detailed understanding has been achieved for vibrationally coupled electron transport in the off-resonant low-voltage regime, e.g., for the interpretation of IETS spectra. On the other hand, transport in the resonant regime at higher bias voltages is much less well understood. This is due to instabilities that make experimental studies of molecular junction transport at higher voltages challenging. Similarly, accurate theoretical methods, which allow a first-principles treatment of vibrationally coupled electron transport in the resonant regime beyond perturbation theory are still to be developed. On the other hand, many interesting transport phenomena occur in the resonant transport regime and remain to be elucidated. Interesting processes and phenomena for future research include switching and multistability, current induced chemical reactions as well as time-dependent (non-stationary) transport.

Acknowledgements

We thank K. F. Albrecht, S. Ballmann, C. Benesch, D. Brisker-Klaiman, M. Bockstedte, M. Butzin, M. Cizek, P. B. Coto, W. Domcke, A. Erpenbeck, C. Hofmeister, A. Komnik, B. Kubala, S. Kurz, F. Langer, S. Leitherer, W. H. Miller, L. Mühlbacher, A. Nitzan, U. Peskin, I. Pshenichnyuk, E. Rabani, O. Rubio-Pons, C. Schinabeck, D. Secker, R. Volkovich, S. Wagner, H. Wang, H. B. Weber, E. Y. Wilner, and

A. Ziegler for fruitful discussions and collaborations. This work has been supported by the German Science Foundation (DFG) and the German-Israeli Foundation for Scientific Development (GIF). R. H. acknowledges financial support by the European Cooperation in Science and Technology (COST) and the Alexander-von-Humboldt Foundation via a Feodor-Lynen research fellowship. The generous allocation of computing time by the computing centers in Erlangen (RRZE) and Munich (LRZ) is gratefully acknowledged.

References

Agrait, N., Untiedt, C., Rubio-Bollinger, G., and Vieira, S. (2002a). Electron transport and phonons in atomic wires, *Chem. Phys.*, **281**(2–3), 231–234.

Agrait, N., Untiedt, C., Rubio-Bollinger, G., and Vieira, S. (2002b). Onset of energy dissipation in ballistic atomic wires, *Phys. Rev. Lett.*, **88**(21), 216803.

Albrecht, K. F., Wang, H., Mühlbacher, L., Thoss, M., and Komnik, A. (2012). Bistability signatures in non-equilibrium charge transport through molecular quantum dots, *Phys. Rev. B*, **86**(8), 081412(R).

Arroyo, C. R., Frederiksen, T., Rubio-Bollinger, G., Velez, M., Arnau, A., Sanchez-Portal, D., and Agrait, N. (2010). Characterization of single-molecule pentanedithiol junctions by inelastic electron tunneling spectroscopy and first-principles calculations, *Phys. Rev. B*, **81**(7), 075405.

Averin, D. V., and Nazarov, Y. V. (1990). Virtual electron diffusion during quantum tunneling of the electric charge, *Phys. Rev. Lett.*, **65**(19), 2446.

Avriller, R., and Levy Yeyati, A. (2009). Electron-phonon interaction and full counting statistics in molecular junctions, *Phys. Rev. B*, **80**(4), 041309.

Ballmann, S., Härtle, R., Coto, P. B., Elbing, M., Mayor, M., Bryce, M. R., Thoss, M., and Weber, H. B. (2012). Experimental evidence for quantum interference and vibrationally induced decoherence in single-molecule junctions, *Phys. Rev. Lett.*, **109**(5), 056801.

Ballmann, S., Hieringer, W., Härtle, R., Coto, P. B., Bryce, M. R., Görling, A., Thoss, M., and Weber, H. B. (2013). The role of vibrations in single-molecule charge transport: A case study of oligoynes with pyridine anchor groups, *Phys. Status Solidi B*, **250**(11), 2452–2457.

Ballmann, S., Hieringer, W., Secker, D., Zheng, Q., Gladysz, J. A., Görling, A., and Weber, H. B. (2010). Molecular wires in single-molecule junctions: Charge transport and vibrational excitations, *Chem. Phys. Chem.*, **11**(10), 2256–2260.

Bedkihal, S., Bandyopadhyay, M., and Segal, D. (2013). Magnetic field symmetries of nonlinear transport with elastic and inelastic scattering, *Phys. Rev. B*, **88**(15), 155407.

Begemann, G., Darau, D., Donarini, A., and Grifoni, M. (2008). Symmetry fingerprints of a benzene single-electron transistor: Interplay between coulomb interaction and orbital symmetry, *Phys. Rev. B*, **77**(20), 201406.

Benesch, C. (2009). Charge transport in single molecule junctions: Vi-bronic effects and conductance switching, Ph.D. thesis, Technische Universität München, Garching, Germany.

Benesch, C., Cizek, M., Klimes, J., Thoss, M., and Domcke, W. (2008). Vibronic effects in single molecule conductance: First-principles description and application to benzenealkanethiolates between gold electrodes, *J. Phys. Chem. C*, **112**(26), 9880–9890.

Benesch, C., Cizek, M., Thoss, M., and Domcke, W. (2006). Vibronic effects on resonant electron conduction through single molecule junctions, *Chem. Phys. Lett.*, **430**(4–6), 355–360.

Benesch, C., Rode, M. F., Cizek, M., Härtle, R., Rubio-Pons, O., Thoss, M., and Sobolewski, A. L. (2009). Switching the conductance of a single molecule by photoinduced hydrogen transfer, *J. Phys. Chem. C*, **113**(24), 10315–10318.

Bergfield, J. P., and Stafford, C. A. (2009). Many-body theory of electronic transport in single-molecule heterojunctions, *Phys. Rev. B*, **79**(24), 245125.

Blum, A. S., Kushmerick, J. G., Long, D. P., Patterson, C. H., Jang, J. C., Henderson, J. C., Yao, Y., Tour, J. M., Shashidhar, R., and Ratna, B. R. (2005). Molecularly inherent voltage-controlled conductance switching, *Nat. Mater.*, **4**(2), 167–172.

Böhler, T., Edtbauer, A., and Scheer, E. (2007). Conductance of individual C_{60} molecules measured with controllable gold electrodes, *Phys. Rev. B*, **76**(12), 125432.

Böhler, T., Grebing, J., Mayer-Gindner, A., Löhneysen, H., and Scheer, E. (2004). Mechanically controllable break-junctions for use as electrodes for molecular electronics, *Nanotechnology*, **15**(7), 465–471.

Bonitz, M. (1998). *Quantum Kinetic Theory* (Teubner, Stuttgart).

Braig, S., and Flensberg, K. (2003). Vibrational sidebands and dissipative tunneling in molecular transistors, *Phys. Rev. B*, **68**(20), 205324.

Brandbyge, M., Mozos, J.-L., Ordejon, P., Taylor, J., and Stokbro, K. (2002). Density-functional method for non-equilibrium electron transport, *Phys. Rev. B*, **65**(16), 165401.

Brisker, D., Cherkes, I., Gnodtke, C., Jarukanont, D., Klaiman, S., Koch, W., Weissmann, S., Volkovich, R., Caspary Toroker, M., and Peskin, U. (2008). Controlled electronic transport through branched molecular conductors, *Mol. Phys.*, **106**(2–4), 281.

Brumer, P. W., and Shapiro, M. (2003). *Principles of the Quantum Control of Molecular Processes* (Wiley, New Jersey).

Brumme, T., Neucheva, O. A., Toher, C., Gutiérrez, R., Weiss, C., Temirov, R., Greuling, A., Kaczmarski, M., Rohlfing, M., Tautz, F. S., and Cuniberti, G. (2011). Dynamical bistability of single-molecule junctions: A combined experimental and theoretical study of ptcda on Ag(111), *Phys. Rev. B*, **84**(11), 115449.

Cederbaum, L. S., and Domcke, W. (1974). On the vibrational structure in photoelectron spectra by the method of Green's functions, *J. Chem. Phys.*, **60**(7), 2878.

Cederbaum, L. S., and Domcke, W. (1976). A many-body approach to the vibrational structure in molecular electronic spectra. i. theory, *J. Chem. Phys.*, **64**(2), 603.

Chapline, G., and Wang, S. X. (2007). Analytical formula for the tunneling current versus voltage for multilayer barrier structures, *J. Appl. Phys.*, **101**(8), 083706.

Chen, Y., Prociuk, A., Perrine, T., and Dunietz, B. D. (2009). Spin-dependent electronic transport through a porphyrin ring ligating an Fe(ii) atom: An ab initio study, *Phys. Rev. B*, **74**(24), 245320.

Chen, Y., Zwolak, M., and Ventra, M. D. (2005). Inelastic effects on the transport properties of alkanethiols, *Nano Lett.*, **5**(4), 621–624.

Choi, B. Y., Kahng, S. J., Kim, S., Kim, H., Kim, H. W., Song, Y. J., Ihm, J., and Kuk, Y. (2006). Conformational molecular switch of the azobenzene molecule: A scanning tunneling microscopy study, *Phys. Rev. Lett.*, **96**(15), 156106.

Choi, S. H., Kim, B., and Frisbie, C. D. (2008). Electrical resistance of long conjugated molecular wires, *Science*, **320**(5882), 1482.

Cizek, M., Thoss, M., and Domcke, W. (2004). Theory of vibrationally inelastic electron transport through molecular bridges, *Phys. Rev. B*, **70**(12), 125406.

Cizek, M., Thoss, M., and Domcke, W. (2005). Charge transport through a flexible molecular junction, *Czech. J. Phys.*, **55**(2), 189–202.

Cohen, G., and Rabani, E. (2011). Memory effects in non-equilibrium quantum impurity models, *Phys. Rev. B*, **84**(7), 075150.

Cuevas, J. C., and Scheer, E. (2010). *Molecular Electronics: An Introduction to Theory and Experiment* (World Scientific, Singapore).

Cuniberti, G., Fagas, G., and Richter, K. (2005). *Introducing Molecular Electronics* (Springer, Heidelberg).

D'Agosta, R., Sai, N., and Ventra, M. D. (2006). Local electron heating in nanoscale conductors, *Nano Lett.*, **6**(12), 2935–2938.

Darau, D., Begemann, G., Donarini, A., and Grifoni, M. (2009). Interference effects on the transport characteristics of a benzene single-electron transistor, *Phys. Rev. B*, **79**(23), 235404.

Datta, S., Tian, W., Hong, S., Reifenberger, R., Henderson, J. I., and Kubiak, C. P. (1997). Current–voltage characteristics of self-assembled monolayers by scanning tunneling microscopy, *Phys. Rev. Lett.*, **79**(13), 2530.

Davies, J. H., Hershfield, S., Hyldgaard, P., and Wilkins, J. W. (1993). Current and rate equation for resonant tunneling, *Phys. Rev. B*, **47**(8), 4603.

Davies, J. H., Hershfield, S., Hyldgaard, P., and Wilkins, J. W. (1994). Resonant tunneling with an electron-phonon interaction, *Ann. Phys.* (NY), **236**(1), 1–42.

de Leon, N. P., Liang, W., Gu, Q., and Park, H. (2008). Vibrational excitation in single-molecule transistors: Deviation from the simple franckcondon prediction, *Nano Lett.*, **8**(9), 2963–2967.

Dhirani, A., Lin, P.-H., Guyot-Sionnest, P., Zehner, R. W., and Sita, L. R. (1997). Self-assembled molecular rectifiers, *J. Chem. Phys.*, **106**(12), 5249.

Domcke, W., Yarkony, D. R., and Köppel, H. (2004). *Conical Intersections: Electronic Structure, Dynamics and Spectroscopy* (World Scientific, Singapore).

Donarini, A., Begemann, G., and Grifoni, M. (2009). All-electric spin control in interference single electron transistors, *Nano Lett.*, **9**(8), 2897–2902.

Dzhioev, A. A., and Kosov, D. S. (2011). Kramers problem for non-equilibrium current-induced chemical reactions, *J. Chem. Phys.*, **135**(7), 074701.

Eckel, J., Heidrich-Meisner, F., Jakobs, S. G., Thorwart, M., Pletyukhov, M., and Egger, R. (2010). Comparative study of theoretical methods for non-equilibrium quantum transport, *New J. Phys.*, **12**(4), 043042.

Eigler, D. M., Lutz, C. P., and Rudge, W. E. (1991). An atomic switch realized with the scanning tunnelling microscope, *Nature*, **352**(6336), 600–603.

Eigler, D. M., and Schweizer, E. K. (1991). Positioning single atoms with a scanning tunnelling microscope, *Nature*, **344**, 524–526.

Elbing, M., Ochs, R., Koentopp, M., Fischer, M., von Hänisch, C., Weigend, F., Evers, F., Weber, H., and Mayor, M. (2005). A single-molecule diode, *PNAS*, **102**(25), 8815–8820.

Entin-Wohlman, O., Imry, Y., and Aharony, A. (2009). Voltage-induced singularities in transport through molecular junctions, *Phys. Rev. B*, **80**(3), 035417.

Entin-Wohlman, O., Imry, Y., and Aharony, A. (2010). Transport through molecular junctions with a non-equilibrium phonon population, *Phys. Rev. B*, **81**(11), 113408.

Esposito, M., and Galperin, M. (2009). Transport in molecular states language: Generalized quantum master equation approach, *Phys. Rev. B*, **79**(20), 205303.

Esposito, M., and Galperin, M. (2010). Self-consistent quantum master equation approach to molecular transport, *J. Phys. Chem. C*, **114**(48), 20362–20369.

Evers, F., Weigend, F., and Koentopp, M. (2004). Conductance of molecular wires and transport calculations based on density-functional theory, *Phys. Rev. B*, **69**(23), 235411.

Fainberg, B. D., Jouravlev, M., and Nitzan, A. (2007). Light-induced current in molecular tunneling junctions excited with intense shaped pulses, *Phys. Rev. B*, **76**(24), 245329.

Flensberg, K. (2003). Tunneling broadening of vibrational sidebands in molecular transistors, *Phys. Rev. B*, **68**(20), 205323.

Flindt, C., Novotny, T., Braggio, A., Sassetti, M., and Jauho, A.-P. (2008). Counting statistics of non-markovian quantum stochastic processes, *Phys. Rev. Lett.*, **100**(15), 150601.

Frederiksen, T., Brandbyge, M., Lorente, N., and Jauho, A. P. (2004). Inelastic scattering and local heating in atomic gold wires, *Phys. Rev. Lett.*, **93**(25), 256601.

Frederiksen, T., Lorente, N., Paulsson, M., and Brandbyge, M. (2007a). From tunneling to contact: Inelastic signals in an atomic gold junction from first principles, *Phys. Rev. B*, **75**(23), 235441.

Frederiksen, T., Paulsson, M., Brandbyge, M., and Jauho, A.-P. (2007b). Inelastic transport theory from first principles: Methodology and application to nanoscale devices, *Phys. Rev. B*, **75**(20), 205413.

Gagliardi, A., Solomon, G. C., Pecchia, A., Frauenheim, T., Di Carlo, A., Reimers, J. R., and Hush, N. S. (2007). A priori method for propensity rules for inelastic electron tunneling spectroscopy of single-molecule conduction, *Phys. Rev. B*, **75**(17), 174306.

Galperin, M., Nitzan, A., and Ratner, M. A. (2006a). Inelastic tunneling effects on noise properties of molecular junctions, *Phys. Rev. B*, **74**(7), 075326.

Galperin, M., Nitzan, A., and Ratner, M. A. (2006b). Molecular transport junctions: Current from electronic excitations in the leads, *Phys. Rev. Lett.*, **96**(16), 166803.

Galperin, M., Nitzan, A., and Ratner, M. A. (2006c). Resonant inelastic tunneling in molecular junctions, *Phys. Rev. B*, **73**(4), 045314.

Galperin, M., Nitzan, A., and Ratner, M. A. (2007a). Inelastic effects in molecular junctions in the coulomb and kondo regimes: Non-equilibrium equation-of-motion approach, *Phys. Rev. B*, **76**(3), 035301.

Galperin, M., Nitzan, A., and Ratner, M. A. (2008). Inelastic transport in the coulomb blockade regime within a non-equilibrium atomic limit, *Phys. Rev. B*, **78**(12), 125320.

Galperin, M., Ratner, M. A., and Nitzan, A. (2004). Inelastic electron tunneling spectroscopy in molecular junctions: Peaks and dips, *J. Chem. Phys.*, **121**(23), 11965.

Galperin, M., Ratner, M. A., and Nitzan, A. (2005). Hysteresis, switching, and negative differential resistance in molecular junctions: A polaron model, *Nano Lett.*, **5**(1), 125–130.

Galperin, M., Ratner, M. A., and Nitzan, A. (2007b). Molecular transport junctions: Vibrational effects, *J. Phys. Cond. Matter*, **19**(10), 103201.

Galperin, M., Saito, K., Balatsky, A. V., and Nitzan, A. (2009). Cooling mechanisms in molecular conduction junctions, *Phys. Rev. B*, **80**(11), 115427.

Galperin, M., and Tretiak, S. (2008). Linear optical response of current-carrying molecular junction: A non-equilibrium greens function time-dependent density functional theory approach, *J. Chem. Phys.*, **128**(12), 124705.

Gao, S., Persson, M., and Lundqvist, B. (1992). Atomic switch proves importance of electron–hole pair mechanism in processes on metal surfaces, *Sol. State Commun.*, **84**(3), 271–273.

Gaudioso, J., Lauhon, L. J., and Ho, W. (2000). Vibrationally mediated negative differential resistance in a single molecule, *Phys. Rev. Lett.*, **85**(9), 1918–1921.

Groshev, A., Ivanov, T., and Valtchinov, V. (1991). Charging effects of a single quantum level in a box, *Phys. Rev. Lett.*, **66**(8), 1082.

Hanke, U., Galperin, Y., Chao, K. A., Gisselfält, M., Jonson, M., and Shekhter, R. I. (1995). Static and dynamic transport in parity-sensitive systems, *Phys. Rev. B*, **51**(14), 9084–9095.

Hanna, A. E., and Tinkham, M. (1991). Variation of the coulomb staircase in a two-junction system by fractional electron charge, *Phys. Rev. B*, **44**(11), 5919–5922.

Harbola, U., Esposito, M., and Mukamel, S. (2006). Quantum master equation for electron transport through quantum dots and single molecules, *Phys. Rev. B*, **74**(23), 235309.

Härtle, R. (2012). Vibrationally coupled electron transport through single-molecule junctions, Ph.D. thesis, Friedrich-Alexander-Universität Erlangen-Nürnberg, Erlangen, Germany.

Härtle, R., Benesch, C., and Thoss, M. (2008). Multimode vibrational effects in single-molecule conductance: A non-equilibrium Green's function approach, *Phys. Rev. B*, **77**(20), 205314.

Härtle, R., Benesch, C., and Thoss, M. (2009). Vibrational non-equilibrium effects in the conductance of single molecules with multiple electronic states, *Phys. Rev. Lett.*, **102**(14), 146801.

Härtle, R., Butzin, M., Rubio-Pons, O., and Thoss, M. (2011). Quantum interference and decoherence in single-molecule junctions: How vibrations induce electrical current, *Phys. Rev. Lett.*, **107**(4), 046802.

Härtle, R., Butzin, M., and Thoss, M. (2013a). Vibrationally induced decoherence in single-molecule junctions, *Phys. Rev. B*, **87**(8), 085422.

Härtle, R., Cohen, G., Reichman, D. R., and Millis, A. J. (2013b). De-coherence and lead induced inter-dot coupling in non-equilibrium electron transport through interacting quantum dots: A hierarchical quantum master equation approach, *Phys. Rev. B*, **88**(23), 235426.

Härtle, R., Peskin, U., and Thoss, M. (2013c). Vibrationally coupled electron transport in single-molecule junctions: The importance of electronhole pair creation processes, *Phys. Status Solidi B*, **250**(11), 2365–2377.

Härtle, R., and Thoss, M. (2011a). Resonant electron transport in single-molecule junctions: Vibrational excitation, rectification, negative differential resistance, and local cooling, *Phys. Rev. B*, **83**(11), 115414.

Härtle, R., and Thoss, M. (2011b). Vibrational instabilities in resonant electron transport through single-molecule junctions, *Phys. Rev. B*, **83**(12), 125419.

Härtle, R., Volkovich, R., Thoss, M., and Peskin, U. (2010). Communications: Mode-selective vibrational excitation induced by non-equilibrium transport processes in single-molecule junctions, *J. Chem. Phys.*, **133**(8), 081102.

Haug, H., and Jauho, A.-P. (1996). *Quantum Kinetics in Transport and Optics of Semiconductors* (Springer, Berlin).

Haupt, F., Novotny, T., and Belzig, W. (2009). Phonon-assisted current noise in molecular junctions, *Phys. Rev. Lett.*, **103**(13), 136601.

Hettler, M. H., Schoeller, H., and Wenzel, W. (2002). Non-linear transport through a molecular nanojunction, *Europhys. Lett.*, **57**(4), 571.

Hettler, M. H., Wenzel, W., Wegewijs, M. R., and Schoeller, H. (2003). Current collapse in tunneling transport through benzene, *Phys. Rev. Lett.*, **90**(7), 076805.

Hihath, J., Arroyo, C. R., Rubio-Bollinger, G., Tao, N. J., and Agrait, N. (2008). Study of electron-phonon interactions in a single molecule covalently connected to two electrodes, *Nano Lett.*, **8**(6), 1673–1678.

Hihath, J., Bruot, C., and Tao, N. (2010). Electronphonon interactions in single octanedithiol molecular junctions, *ACS Nano*, **4**(7), 3823–3830.

Ho, W. (2002). Single-molecule chemistry, *J. Chem. Phys.*, **117**(24), 11033.

Huang, Z., Xu, B., Chen, Y., Ventra, M. D., and Tao, N. (2006). Measurement of current-induced local heating in a single molecule junction, *Nano Lett.*, **6**(6), 1240–1244.

Huisman, E. H., Gudon, C. M., van Wees, B. J., and van der Molen, S. J. (2009). Interpretation of transition voltage spectroscopy, *Nano Lett.*, **9**(11), 3909–3913.

Hüttel, A. K., Witkamp, B., Leijnse, M., Wegewijs, M. R., and van der Zant, H. S. J. (2009). Pumping of vibrational excitations in the coulomb-blockade regime in a suspended carbon nanotube, *Phys. Rev. Lett.*, **102**(22), 225501.

Hützen, R., Weiss, S., Thorwart, M., and Egger, R. (2012). Iterative summation of path integrals for non-equilibrium molecular quantum transport, *Phys. Rev. B*, **85**(12), 121408(R).

Ingold, G.-L., and Nazarov, Y. V. (1992). *Charge Tunneling Rates in Ultrasmall Junctions*, NATO ASI Series B, vol. 294 (Plenum Press), pp. 21–107.

Ioffe, Z., Shamai, T., Ophir, A., Noy, G., Yutsis, I., Kfir, K., Chesh-novsky, O., and Selzer, Y. (2008). Detection of heating in current-carrying molecular junctions by raman scattering, *Nat. Nanotech.*, **3**(12), 727–732.

Jaklevic, R. C., and Lambe, J. (1966). Molecular vibration spectra by electron tunneling, *Phys. Rev. Lett.*, **17**(22), 1139–1140.

Jewell, A. D., Tierney, H. L., Baber, A. E., Iski, E. V., Laha, M. M., and Sykes, E. C. H. (2010). Time-resolved studies of individual molecular rotors, *J. Phys.: Condens. Matter*, **22**(26), 264006.

Jorn, R., and Seidemann, T. (2009). Competition between current-induced excitation and bath-induced decoherence in molecular junctions, *J. Chem. Phys.*, **131**(24), 244114.

Jortner, J., Levine, R. D., and Pullman, B. (1991). *Mode-Selective Chemistry* (Kluwer, Amsterdam).

Jovchev, A., and Anders, F. (2013). Influence of vibrational modes on quantum transport through a nanodevice, *Phys. Rev. B*, **87**(19), 195112.

Kast, D., Kecke, L., and Ankerhold, J. (2011). Charge transfer through single molecule contacts: How reliable are rate descriptions? *Beilstein J. Nanotechnol.*, **2**, 416–426.

Ke, S.-H., Baranger, H. U., and Yang, W. (2004). Electron transport through molecules:self-consistent and non-self-consistent approaches, *Phys. Rev. B*, **70**(8), 085410.

Keane, Z. K., Ciszek, J. W., Tour, J. M., and Natelson, D. (2006). Three-terminal devices to examine single-molecule conductance switching, *Nano Lett.*, **6**(7), 1518–1521.

Kiguchi, M., Tal, O., Wohlthat, S., Pauly, F., Krieger, M., Djukic, D., Cuevas, J. C., and van Ruitenbeek, J. M. (2008). Highly conductive molecular junctions based on direct binding of benzene to platinum electrodes, *Phys. Rev. Lett.*, **101**(4), 046801.

Kim, Y., Song, H., Strigl, F., Pernau, H.-F., Lee, T., and Scheer, E. (2011). Conductance and vibrational states of single-molecule junctions controlled by mechanical stretching and material variation, *Phys. Rev. Lett.*, **106**(19), 196804.

Koch, J., Raikh, M. E., and von Oppen, F. (2005). Full counting statistics of strongly non-ohmic transport through single molecules, *Phys. Rev. Lett.*, **95**(5), 056801.

Koch, J., Semmelhack, M., von Oppen, F., and Nitzan, A. (2006a). Current-induced non-equilibrium vibrations in single-molecule devices, *Phys. Rev. B*, **73**(15), 155306.

Koch, J., and von Oppen, F. (2005). Franck–Condon blockade and giant fano factors in transport through single molecules, *Phys. Rev. Lett.*, **94**(20), 206804.

Koch, J., von Oppen, F., and Andreev, A. V. (2006b). Theory of the Franck–Condon blockade regime, *Phys. Rev. B*, **74**(20), 205438.

Koch, T., Loos, J., Alvermann, A., Bishop, A. R., and Fehske, H. (2010). Transport through a vibrating quantum dot: Polaronic effects, *J. Phys. Conf. Ser.*, **220**(1), 012014.

Kondov, I., Cizek, M., Benesch, C., Thoss, M., and Wang, H. (2007). Quantum dynamics of photoinduced electron-transfer reactions in dyesemiconductor systems: First-principles description and application to coumarin 343-TiO_2, *J. Chem. Phys. C*, **111**(32), 11970–11981.

König, J., Schoeller, H., and Schön, G. (1996). Zero-bias anomalies and boson-assisted tunneling through quantum dots, *Phys. Rev. Lett.*, **76**(10), 1715.

Kral, P. (1997). Non-equilibrium linked cluster expansion for steady-state quantum transport, *Phys. Rev. B*, **56**(12), 7293.

Kushmerick, J. G., Lazorcik, J., Patterson, C. H., Shashidhar, R., Se-feros, D. S., and Bazan, G. C. (2004). Vibronic contributions to charge transport across molecular junctions, *Nano Lett.*, **4**(4), 639–642.

Lafferentz, L., Ample, F., Yu, H., Hecht, S., Joachim, C., Grill, L., and Reed, M. A. (2009). Conductance of a single conjugated polymer as a continuous function of its length, *Science*, **27**(5918), 1193–1197.

Lang, I. G., and Firsov, Y. A. (1963). Kinetic theory of semiconductors with low mobility, *Sov. Phys. JETP*, **16**(5), 1301.

Langreth, D. C. (1976). *Linear and Nonlinear Response Theory with Applications*, NATO ASI Series B, vol. 17 (Plenum Press), pp. 3–32.

Lehmann, J., Kohler, S., May, V., and Hänggi, P. (2004). Vibrational effects in laser-driven molecular wires, *J. Chem. Phys.*, **121**(5), 2278.

Leijnse, M., and Wegewijs, M. R. (2008). Kinetic equations for transport through single-molecule transistors, *Phys. Rev. B*, **78**(23), 235424.

LeRoy, B. J., Lemay, S. G., Kong, J., and Dekker, C. (2004). Electrical generation and absorption of phonons in carbon nanotubes, *Nature*, **432**(7015), 371–374.

Li, B., Levy, T. J., Swenson, D. W. H., Rabani, E., and Miller, W. H. (2013). A cartesian quasi-classical model to non-equilibrium quantum transport: The anderson impurity model, *J. Chem. Phys.*, **138**(10), 104110.

Li, B., Wilner, E. Y., Thoss, M., Rabani, E., and Miller, W. H. (2014). A quasi-classical mapping approach to vibrationally coupled electron transport in molecular junctions, *J. Chem. Phys.*, **140**(10), 104110.

Liang, W., Shores, M. P., Bockrath, M., Long, J. R., and Park, H. (2002). Kondo resonance in a single-molecule transistor, *Nature*, **417**(6890), 725.

Lorente, N., and Persson, M. (2000). Theory of single molecule vibrational spectroscopy and microscopy, *Phys. Rev. Lett.*, **85**(14), 2997–3000.

Lörtscher, E., Ciszek, J. W., Tour, J., and Riel, H. (2006). Reversible and controllable switching of a single-molecule junction, *Small*, **2**(8–9), 973–977.

Lüffe, M. C., Koch, J., and von Oppen, F. (2008). Theory of vibrational absorption sidebands in the coulomb-blockade regime of single-molecule transistors, *Phys. Rev. B*, **77**(12), 125306.

Mahan, G. D. (1981). *Many Particle Physics* (Plenum Press, New York).

Markussen, T., Stadler, R., and Thygesen, K. S. (2010). The relation between structure and quantum interference in single molecule junctions, *Nano Lett.*, **10**(10), 4260–4265.

Martin, C. A., van Ruitenbeek, J. M., and van der Zant, H. S. J. (2010). Sandwich-type gated mechanical break junctions, *Nanotechnology*, **21**(26), 265201.

May, V. (2002). Electron transfer through a molecular wire: Consideration of electron-vibrational coupling within the liouville space pathway technique, *Phys. Rev. B*, **66**(24), 245411.

May, V., and Kühn, O. (2008a). Optical field control of charge transmission through a molecular wire. i. generalized master equation description, *Phys. Rev. B*, **77**(11), 115439.

May, V., and Kühn, O. (2008b). Optical field control of charge transmission through a molecular wire. ii. photoinduced removal of the Franck–Condon blockade, *Phys. Rev. B*, **77**(11), 115440.

McEniry, E. J., Bowler, D. R., Dundas, D., Horsfield, A. P., Sánchez, C. G., and Todorov, T. N. (2007). Dynamical simulation of inelastic quantum transport, *J. Phys. Cond. Matter*, **19**(19), 196201.

McEniry, E. J., Frederiksen, T., Todorov, T. N., Dundas, D., and Horsfield, A. P. (2008). Inelastic quantum transport in nanostructures: The self-consistent born approximation and correlated electron-ion dynamics, *Phys. Rev. B*, **78**(3), 035446.

McEniry, E. J., Todorov, T. N., and Dundas, D. (2009). Current-assisted cooling in atomic wires, *J. Phys.: Cond. Matter*, **21**(19), 195304.

Meded, V., Bagrets, A., Arnold, A., and Evers, F. (2009). Molecular switch controlled by pulsed bias voltages, *Small*, **5**(19), 2218–2223.

Mii, T., Tikhodeev, S., and Ueba, H. (2002). Theory of vibrational tunneling spectroscopy of adsorbates on metal surfaces, *Surf. Sci.*, **502–503**, 26–33.

Mitra, A., Aleiner, I., and Millis, A. J. (2004). Phonon effects in molecular transistors:Quantal and classical treatment, *Phys. Rev. B*, **69**(24), 245302.

Mohn, F., Repp, J., Gross, L., Meyer, G., Dyer, M. S., and Persson, M. (2010). Reversible bond formation in a gold-atom-organic-molecule complex as a molecular switch, *Phys. Rev. Lett.*, **105**(26), 266102.

Monturet, S., Alducin, M., and Lorente, N. (2010). Role of molecular electronic structure in inelastic electron tunneling spectroscopy: O_2 on Ag(110), *Phys. Rev. B*, **82**(8), 085447.

Mühlbacher, L., and Rabani, E. (2008). Real-time path integral approach to non-equilibrium many-body quantum systems, *Phys. Rev. Lett.*, **100**(17), 176403.

Mujica, V., Ratner, M. A., and Nitzan, A. (2002). Molecular rectification: Why is it so rare? *Chem. Phys.*, **281**(2), 147–150.

Muralidharan, B., and Datta, S. (2007). Generic model for current collapse in spin-blockaded transport, *Phys. Rev. B*, **76**, 035432.

Nakajima, S. (1958). On quantum theory of transport phenomena, *Prog. Theor. Phys.*, **20**(6), 948–959.

Nitzan, A. (2001). Electron transmission through molecules and molecular interfaces, *Ann. Rev. Phys. Chem.*, **52**(1), 681–725.

Ogawa, N., Mikaelian, G., and Ho, W. (2007). Spatial variations in sub-molecular vibronic spectroscopy on a thin insulating film, *Phys. Rev. Lett.*, **98**(16), 166103.

Osorio, E. A., Ruben, M., Seldenthuis, J. S., Lehn, J. M., and van der Zant, H. S. J. (2010). Conductance switching and vibrational fine structure of a [22] coii4 gridlike single molecule measured in a three-terminal device, *Small*, **6**(2), 174–178.

Park, J., Pasupathy, A. N., Goldsmith, J. I., Chang, C., Yaish, Y., Petta, J. R., Rinkoski, M., Sethna, J. P., Abruna, H. D., McEuen, P. L., and Ralph, D. C. (2002). Coulomb blockade and the kondo effect in single-atom transistors, *Nature*, **417**(6890), 722–725.

Parks, J. J., Champagne, A. R., Hutchison, G. R., Flores-Torres, S., Abruna, H. D., and Ralph, D. C. (2007). Tuning the kondo effect with a mechanically controllable break junction, *Phys. Rev. Lett.*, **99**(2), 026601.

Pascual, J. I., Lorente, N., Song, Z., Conrad, H., and Rust, H. P. (2003). Selectivity in vibrationally mediated single-molecule chemistry, *Nature*, **423**(6939), 525–528.

Pasupathy, A. N., Park, J., Chang, C., Soldatov, A. V., Lebedkin, S., Bialczak, R. C., Grose, J. E., Donev, L. A. K., Sethna, J. P., Ralph, D. C., and McEuen,

P. L. (2005). Vibration-assisted electron tunneling in c140 transistors, *Nano Lett.*, **5**(2), 203–207.

Pedersen, J. N., and Wacker, A. (2005). Tunneling through nanosystems: Combining broadening with many-particle states, *Phys. Rev. B*, **72**(19), 195330.

Persson, B. N. J., and Ueba, H. (2002). Theory of inelastic tunneling induced motion of adsorbates on metal surfaces, *Surf. Sci.*, **502–503**, 18–25.

Piva, P. G., DiLabio, G. A., Pitters, J. L., Zikovsky, J., Rezeq, M., Dogel, S., Hofer, W. A., and Wolkow, R. A. (2005). Field regulation of single-molecule conductivity by a charged surface atom, *Nature*, **435**(7042), 658–661.

Pshenichnyuk, I. A., and Cizek, M. (2011). Motor effect in electron transport through a molecular junction with torsional vibrations, *Phys. Rev. B*, **83**(16), 165446.

Pump, F., Temirov, R., Neucheva, O., Soubatch, S., Tautz, S., Rohlfing, M., and Cuniberti, G. (2008). Quantum transport through STM-lifted single ptcda molecules, *Appl. Phys. A*, **93**(2), 335–343.

Quek, S. Y., Kamenetska, M., Steigerwald, M. L., Choi, H. J., Louie, S. G., Hybertsen, M. S., Neaton, J. B., and Venkataraman, L. (2009). Mechanically-controlled binary conductance switching of a single-molecule junction, *Nat. Nanotech.*, **4**(4), 230–234.

Reed, M. A., Zhou, C., Muller, C. J., Burgin, T. P., and Tour, J. M. (1997). Conductance of a molecular junction, *Science*, **278**(5336), 252–254.

Repp, J., Liljeroth, P., and Meyer, G. (2010). Coherent electronnuclear coupling in oligothiophene molecular wires, *Nat. Phys.*, **6**(12), 975–979.

Romano, G., Gagliardi, A., Pecchia, A., and Di Carlo, A. (2010). Heating and cooling mechanisms in single-molecule junctions, *Phys. Rev. B*, **81**(11), 115438.

Ryndyk, D. A., Hartung, M., and Cuniberti, G. (2006). Non-equilibrium molecular vibrons: An approach based on the non-equilibrium green function technique and the self-consistent born approximation, *Phys. Rev. B*, **73**(4), 045420.

Sapmaz, S., Jarillo-Herrero, P., Blanter, Y. M., Dekker, C., and van der Zant, H. S. J. (2006). Tunneling in suspended carbon nanotubes assisted by longitudinal phonons, *Phys. Rev. Lett.*, **96**(2), 026801.

Sapmaz, S., Jarillo-Herrero, P., Blanter, Y. M., and van der Zant, H. S. J. (2005). Coupling between electronic transport and longitudinal phonons in suspended nanotubes, *New J. Phys.*, **7**(1), 243.

Schiff, P. R., and Nitzan, A. (2010). Kramers barrier crossing as a cooling machine, *Chem. Phys.*, **375**(2–3), 399.

Schmidt, T. L., and Komnik, A. (2009). Charge transfer statistics of a molecular quantum dot with a vibrational degree of freedom, *Phys. Rev. B*, **80**(4), 041307.

Schultz, M. G. (2010). Quantum transport through single-molecule junctions with orbital degeneracies, *Phys. Rev. B*, **82**(15), 155408.

Schultz, M. G., and von Oppen, F. (2009). Quantum transport through nanostructures in the singular-coupling limit, *Phys. Rev. B*, **80**(3), 033302.

Schulze, G., Franke, K. J., Gagliardi, A., Romano, G., Lin, C. S., Rosa, A. D., Niehaus, T. A., Frauenheim, T., Carlo, A. D., Pecchia, A., and Pascual, J. I. (2008). Resonant electron heating and molecular phonon cooling in single c60 junctions, *Phys. Rev. Lett.*, **100**(13), 136801.

Secker, D. (2008). Dynamisches Verhalten beim Ladungstransport durch organische Moleküle, Ph.D. thesis, Friedrich-Alexander-Universität Erlangen-Nürnberg, Erlangen, Germany.

Secker, D., Wagner, S., Härtle, S. B. R., Thoss, M., and Weber, H. B. (2011). Resonant vibrations, peak broadening, and noise in single molecule contacts: The nature of the first conductance peak, *Phys. Rev. Lett.*, **106**(13), 136807.

Shen, X. Y., Dong, B., Lei, X. L., and Horing, N. J. M. (2007). Vibration-mediated resonant tunneling and shot noise through a molecular quantum dot, *Phys. Rev. B*, **76**(11), 115308.

Siddiqui, L., Ghosh, A. W., and Datta, S. (2007). Phonon runaway in carbon nanotube quantum dots, *Phys. Rev. B*, **76**(8), 085433.

Simine, L., and Segal, D. (2013). Path-integral simulations with fermionic and bosonic reservoirs: Transport and dissipation in molecular electronic junctions, *J. Chem. Phys.* **138**(21), 214111.

Simmons, J. (1963). Generalized formula for the electric tunnel effect between similar electrodes separated by a thin insulating film, *J. Appl. Phys.*, **34**(6), 1793.

Smit, R. H. M., Noat, Y., Untiedt, C., Lang, N. D., van Hemert, M. C., and van Ruitenbeek, J. M. (2002). Measurement of the conductance of a hydrogen molecule, *Nature*, **419**(6910), 906–909.

Solomon, G. C., Gagliardi, A., Pecchia, A., Frauenheim, T., Carlo, A. D., Hush, N. S., and Reimers, J. R. (2006a). The symmetry of single-molecule conduction, *J. Chem. Phys.*, **125**(18), 184702.

Solomon, G. C., Gagliardi, A., Pecchia, A., Frauenheim, T., Di Carlo, A., Reimers, J. R., and Hush, N. S. (2006b). Molecular origins of conduction channels observed in shot-noise measurments, *Nano Lett.*, **6**(11), 2431–2437.

Song, H., Kim, Y., Jang, Y. H., Jeong, H., Reed, M. A., and Lee, T. (2009). Observation of molecular orbital gating, *Nature*, **462**(7276), 1039–1043.

Stafford, C. A., Cardamone, D. M., and Mazumdar, S. (2007). The quantum interference effect transistor, *Nanotechnology*, **18**(42), 424014.

Stafford, C. A., Cardamone, D. M., and Mazumdar, S. (issued August 31, 2010). Quantum interference effect transistor (QuIET) U.S. Patent No. 7,786,472.

Stipe, B. C., Rezai, M. A., and Ho, W. (1998). Single-molecule vibrational spectroscopy and microscopy, *Science*, **280**(5370), 1732.

Swenson, D. W. H., Cohen, G., and Rabani, E. (2012). A semiclassical model for the non-equilibrium quantum transport of a many-electron Hamiltonian coupled to phonons, *Mol. Phys.*, **110**(9–10), 743.

Tahir, M., and MacKinnon, A. (2008). Quantum transport in a resonant tunnel junction coupled to a nanomechanical oscillator, *Phys. Rev. B*, **77**(22), 224305.

Tal, O., Krieger, M., Leerink, B., and van Ruitenbeek, J. M. (2008). Electron-vibration interaction in single-molecule junctions: From contact to tunneling regimes, *Phys. Rev. Lett.*, **100**(19), 196804.

Tao, N. J. (2006). Electron transport in molecular junctions, *Nat. Nanotech.*, **1**(3), 173–181.

Thijssen, W. H. A., Djukic, D., Otte, A. F., Bremmer, R. H., and van Ruitenbeek, J. M. (2006). Vibrationally induced two-level systems in single-molecule junctions, *Phys. Rev. Lett.*, **97**(22), 226806.

Timm, C. (2008). Tunneling through molecules and quantum dots: Master-equation approaches, *Phys. Rev. B*, **77**(19), 195416.

Toroker, M. C., and Peskin, U. (2007). Electronic transport through molecular junctions with nonrigid molecule–leads coupling, *J. Chem. Phys.*, **127**(15), 154706.

Troisi, A., and Ratner, M. A. (2006). Molecular transport junctions: Propensity rules for inelastic electron tunneling spectra, *Nano Lett.*, **6**(8), 1784–1788.

Trouwborst, M. L., Martin, C. A., Smit, R. H. M., Gudon, C. M., Baart, T. A., van der Molen, S. J., and van Ruitenbeek, J. M. (2011). Transition voltage spectroscopy and the nature of vacuum tunneling, *Nano Lett.*, **11**(2), 614–617.

van der Molen, S. J., Liao, J., Kudernac, T., Agustsson, J. S., Bernard, L., Calame, M., van Wees, B. J., Feringa, B. L., and Schönenberger, C. (2009). Light-controlled conductance switching of ordered metal-molecule-metal devices, *Nano Lett.*, **9**(1), 76–80.

van der Molen, S. J., and Liljeroth, P. (2010). Charge transport through molecular switches, *J. Phys.: Cond. Matter*, **22**(13), 133001.

Volkovich, R., Härtle, R., Thoss, M., and Peskin, U. (2011). Bias-controlled selective excitation of vibrational modes in molecular junctions: A route towards mode-selective chemistry, *Phys. Chem. Chem. Phys.*, **13**(32), 14333–14349.

Volkovich, R., and Peskin, U. (2011). Transient dynamics in molecular junctions: Coherent bichromophoric molecular electron pumps, *Phys. Rev. B*, **83**(3), 033403.

Volkovich, R., Toroker, M. C., and Peskin, U. (2008). Site-directed electronic tunneling in a dissipative molecular environment, *J. Chem. Phys.*, **129**(3), 034501.

Wang, H., Pshenichnyuk, I., Härtle, R., and Thoss, M. (2011). Numerically exact, time-dependent treatment of vibrationally coupled electron transport in single-molecule junctions, *J. Chem. Phys.*, **135**(24), 244506.

Wang, H., and Thoss, M. (2009). Numerically exact quantum dynamics for indistinguishable particles: The multilayer multiconfiguration time-dependent hartree theory in second quantization representation, *J. Chem. Phys.*, **131**(2), 024114.

Wang, Y., Kröger, J., Berndt, R., and Hofer, W. A. (2009). Pushing and pulling a SN-ion through an adsorbed phthalocyanine molecule, *J. Am. Chem. Soc.*, **131**(10), 3639–3643.

Ward, D. R., Corley, D. A., Tour, J. M., and Natelson, D. (2011). Vibrational and electronic heating in nanoscale junctions, *Nat. Nanotechnol.*, **6**(1), 33–38.

Ward, D. R., Halas, N. J., Ciszek, J. W., Tour, J. M., Wu, Y., Nordlander, P., and Natelson, D. (2008). Simultaneous measurements of electronic conduction and raman response in molecular junctions, *Nano Lett.*, **8**(3), 919–924.

Weiss, S., Eckel, J., Thorwart, M., and Egger, R. (2008). Iterative realtime path integral approach to non-equilibrium quantum transport, *Phys. Rev. B*, **77**(19), 195316.

Wheeler, W. D., and Dahnovsky, Y. (2008). Molecular transistors based on BDT-type molecular bridges, *J. Chem. Phys.*, **129**(15), 154112.

Wilner, E. Y., Wang, H., Cohen, G., Thoss, M., and Rabani, E. (2013). Bistability in a non-equilibrium quantum system with electron-phonon interactions, *Phys. Rev. B*, **88**(4), 045137.

Wilner, E. Y., Wang, H., Thoss, M., and Rabani, E. (2014). Non-equilibrium quantum systems with electron-phonon interactions: Transient dynamics and approach to steady state, *Phys. Rev. B*, **89**(20), 205129.

Wu, S. W., Nazin, G. V., Chen, X., Qiu, X. H., and Ho, W. (2004). Control of relative tunneling rates in single molecule bipolar electron transport, *Phys. Rev. Lett.*, **93**(23), 236802.

Xue, Y., and Ratner, M. A. (2003a). Microscopic study of electrical transport through individual molecules with metallic contacts. i. band lineup, voltage drop, and high-field transport, *Phys. Rev. B*, **68**(11), 115407.

Xue, Y., and Ratner, M. A. (2003b). Microscopic study of electrical transport through individual molecules with metallic contacts. ii. effect of the interface structure, *Phys. Rev. B*, **68**(11), 115406.

Yeganeh, S., Galperin, M., and Ratner, M. A. (2007). Switching in molecular transport junctions: Polarization response, *J. Am. Chem. Soc.*, **129**(43), 13313–13320.

Yu, L. H., Keane, Z. K., Ciszek, J. W., Cheng, L., Stewart, M. P., Tour, J. M., and Natelson, D. (2004). Inelastic electron tunneling via molecular vibrations in single-molecule transistors, *Phys. Rev. Lett.*, **93**(26), 266802.

Zazunov, A., Feinberg, D., and Martin, T. (2006). Phonon-mediated negative differential conductance in molecular quantum dots, *Phys. Rev. B*, **73**(11), 115405.

Zimbovskaya, N. A., and Kuklja, M. M. (2009). Vibration-induced inelastic effects in the electron transport through multisite molecular bridges, *J. Chem. Phys.*, **131**(11), 114703.

Zwanzig, R. (1960). Ensemble method in the theory of irreversibility, *J. Chem. Phys.*, **33**(5), 1338.

Chapter 6

Vibration Spectroscopy of Single Molecular Junctions

Manabu Kiguchi and Ryuji Matsushita

Department of Chemistry, Tokyo Institute of Technology, Ookayama, Meguro-ku, Tokyo 152-8551, Japan

kiguti@chem.titech.ac.jp

The characterization of single molecular junctions is essential to understanding their properties, including their electronic, magnetic, and optical properties. Vibration spectroscopy is a powerful technique for characterizing the atomic structure of single molecular junctions. In this chapter, we describe the point-contact spectroscopy (PCS), inelastic electron tunneling spectroscopy (IETS), and surface-enhanced Raman scattering (SERS) as a vibration spectroscopy of single molecular junctions. Both PCS and IETS make use of electron–vibration interaction. Above a threshold voltage, the electron can excite a vibration mode of the single molecular junctions. This excitation of the vibration modes leads to an increase in differential conductance for IETS, and a decrease in differential conductance for PCS. The threshold voltage provides us with the energy of the vibration mode. In contrast with PCS and IETS, SERS does not require low temperature.

Molecular Electronics: An Experimental and Theoretical Approach
Edited by Ioan Bâldea

ISBN 978-981-4613-90-3 (Hardcover), 978-981-4613-91-0 (eBook)
www.panstanford.com

A strong electric field is formed in the nanogap in the single molecular junctions, which enhances the intensity of the Raman signal, and thus the Raman spectrum from a single molecule in the nanogap is selectively observed. Simultaneous SERS and conductance measurements can give us information on distinct states of the electronic and geometric structure of the single molecular junctions, which makes the single molecule dynamics clear.

6.1 Introduction

The single molecular junction, where a single molecule is bridged between metal electrodes, has attracted great attention owing to its potential application to molecular electronics [1–3]. The single molecular junctions have several key characteristics: (1) They have two metal–molecule interfaces, (2) they are low-dimensional nanomaterials, and (3) molecules are trapped between nanogap electrodes. Thanks to these characteristics, the atomic and electronic structure of the molecule in the single molecular junction is different from those of an isolated molecule and bulk crystal. We can thus expect the appearance of novel properties that are not observed in other phases. Based on these features, the single molecular junction is an important current topic in materials science. The single molecular junctions can be fabricated using the break junction technique, such as the mechanically controllable break junction (MCBJ) [4] and the scanning tunneling microscope break junction (STM-BJ) [5]. Using the STM-BJ and MCBJ, various single molecular junctions have been fabricated and their electron transport properties have been investigated. Novel properties including diode and transistor properties and switching behaviors have been revealed for single molecular junctions [6].

Characterization of the atomic and electronic structures of the single molecular junctions is essential for discussing and understanding their properties. Vibration spectroscopy is a powerful technique for characterizing the atomic configuration of single molecular junctions. Various techniques for vibration spectroscopy of single molecular junctions, including IETS, PCS, and SERS, have been developed to characterize single molecular junctions. The electronic structure of single molecular junctions can be characterized via transition voltage spectroscopy, shot noise, and thermopower measurements.

6.2 IETS and PCS

6.2.1 Principles of IETS and PCS

Both IETS and PCS are the vibration spectroscopy of single molecular junctions and are based on electron–vibration interaction. The differential conductance of a single molecular junction (dI/dV) is measured as a function of the bias voltage. Electron–vibration interaction increases the junction differential conductance in the tunneling regime (for IETS) and decreases junction differential conductance in the contact regime (for PCS). IETS was first applied to molecules buried in a metal–oxide interface for a metal–oxide–metal tunneling junction [7], while PCS has been applied to metal contacts [8]. The first IETS measurement for a single molecule adsorbed on a metal substrate (not a molecular junction) was reported by Ho's group with a scanning tunneling microscope (STM) [9]. For molecular junction, IETS and PCS were first applied to the single hydrogen molecular junction [10], and have since been applied to various single molecular junctions, including benzene, alkanedithiol, ethylene, and acetylene [11–14].

Figure 6.1 shows a schematic representation of IETS and PCS for single molecular junctions. In the tunneling regime, as shown in Fig. 6.1a, when the bias voltage is less than the vibrational energy $h\nu$, where h and ν are the Planck constant and the frequency of the vibration mode, the electrons can tunnel elastically. Above a threshold voltage, where $h\nu = eV$, an electron can excite a vibration mode of the single molecular junction. This means that a second inelastic channel opens, in addition to the elastic channel. The opening of this inelastic channel is accompanied by an increase of the differential conductance (dI/dV) at $eV = \pm h\nu$. This change is clearly observed in the derivative of the differential conductance, d^2I/dV^2, where a peak and a dip are observed for the positive and negative biases, respectively. In the contact regime, where the transmission probability is close to one, electrons are delocalized over the two electrodes. We can thus consider the momentum space as shown in Fig. 6.1b. The right-moving electrons occupy higher states than the left-moving electrons when the bias voltage is applied to the single molecular junction. Above a threshold voltage, the electron can excite a vibration mode of the single molecular junction, as with the case of the tunneling regime. Here, electrons around the top of the right-moving states lose energy via excitation of the vibration mode. Electrons should scatter backwards, because the right-moving states are occupied at a lower energy. This backscattering leads to the decrease in the differential conductance at $eV = \pm h\nu$.

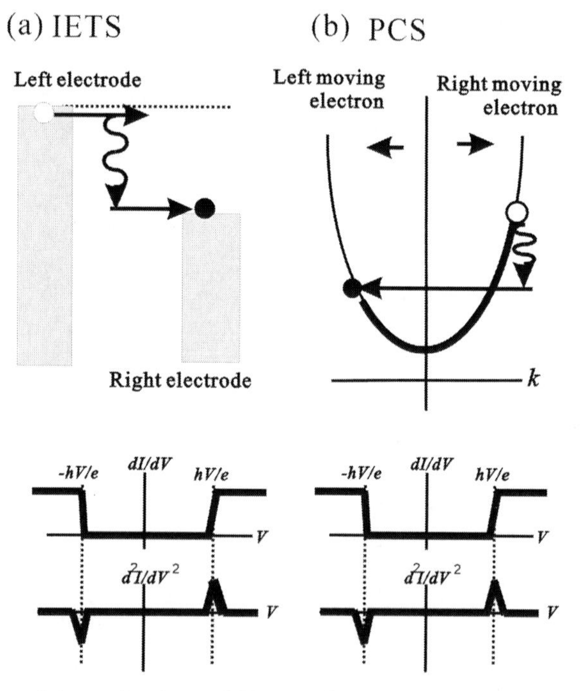

Figure 6.1 Schematic view of IETS and PCS, together with dI/dV and d^2I/dV^2. (a) An additional inelastic channel is opened by the excitation of a vibration mode. The differential conductance (dI/dV) increases above a threshold voltage. (b) The right-moving electrons are scattered backwards by the excitation of a vibration mode. The differential conductance decreases above the threshold voltage.

6.2.2 Application of IETS and PCS to Single Molecular Junctions

Figure 6.2a shows an example of IETS spectra for a single benzene/Pt junction [12]. The differential conductance (dI/dV) is measured using a lock-in technique. The differential conductance is monitored for a fixed contact configuration during sweeping of the DC bias. A symmetric upward step in the differential conductance is observed around 40 mV, and peaks are observed in its derivative (d^2I/dV^2: IETS). The upward steps in the dI/dV spectrum and the peaks in the d^2I/dV^2 spectrum indicate the excitation of the vibrational mode by conduction electrons with energies of 40 meV. The isotope effect is observed in the IETS for the single benzene/Pt junction. Figure 6.2b shows the distribution of vibrational energies for single $^{12}C_6H_6$ and $^{13}C_6H_6$/Pt junctions. The histograms for the single $^{12}C_6H_6$ and $^{13}C_6H_6$/Pt junctions show peaks at 42 meV with a 5 meV width and at 40 meV with a 5 meV width, respectively. The mass ratio of $^{13}C_6H_6$ and $^{12}C_6H_6$ is 84/78. The vibrational mode for the single $^{13}C_6H_6$/Pt junction is predicted to shift from 42 meV ($^{12}C_6H_6$/Pt junction) to 40 meV, assuming the harmonic oscillator model. The good agreement between the predicted and experimental values for vibration energies indicates that the benzene molecule bridges between the Pt electrodes. The theoretical calculation shows that the observed vibration mode is attributed to the hindered rotation mode of the benzene molecule bridging between the Pt electrodes. Here, we note that the single benzene/Pt junction exhibits a conductance value close to 1 G_0 ($2e^2/h$). In the conventional single molecular junctions, where molecules are attached to metal electrodes via an anchoring group (e.g., thiol, amine), the conductance is below 0.01 G_0 [1, 3]. A highly conductive molecular junction can be fabricated by the direct binding of a π-conjugated molecule to the metal electrodes without the anchoring group. Currently, the direct binding technique has been applied to other π-conjugated molecules, including ethylene, acetylene, and C_{60} [14, 15].

Figure 6.3 shows another example of IETS of a single hexanedithiol molecular junction as the electrode distance increases [16]. In contrast with the single benzene molecular junction, intra-molecule vibration modes are clearly observed, in addition to metal–molecule vibration modes. The vibrational peaks in the spectra are assigned as follows: Z: longitudinal metal phonon; I: Au-S stretching [ν(Au-S)]; II: C-S stretching [ν(C-S)]; III: C-H rocking [$\delta_r(CH_2)$]; IV: C-C stretching [ν_w(C-C)]; V: C-H wagging [$\gamma_w(CH_2)$]; and VI: C-H scissoring [$\delta_s(CH_2)$]. With an increase in the electrode distance, the intensity of the $\delta_r(CH_2)$ and $\gamma_w(CH_2)$ peaks increases, which is caused by the introduction of gauche defects in the hexanedithiol. In the gauche conformation, the C-H bonds become nearly perpendicular to the metal surface and, thus, it becomes easier for the conduction electron to excite the vibration modes, leading to the enhancement of these peaks.

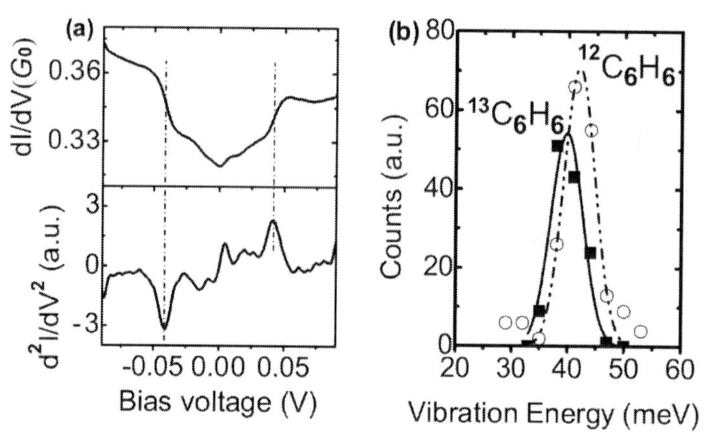

Figure 6.2 (a) Differential conductance (top) and its derivative (bottom) for a single benzene/Pt junction taken at a zero bias conductance of 0.3 G_0. The differential conductance is measured using a lock-in technique with AC modulation at 1 mV and 7.777 kHz. The differential conductance is monitored for a fixed contact configuration during sweeping of the DC bias from −100 mV to +100 mV. (b) Distribution of vibration energy [12].

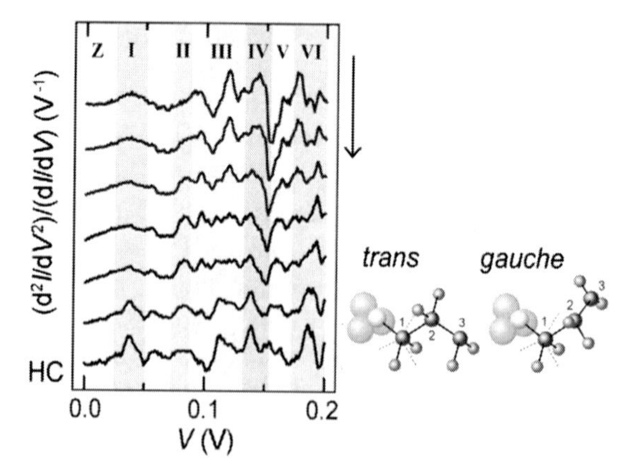

Figure 6.3 IETS of a single hexanedithiol molecular junction as the electrode distance increases. Schematic diagrams of both trans and gauche conformations [16].

The full-width at half-maximum (FWHM) of the peaks in the IETS is given by $W = [(1.7V_m)^2 + (5.4k_BT/e)^2 + W_I^2]^{1/2}$, where V_m is the modulation voltage in the lock-in technique, k_B is the Boltzmann constant, T is the temperature, and W_I is the intrinsic width [1]. The thermal broadening is caused by the broadening of the Fermi distribution, and the modulation broadening is caused by the dynamic detection technique used to obtain the second-harmonic signals. The peaks in the IETS are not clear above 100 K in most single molecular junctions. IETS and PCS thus require low temperatures.

The boundary between the contact and tunneling regimes, that is, those employed by PCS and IETS, is discussed by Paulsson et al. [17]. Figure 6.4 shows the phase diagram for a one-level model illustrating the sign of the change in differential conductance at the onset of the vibration mode. It shows that the change in differential conductance due to the excitation of a vibration mode depends on the conductance of the single molecular junctions and on the symmetry of the coupling of the molecule to the metal electrodes. In the case of symmetric coupling, theoretical calculations predict that the junction differential conductance changes from increasing to decreasing at a transition probability of 0.5 [17]. To confirm this theoretical prediction, there have been some experimental investigations for H_2, H_2O, ethylene, and benzenedithiol (BDT) junctions [18–20].

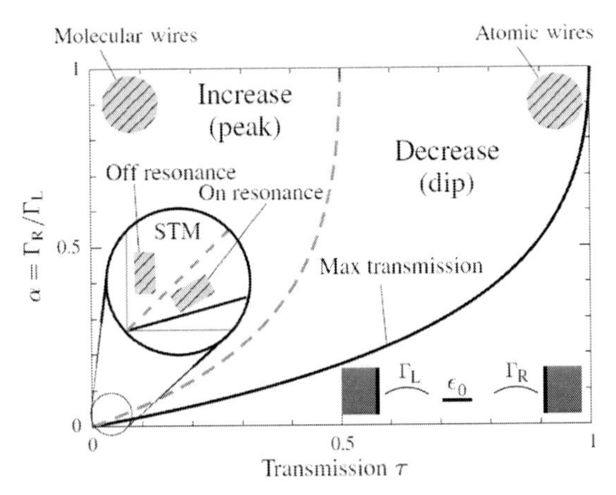

Figure 6.4 Phase diagram for a one-level model (inset) illustrating the sign of the change in differential conductance at the onset of excitation of the vibration mode [17].

Figure 6.5 shows the histogram of step-up (increase in differential conductance) and step-down (decrease in differential conductance) features in the dI/dV spectra for a Pt/H_2O junction as a function of zero-bias conductance [19]. Tal et al. [19] fabricated various Pt/H_2O junctions displaying conductance values of 0.2–1.2 G_0 with a MCBJ at 5 K. The vibration modes are observed around 42 meV as step-up or step-down features in the dI/dV spectra. The curves with step-up appear below 0.57 G_0 and curves with step-down are detected only above 0.72 G_0, which is consistent with the theoretical prediction. The small deviation from the expected value of 0.5 originates from the contribution of the second conductance channel with small transmission probability.

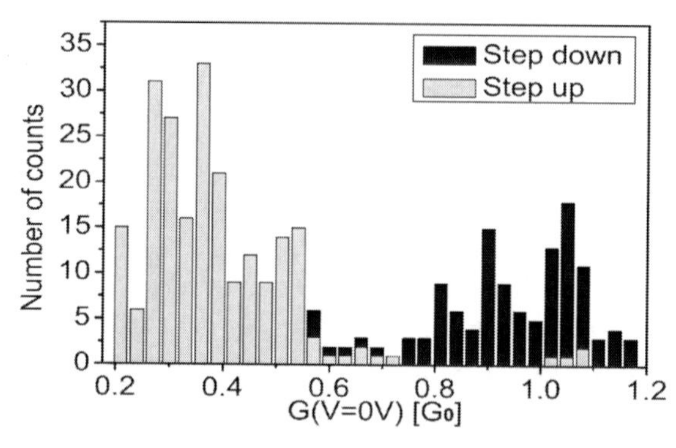

Figure 6.5 Histogram of step-up (gray) and step-down (dark) features in dI/dV spectra for Pt/H_2O junctions as a function of zero-bias conductance [19].

6.2.3 Atomic Motion Induced by Excitation of Vibration Mode

The excitation of the vibration mode by the conduction electron can induce a structural change in single molecular junctions. Symmetric peaks are observed in the dI/dV spectra. The energy of the peak indicates the vibration energy of the single molecular junction, and, thus, this abrupt change in differential conductance is utilized as "action spectroscopy" of the single molecular junctions. In conventional action spectroscopy, the reaction yield is measured as a function of energy. Here, we call this form of spectroscopy "action spectroscopy" because the action of single molecular junctions is utilized to determine the energy of the vibration mode. Action spectroscopy has some advantages compared to IETS. First, it can detect vibration modes that cannot be detected by IETS. Second, action spectroscopy can be applied to system in which the IETS signal is too weak to detect, since the signal of the action spectra is much larger than that of IETS.

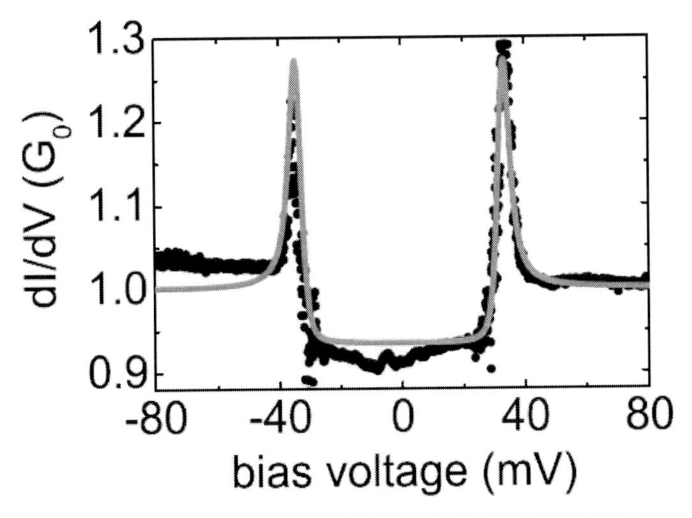

Figure 6.6 The *dI/dV* spectrum of the CO/Pt junction, showing peak features [21].

Action spectroscopy of the single molecular junctions was first applied to the CO/Pt and H_2/Pt junctions [21]. Figure 6.6 shows an example of the *dI/dV* spectrum of a single CO/Pt junction. Symmetric peaks are observed around 40 meV. The abrupt differential conductance change in the *dI/dV* spectrum is explained using a model of vibration-induced two-level systems [21]. In this model, the potential curve of the molecular junction is represented as a double-well potential with a ground state Ψ_1 and a metastable state Ψ_2 in the two energy minima. The two energy minima are separated by the activation barrier (E_{AC}). The molecular junction can be vibrationally excited by the conduction electron. If the junction is fully excited, then the junction with ground state Ψ_1 can overcome the activation barrier, and change into a junction with metastable Ψ_2, which leads to the abrupt change in differential conductance (the peak in the *dI/dV* spectrum). The Ψ_1 and Ψ_2 states correspond to junctions with slightly different local geometric configurations, such as the adsorption site of the molecule on the metal electrodes, the molecule tilt angle, and the configuration of the metal electrodes.

An interesting isotope effect is observed for the action spectroscopy of the single hydrogen/Au junctions [22]. While "peak" spectra (action spectra) are frequently observed for the single H_2/Au junctions, they are not observed for the single D_2/Au junctions. The difference between the two single molecular junctions can be explained by considering the zero-point energy of the single molecular junctions as shown in Fig. 6.7. Since a H_2 molecule has half the mass of a D_2 molecule, the energy of the Au-H_2 vibration mode and the zero-point motion are larger than those for D_2. Therefore, E_{AC} is smaller for the single H_2/Au junctions than for the D_2/Au junctions. A structural change could thus easily occur for the single H_2/Au junctions owing to smaller E_{AC}. Another possible explanation is the difference in the number of excitation processes in the multiple excitation processes. When E_{AC} is larger than the energy of the Au-H_2 (D_2) vibration mode, the system should be excited into states with higher vibrational quanta in a ladder-climbing manner in order to overcome the activation barrier for the structural change. In the multiple excitation process, exciting the system becomes getting harder with the number of excitation processes. The number of excitation processes required to overcome the activation barrier would be smaller for the single H_2/Au junctions than for the single D_2/Au junctions, owing to the higher energy of the single Au-H_2 vibration mode.

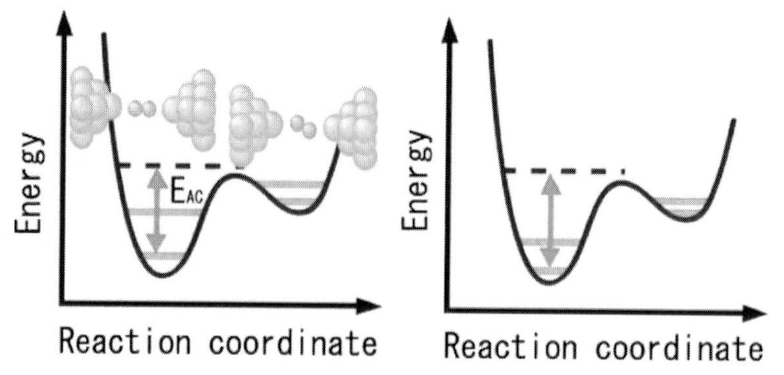

Figure 6.7 The potential energies of (left) single H_2/Au and (right) D_2/Au junctions [22].

6.3 SERS

6.3.1 Principles of SERS

While IETS, PCS, and action spectroscopy are powerful techniques for characterizing single molecular junctions, these techniques require low temperatures. From an application standpoint, vibration spectroscopy at room temperature is desirable for *in situ* characterization of single molecular devices working at room temperature. Conventional optical spectroscopic techniques, such as IR and Raman spectroscopy, are the most promising for vibration spectroscopy of the single molecular junctions at room temperature. However, it appears very difficult to use conventional optical spectroscopes to observe a single molecule in a single molecular junction. First, it is not easy to focus light to single-molecule size; second, the signal from a single molecule is too weak to detect; and third, the excited states may rapidly relax because of the proximity to a metal surface. Fortunately, SERS can overcome these difficulties. In the single molecular junction, a single molecule is trapped in the nanogap. A strong electric field is formed between the nanogap electrodes, which enhances the intensity of the Raman signal, and thus the Raman spectrum from a single molecule in the nanogap is selectively observed.

SERS measurement of a single molecule was first demonstrated using random aggregates of colloidal Ag nanoparticles with crystal violet, and rhodamine 6G molecules [23, 24]. The enhancement of Raman scattering intensity is attributed to electromagnetic and chemical effects. The electromagnetic enhancement arises from the enhanced electric field due to the excitation of plasmon resonance on the metallic nanostructures. Since Raman scattering is a second-order optical process, the metallic nanostructure acts as an amplifier for both the incoming (ω) and scattered (ω') wave fields, where ω and ω' are the frequencies of the incident and scattered waves, respectively. The Raman cross section is enhanced by $f(\omega)^2 f(\omega')^2$, where f is the ratio of the total electric field to the incident electric field. Thus, the Raman cross section depends on the fourth power of f. Let us say that the electric field is enhanced by 100; then the Raman cross section is enhanced by 10^8. The chemical effect results from a metal electron-mediated resonance Raman effect via a charge transfer intermediate state.

6.3.2 SERS of Molecular Junctions

The first SERS measurement of molecular junctions was reported by Tian et al. [25]. They fabricated BDT molecular junctions using an MCBJ and microfabricated Au electrodes on a Si substrate. In this setup, the strong SERS signal is observed from the gap, and it depends critically on the polarization of incident light. When the electrodes are parallel to the polarization direction, the SERS intensity increases dramatically. Since SERS from molecules other than the nanogap region should be depolarized, the observed polarization dependence indicates that the SERS signal originates from the molecules in the nanogap. Figure 6.8 shows the SERS of the BDT in the nanogap as a function of the gap width. The SERS intensity increases considerably with the decrease in the gap width from 0.8 nm to 0.4 nm. The enhancement of SERS intensity is explained by the increase in the intensity of the electric field formed in the nanogap.

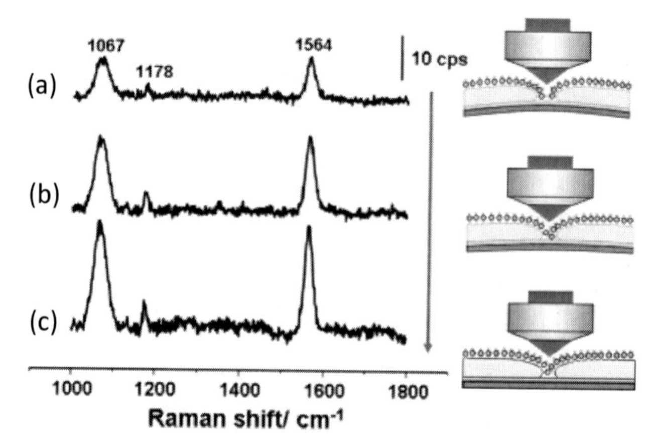

Figure 6.8 SERS of 1,4-benzenedithiol in the nanogap with the process of bending the pair of metallic electrodes. The gap width is (a) 0.8 nm, (b) 0.6 nm, and (c) 0.4 nm [25].

The first simultaneous SERS and conductance measurements were performed for *p*-mercaptoaniline (pMA) and fluorinated oligophenylyne ethynylene molecular junctions using the nanogap electrodes fabricated with electron migration technique [26, 27]. Figure 6.9 shows a scanning electron image of Au constriction with a nanogap, a map of the Si substrate at the 520 cm^{-1} peak, and a map of the pMA SERS signal from the a1 symmetry mode at 1590 cm^{-1}. The Au pads that attenuate the Si signal are clearly visible. The SERS signal is enhanced around the nanogap region. Temporal fluctuations of the SERS intensity (blinking) are observed, suggesting that a few molecules are detected. In some junctions, there are temporal correlations between the fluctuations in the conductance and change in the SERS spectra. Figure 6.10 shows a waterfall plot of the Raman spectrum (at 1-s integrations) and the conduction measurements for the pMA molecular junction. At the boundary shown by the dotted line, the sudden changes in the Raman spectrum are correlated in time with changes in the conductance. However, the correlation between conductance and SERS is complex.

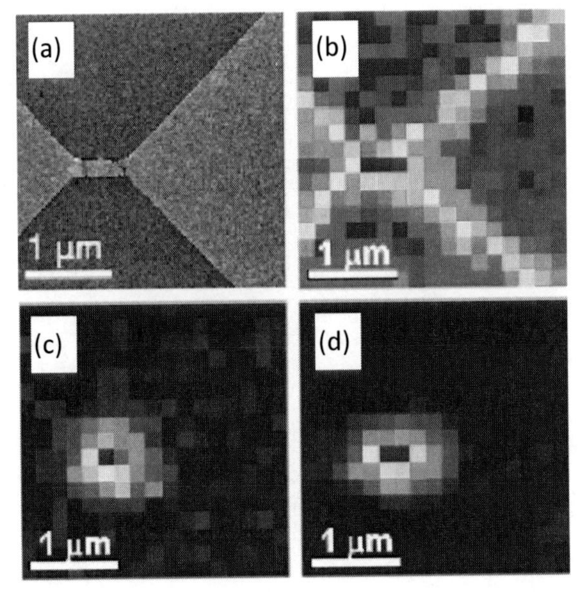

Figure 6.9 (a) Scanning electron image of Au constriction with a nanogap. (b) Map of the Si substrate at the 520 cm^{-1} peak. (c) Map of the pMA SERS signal from the a1 symmetry mode at 1590 cm^{-1}. (d) Map of the integrated continuum signal (integrated from 50 to 300 cm^{-1}). The Raman spectra were captured with 1–2 s integration at an incident 785 nm wavelength laser intensity of 0.5 mW [27].

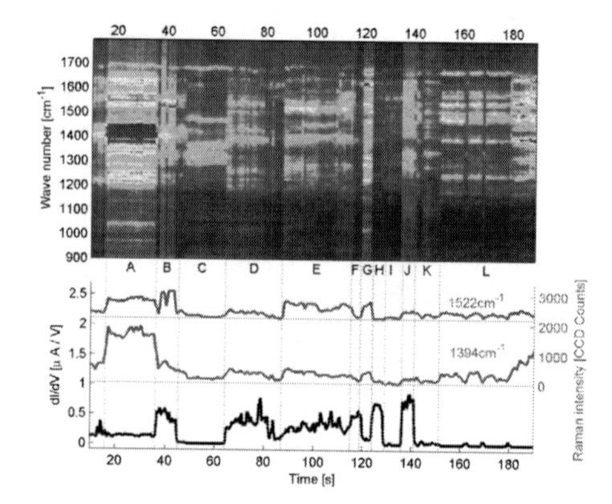

Figure 6.10 Waterfall plot of the Raman spectrum (1-s integrations) and conduction measurements for the pMA molecular junction. Distinct changes in conduction are observed with every significant change in the Raman spectrum and are indicated by vertical lines [27].

Simultaneous SERS and conductance measurements have been performed for various molecular junctions. Ioffe et al. and Ward et al. succeeded in evaluating the effective temperature of 4,4-biphenyldithiol, oligophenylene vinylene (OPV) and 1-dodecanethiol molecular junctions by measuring both the Stokes and anti-Stokes components of the Raman scattering with fixed-nanogap electrodes [28, 29]. The effective temperature of a particular vibration mode, T_ν^{eff}, is defined as

$$\frac{I_\nu^{\mathrm{AS}}}{I_\nu^{\mathrm{S}}} = A_\nu \frac{(\varpi_{\mathrm{L}} + \varpi_\nu)^4}{(\varpi_{\mathrm{L}} - \varpi_\nu)^4} \exp(-\hbar\varpi_\nu / k_{\mathrm{B}} T_\nu^{\mathrm{eff}}),$$

where I_ν^{S} and I_ν^{AS} are the Stokes and anti-stokes Raman intensities for that mode, ϖ_{L} is the incident laser frequency, ϖ_ν is the frequency of Raman scattering, and A_ν is a correction factor that accounts for the ratio of the anti-Stokes and Stokes cross sections. We note that this equation is only valid under equilibrium conditions, that is, at zero bias. In the case of the OPV molecular junction, the effective temperature linearly increases with the bias voltage for two modes: 1317 cm^{-1} and 1625 cm^{-1}. For the vibration mode at 1815 cm^{-1}, the effective temperature does not increase until the bias voltage (V) exceeds $\hbar\varpi_\nu$. Once V exceeds this threshold voltage, the effective temperature linearly increases with the bias voltage. This observation indicates that heating results from the pumping of vibration modes by the conduction electrons. Although the effective temperature varies with the vibration mode, it can increase by more than 100 K by applying a bias voltage of 400 mV.

6.3.3 SERS of Single Molecular Junctions

The next step is the SERS measurement of single molecular junctions. The single molecular junctions should be fabricated with the break junction technique, instead of a fixed nanogap electrode. Simultaneous SERS and conductance measurements are performed for a single 4,4′-bipyridine molecular junction fabricated with an MCBJ using the setup as shown in Fig. 6.11 [31]. An oxide film (Al_2O_3) is used as an insulating film, because it can reduce the background of the Raman spectrum more than a system using organic insulating films can. The Au nanoelectrode is patterned on a substrate covered with a thin polyimide layer using electron beam lithography and a lift-off technique. Subsequently, the polyimide underneath the Au nanobridge and uncovered polyimide layer is completely removed by reactive ion etching using O_2/CF_4 plasma. A free- standing Au nanobridge is obtained. Figure 6.11c shows the Raman imaging of the Au nanogap at 1015 cm^{-1}, which corresponds to the 4,4′-bipyridine ring breathing vibrational mode. Intense Raman scattering is observed at the gap. The intensity of this scattering is very sensitive to the polarization direction of the laser light. Polarization parallel to the Au nanogap results in the most intense scattering signal at the nanogap, which indicates that the 4,4′-bipyridine signal is due largely to the localized surface plasmon excitation of the Au nanoelectrodes.

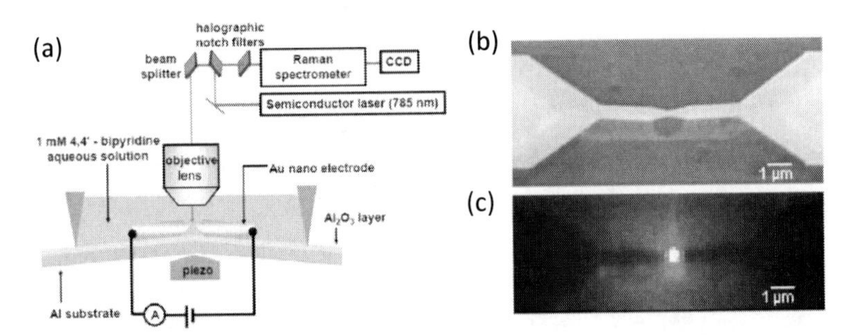

Figure 6.11 (a) Schematic view of the MCBJ-SERS system. (b) SEM image of an Au nanobridge in the MCBJ sample. (c) Raman imaging of the Au nanobridge observed at 1015 cm^{-1} [31].

Figure 6.12 shows the typical time course of the conductivity and the SERS spectrum observed during self-breaking of the junction. Here, the Au nanocontacts, having a conductance of 3 G_0, are fabricated by controlling substrate bending. The contacts spontaneously break as a result of thermal fluctuations and current-induced forces. Since the substrate bending is fixed, the imaging and spectra are in focus during the SERS measurements. The conductance has a plateau at 1 G_0, which corresponds to the Au atomic contact (Region A). Another conductance plateau appears at 0.01 G_0, which corresponds to the formation of a single 4,4′-bipyridine molecular junction (Region B). During these conductance changes, the SERS spectra exhibit very characteristic behavior. Intense bands appear during the formation of the single molecular junction (Region B). Figure 6.13 shows that the SERS intensities of the totally and non-totally symmetric modes are dependent on the conductance of the junction. The vibration modes are strongly enhanced around 0.01 G_0. The close agreement between the conductance value of the single 4,4′-bipyridine molecular junction and the conductance value at maximum SERS intensity indicates conclusively that the SERS spectra are induced by the formation of single molecular junctions.

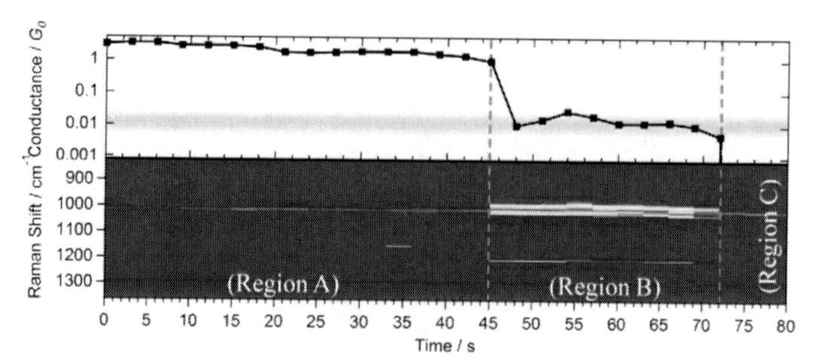

Figure 6.12 Time course of the conductivity and SERS spectrum during breaking of the Au contact in an aqueous solution containing 1 mM of 4,4′-bipyridine. Before (Region A) and after (Region C) the formation of the single molecular junction, SERS bands of 4,4′-bipyridine were observed at 1015, 1074, 1230, and 1298 cm^{-1}. During the presence of the single molecular junction (Region B), additional intense bands were observed at 990, 1020, 1065, and 1200 cm^{-1}. Spectra are obtained with a commercial Raman microprobe spectrometer (λ_{ex} = 785 nm). The DC conductance is measured under an applied bias voltage of 100 mV [31].

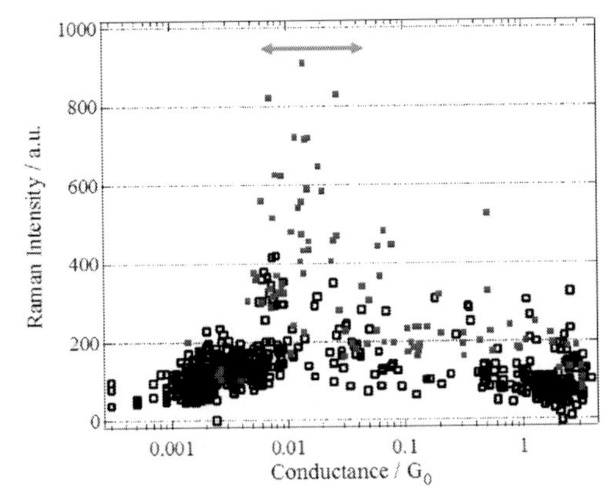

Figure 6.13 The dependence of SERS intensity on the conductance of the 4,4′-bipyridine molecular junction. The arrow indicates the conductance regime of a single 4,4′-bipyridine molecular junction.

Liu et al. reported simultaneous SERS and conductance measurements for single 4,4′-bipyridine molecular junction using "fishing mode" tip-enhanced Raman spectroscopy [30]. In the fishing mode measurements, the distance between the tip and substrate is controlled, while the response of the STM is decreased by lowering the proportional gain and the integral gain to around 0.03% of their normal values. Thermally driven movement of a molecule on the surface leads to the formation and breaking of the molecular junctions in the gap. Interestingly, the 1609 cm^{-1} peak can be reversibly changed to a doublet by increasing the bias voltage. This change is explained by the asymmetric metal–molecule coupling caused by application of the bias voltage. Increasing the bias voltage decreases the Fermi level energy, and thus increases the surface charge density on the STM tip. This leads to stronger chemical bonding between the STM tip and the pyridine ring with which it is in contact. The interaction between the pyridine ring and the STM tip is different from that between the pyridine ring and the substrate. Therefore, the 1609 cm^{-1} peak becomes a doublet as a result of applying the bias voltage.

6.3.4 Single Molecule Dynamics

Further analysis of the simultaneous conductance and SERS measurements reveals the dynamic motion of single molecular junctions. Figure 6.14a shows the time courses of the conductance and the intensity of the a, b_1, and b_2 modes in the SERS spectra, where the 4,4′-bipyridine molecule is assumed to have C_{2v} symmetry. The b_1 mode is observed up to 3 s, and the corresponding conductance value is higher than 0.01 G_0. At 4 s, the b_1 mode disappears along with a sudden drop of conductance below 0.01 G_0. At 7 s, the b_2 mode appears and the conductance returns to the conductance value of the single molecular junction. A simulation of Raman spectra based on density functional theory shows that the b_2 mode is observed when a molecule vertically bridges the gap, while the b_1 mode is observed when a molecule is tilted. Therefore, the SERS spectra (Fig. 6.14a) indicate that the molecule initially bridges the gap with its molecular long axis inclining to the gap direction, after which the molecular junction break occurs (~3 s), and finally the molecule bridges the gap vertically (~7 s).

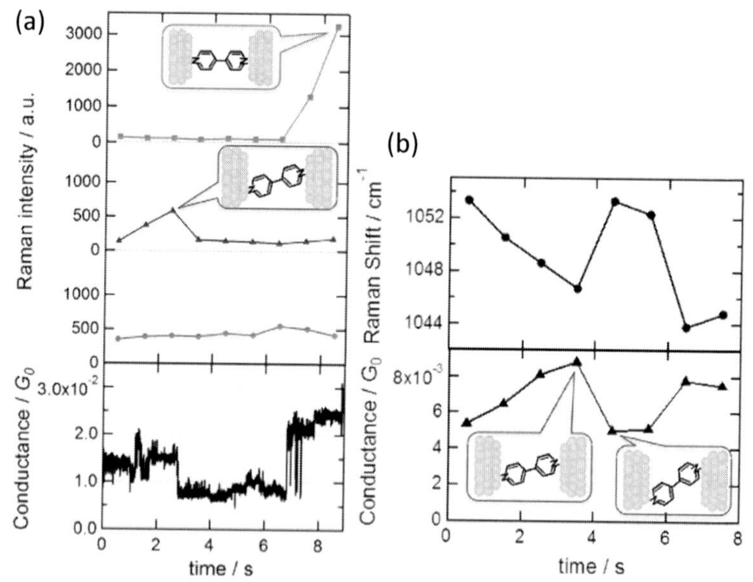

Figure 6.14 (a) Time course of the Raman intensity of the a mode, b_1 mode, and b_2 mode together with the conductance of the 4,4′-bipyridine molecular junction. (b) Time course of the energy of the ring breathing mode around 1050 cm^{-1} (b_1 mode) and the conductance of the single molecular junction. Schematics of the relevant molecular orientation are shown as insets [31].

In addition to the dramatic change in orientation of the molecule in the junction, a slight change in its orientation is observed through a change in peak energy (wavenumber) and conductance. Figure 6.14b shows the time-course of the conductivity and the energy of the ring breathing mode around 1050 cm^{-1} in the single molecular junction regime. The energy of the ring breathing mode becomes higher (lower) as the conductance becomes smaller (larger). This anti-correlation between peak energy and conductance can be explained by the change in the strength of the metal–molecule interaction. When the molecule is adsorbed on the metal surface, the bonding orbitals are formed by the interaction between the highest occupied molecular orbital (HOMO) of the molecule and the unoccupied state of the metal, and between the lowest unoccupied molecular orbital (LUMO) of the molecule and the occupied state of the metal. Electrons are transferred from the HOMO (bonding orbital) to the metal state, and electrons are transferred from the metal to the LUMO (anti-bonding). Both electron transfers lead to a decrease in the bond strength of the molecule, while they increase the metal–molecule interaction. The amount of decrease in the bond strength of the molecule, that is, the energy of the vibration mode, depends on the strength of the metal–molecule interaction. Therefore, an increase in the strength of the metal–molecule interaction leads to a decrease in the energy of the vibration mode of the molecule.

On the other hand, in the single-level tunneling model, the zero bias conductance of the single molecular junction can be represented by:

$$G = \frac{2e^2}{h} \frac{4\Gamma^2}{\Delta^2 + 4\Gamma^2}, \quad (1.1)$$

where Γ and Δ are the coupling between the molecule and the electrode, and the energy difference between the conduction orbital and the Fermi level of the metal electrodes, respectively. The conductance of the single molecular junction increases with the coupling between the molecule and the electrode, that is, the strength of the metal–molecule interaction. Therefore, the increase in the strength of the metal–molecule interaction leads to the increase in the conductance of the single molecular junction, and decrease in the energy of the ring breathing mode of 4,4′-bipyridine. The fluctuation of the vibrational energy synchronized with the change in conductance observed in Fig. 6.14b directly reflects the dynamic motion of the single molecule in the single molecular junction.

6.3.5 SERS Active Mode in Single Molecular Junctions

Here we note that the b_1 and b_2 modes are observed in the SERS of single 4,4′-bipyridine molecular junctions, where the molecule is assumed to have C_{2v} symmetry. Both the b_1 and b_2 modes are non-totally symmetric modes. The non-totally symmetric modes (b_1 and b_2 modes) are inactive in normal Raman scattering at visible excitations (λ_{ex} = 785 nm) for 4,4′-bipyridine, while the totally symmetric mode (a_1 mode) is active in Raman scattering. The b_1 and b_2 modes can be observed for the free molecule only when the electronic transition from the ground state to the LUMO is excited.

The appearance of the non-totally symmetric mode is also observed for other single molecular junctions. Figure 6.15 shows typical SERS spectra of single BDT (C_{2v} symmetry) molecular junctions [32]. The b_2 modes are observed around 1400 cm^{-1} (the upper two spectra). The appearance of the non-totally symmetric mode (b_2 mode) can be explained by considering the charge transfer process. Lombardi et al. extended resonance Raman theory to fit the metal–adsorbate system. According to this model, Raman tensor elements are represented by α = A + B + C [33]. Term A represents a Franck–Condon contribution, and the B and C terms represent Herzberg–Teller contributions. Terms B and C correspond to molecule-to-metal and metal-to-molecule charge transfer transitions, respectively. The totally symmetric modes can be enhanced by all three terms, while the non-totally symmetric modes can be enhanced by terms B and C.

The previously reported thermoelectricity measurement of single BDT molecular junctions showed that the HOMO level and the LUMO level of BDT are respectively 1.2 eV below and 2.5 eV above the Fermi level of the metal electrodes [34]. The wavelength of the incident laser used in this SERS study was 785 nm (~1.6 eV). Therefore, only the B term (the molecule-to-metal charge transfer transition) could contribute to the Raman scattering. There are three requirements for term B to be non-vanishing: (1) there must be at least one allowed transition from the molecular ground state to the excited state; (2) transitions from the molecule to the metal must be allowed; and (3) the direct product of $\Gamma_M \times \Gamma_K \times \Gamma_Q$ must contain the totally symmetric representation, where Γ_M, Γ_K, and Γ_Q are the irreducible representations of charge transfer state, the excited state, and the vibrational mode, respectively. The HOMO of BDT has a b_1 symmetry and the LUMO has an a_2 symmetry, assuming a C_{2v} symmetry. Therefore, the transitions from the molecular ground state to the excited state ($b_1 \times a_2 = b_2$) are allowed. At room temperature, the molecular orientation fluctuates with time in the single molecular junction. In some instances, the molecular axis can be oriented normal to the surface. In this configuration, the charge transfer state has a_1 symmetry. Thus, $\Gamma_M \times \Gamma_K \times \Gamma_Q$ contains the totally symmetric representation for the b_2 mode. Under these conditions, the b_2 mode is enhanced by the B term.

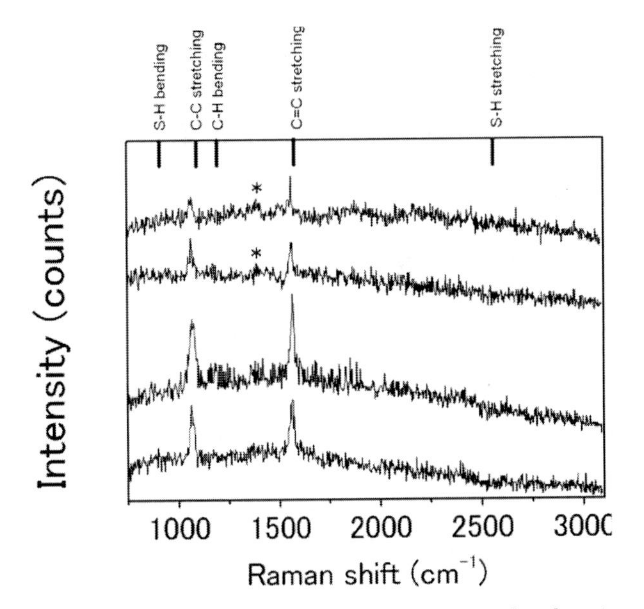

Figure 6.15 Examples of SERS of single BDT molecular junctions. Spectra are obtained with a commercial Raman microprobe spectrometer (λ_{ex} = 785 nm). The vertical lines indicate the vibration modes of the bulk crystal assigned to each mode with the calculation result [32].

In addition to the topics related to the selection rule, the SERS of the single BDT molecular junctions (Fig. 6.15) has two characteristic features. First, the SH stretching and bending modes (908 and 2556 cm^{-1}) are not observed in the BDT molecular junction. The absence of the SH modes indicates that the S-H bonds are broken and that the BDT molecule is bound to both Au electrodes. The SH modes would be observed when the BDT molecule binds to one of the Au electrodes and the other SH anchoring group is free. The second feature is the shift of the peak of the vibration mode of the molecular junction. On average, the C=C (C–C) stretching mode is observed at 1562 cm^{-1} (1063 cm^{-1}) for the molecular junction, while it is observed at 1572 cm^{-1} (1094 cm^{-1}) in the bulk crystal. The red-shift of the C=C and C–C stretching modes originates from the interaction between the BDT molecule and the metal electrode, as discussed in the previous section, which confirms the bridging of the BDT molecule between the Au electrodes.

6.4 Conclusion

In this chapter, we have discussed vibration spectroscopy for single molecular junctions. Both IETS and PCS utilize the change in conductance induced by the excitation of a vibration mode. These spectroscopy techniques provide information about the atomic configuration of the single molecular junctions, which reveals the mechanisms of single molecular devices. Excitation of a vibration mode by a conduction electron can induce atomic motion of the single molecular junctions. This motion can be utilized for use in vibration spectroscopy, as an action spectroscopy technique for the single molecular junctions. Isotope effects have been observed using IETS and action spectroscopy. Simultaneous SERS and conductance measurements provide us with the information about the atomic configuration and dynamics of the single molecular junctions. Raman-inactive vibration modes in bulk and a shifting of the vibration energy are observed for the single molecular junctions, which indicates a molecule–metal interaction. Single molecular junctions can exhibit interesting properties that are not observed in other phases, such as gas, liquid, and solid phases. The mechanism for the appearance of unique properties can be revealed with further development of spectroscopy techniques for the single molecular junctions, which will lead to high-performance electronic devices and new topics in material science.

Acknowledgments

We would like to thank Prof. J. M. van Ruitenbeek, Prof. K. Murakoshi, Dr. T. Konishi, Dr. T. Nakazumi for useful discussions. We gratefully acknowledge the financial support of the Grants-in-Aid for Scientific Research from the Ministry of Education, Culture, Sports, Science and Technology (MEXT) in Japan.

References

1. Cuevas, J. C., and Scheer, E. (2010). *Molecular Electronics: An Introduction to Theory and Experiment* (World Scientific Publishing Co. Pte. Ltd., Singapore).
2. Agraït, N., Yeyati, A. L., and van Ruitenbeek, J. M. (2003). *Phys. Rep.*, **377**, 81–279.
3. Kiguchi, M., and Kaneko, S. (2013). Single molecule bridging between metal electrodes, *Phys. Chem. Chem. Phys.*, **15**, 2253–2267.
4. van Ruitenbeek, J. M., Alvarez, A., Piñeyro, I., Grahmann, C., Joyez, P., Devoret, M. H., Esteve, D., and Urbina, C. (1996). Adjustable nanofabricated atomic size contacts, *Rev. Sci. Instrum.* **67**, 108–111.
5. Xu, B., and Tao, N. J. (2003). Measurement of single-molecule resistance by repeated formation of molecular junctions, *Science*, **301**, 1221–1223.
6. Song, H., Kim, Y., Jang, Y. H., Jeong, H., Reed, M. A., and Lee, T. (2009). Observation of molecular orbital gating, *Nature*, **462**, 1039–1043.
7. Jaklevic R. C., and Lambe, J. (1966). Molecular vibration spectra by electron tunneling, *Phys. Rev. Lett.*, **17**, 1139–1140.
8. Naidyuk, Y., and Yanson, I. (2005). *Point-Contact Spectroscopy* (Springer, New York).
9. Stipe, B. C., Rezaei, M. A., and Ho, W. (1998). Single-molecule vibrational spectroscopy and microscopy, *Science*, **280**, 1732–1735.
10. Smit, R. H. M., Noat, Y., Untiedt, C., Lang, N. D., van Hemert, M. C., and van Ruitenbeek, J. M. (2002). Measurement of the conductance of a hydrogen molecule, *Nature*, **419**, 906–909.
11. Kiguchi, M., Stadler, R., Kristensen, I. S., Djukic, D., and van Ruitenbeek, J. M. (2007). Evidence for a single hydrogen molecule connected by an atomic chain, *Phys. Rev. Lett.*, **98**, 146802/1–146802/4.
12. Kiguchi, M., Tal, O., Wohlthat, S., Pauly, F., Krieger, M., Djukic, D., Cuevas, J. C., and van Ruitenbeek, J. M. (2008). Highly conductive

molecular junctions based on direct binding of benzene to platinum electrodes, *Phys. Rev. Lett.*, **101**, 046801/1–046801/4.

13. Hihath, J., Arroyo, C. R., Rubio-Bollinger, G., Tao, N., and Agraït, N. (2008). Study of electron-phonon interactions in a single molecule covalently connected to two electrodes, *Nano Lett.*, **8**, 1673–1678.
14. Nakazumi, T., Kaneko, S., Matsushita, R., and Kiguchi, M. (2012) Electric conductance of single ethylene and acetylene molecules bridging between Pt electrodes, *J. Phys. Chem. C*, **116**, 18250–18255.
15. Kiguchi, M. (2009). Electrical conductance of single C_{60} and benzene molecules bridging between Pt electrode, *Appl. Phys. Lett.*, **95**, 073301/1–073301/3.
16. Kim, Y., Song, H., Strigl, F., Pernau, H.-F., Lee, T., and Scheer, E. (2011). Conductance and vibrational states of single-molecule junctions controlled by mechanical stretching and material variation, *Phys. Rev. Lett.*, **106**, 196804/1–196804/4.
17. Paulsson, M., Frederiksen, T., Ueba, H., Lorente, N., and Brandbyge, M. (2008). Unified description of inelastic propensity rules for electron Transport through nanoscale junctions, *Phys. Rev. Lett.*, **100**, 226604/1–226604/4.
18. Matsushita, R., Kaneko, S., Nakazumi, T., and Kiguchi, M. (2011). Effect of metal-molecule contact on electron-vibration interaction in single hydrogen molecule junction, *Phys. Rev. B.*, **84**, 245412/1–245412/5.
19. Tal, O., Krieger, M., Leerink, B., and van Ruitenbeek, J. M. (2008). Electron-vibration interaction in single-molecule junctions: From contact to tunneling regimes, *Phys. Rev. Lett.*, **100**, 196804/1–196804/4.
20. Kim, Y., Pietsch, T., Erbe, A., Belzig, W., and Scheer, E. (2011). Benzenedithiol: A broad-range single-channel molecular conductor, *Nano Lett.*, **11**, 3734–3738.
21. Thijssen, W. H. A., Djukic, D., Otte, A. F., Bremmer, R. H., and van Ruitenbeek, J. M. (2006). Vibrationally induced two-level systems in single-molecule junctions, *Phys. Rev. Lett.*, **97**, 226806/1–226806/4.
22. Kiguchi, M., Nakazumi, T., Hashimoto, K., and Murakoshi, K. (2010). Atomic motion in H_2 and D_2 single-molecule junctions induced by phonon excitation, *Phys. Rev. B.*, **81**, 045420/1–045420/4.
23. Nie, S., and Emory, S. R. (1997). Probing single molecules and single nanoparticles by surface-enhanced Raman scattering, *Science*, **275**, 1102–1106.

24. Kneipp, K., Wang, Y., Kneipp, H., Perelman, L. T., and Itzkan, I. (1997). Single molecule detection using surface-enhanced Raman scattering (SERS), *Phys. Rev. Lett.,* **78**, 1667–1670.

25. Tian, J. H., Liu, B., Li, X., Yang, Z. L., Ren. B., Wu, S. T., Tao, N., and Tian, Z. Q. (2006). Study of molecular junction with a combined surface-enhanced Raman and mechanically controllable break junction method, *J. Am. Chem. Soc.,* **128**, 14748–14749.

26. Ward, D. R., Grady, N. K., Levin, C. S., Halas, N. J., Wu, Y., Nordlander, P., and Natelson, D. (2007). Electromigrated nanoscale gaps for surface enhanced Raman spectroscopy, *Nano Lett.,* **7**, 1396–1400.

27. Ward, D. R., Halas, N. J., Ciszek, J. W., Tour, J. M., Wu, Y., Nordlander, P., and Natelson, D. (2008). Simultaneous measurements of electronic conduction and Raman response in molecular junctions, *Nano Lett.,* **8**, 919–924.

28. Ioffe, Z., Shamai, T., Ophir, A., Noy, G., Yutsis, I., Kfir, K., Cheshnovsky, O., and Selzer, Y. (2008). Detection of heating in current-carrying molecular junctions by Raman scattering, *Nat. Nanotechnol.,* **3**, 727–732.

29. Ward, D. R., Corley, D. A., Tour, J. M., and Natelson, D. (2010). Vibrational and electronic heating in nanoscale junctions, *Nat. Nanotechnol.,* **6**, 33–38.

30. Liu, Z., Ding, S. Y., Chen, Z.-B., Wang, X., Tian, J. H., Anema, J. R., Zhou, X. S., Wu, D. Y., Mao, B. W., Xu, X., Ren, B., and Tian, Z. Q. (2011). Revealing the molecular structure of single-molecule junctions in different conductance states by fishing-mode tip-enhanced Raman spectroscopy, *Nature Commun.,* **2**, 305.

31. Konishi, T., Kiguchi, M., Takase, M., Nagasawa, F., Nabika, H., Ikeda, K., Uosaki, K., Ueno, K., Misawa, H., and Murakoshi, K. (2013). Single molecule dynamics at a mechanically controllable break junction in solution at room temperature, *J. Am. Chem. Soc.,* **135**, 1009–1014.

32. Matsushita, R., Horikawa, M., Naitoh, Y., Nakamura, H., and Kiguchi, M. (2013). Conductance and SERS measurement of benzenedithiol molecules bridging between Au electrodes, *J. Phys. Chem. C.,* **117**, 1791–1795.

33. Lombardi, J. R., Brike, R. L., Lu, T., and Xu, J. (1986). Charge-transfer theory of surface enhanced Raman spectroscopy: Heizberg–Teller contributions, *J. Chem. Phys.,* **84**, 4174–4180.

34. Reddy, P., Jang, S.-Y., Segalman, R. A., and Majumdar, A. (2007). Thermoelectricity in molecular junctions, *Science,* **315**, 1568–1571.

Chapter 7

Currents from Pulse-Driven Leads in Molecular Junctions: A Time-Independent Scattering Formulation

Maayan Kuperman and Uri Peskin

Schulich Faculty of Chemistry and the Lise Meitner Center for Computational Quantum Chemistry, Technion—Israel Institute of Technology, Haifa 32000, Israel

uri@tx.technion.ac.il

Electronic transport through molecular wires and junctions has been attracting much attention in past years due to remarkable experimental and theoretical advances. Statistically reliable measurements of single molecule junctions at steady state are being performed and analyzed, revealing a wealth of unique physical characteristics of non-equilibrium transport through molecules. Much less is known about the transient dynamics in molecular junctions following excitation by external time dependent fields. Anticipation for new physical phenomena as well as technological breakthroughs associated with electronic devices operation at fast intra-molecular (terahertz) frequencies drives experimental efforts along this direction, based on irradiation of the junctions with periodic sequences of pulses. In this chapter, we address one of the important challenges in the theory of field-driven

Molecular Electronics: An Experimental and Theoretical Approach
Edited by Ioan Bâldea

ISBN 978-981-4613-90-3 (Hardcover), 978-981-4613-91-0 (eBook)
www.panstanford.com

molecular junctions, which is the need to account for direct excitations of the leads (plasmons, e-h pairs, phonons) and to unveil their role in the charge transport processes. First we consider the long range nature of excitations in the leads and map them onto finite range time-dependent interactions between the molecule and the leads. Then we focus on implementation of quantum scattering theory to calculations of steady state currents in the presence of transient driving fields of general shape and intensity. Generalized transport equations are rigorously derived and approximations to these equations are shown to lead to simpler formulas (generalized Tien–Gordon formulas) for photo-assisted currents. Applications to analytically solvable models are given, demonstrating the basic principles of current enhancement or suppression by time-periodic, but otherwise general, driving fields.

7.1 Introduction

Electronic transport through molecular wires and junctions has been attracting much attention in past years due to remarkable experimental and theoretical advances [1–18]. The field has reached a level where statistically reliable measurements of single molecule junctions can be performed in order to explore the effect of various parameters on their conductance properties [19–25]. Along with advances in the theoretical understanding and the modeling of these systems [26], the unique physics of non-equilibrium transport through molecules is revealed. Yet, our current understanding of molecular junctions is primarily based on steady state measurements under a fixed bias potential, where the intricate many-body electronic and nuclear dynamics is essentially averaged by the measurement process. The dynamical response of molecular junctions to transient perturbations leading to charge and energy transfer within the molecule or the leads, or between the molecule and the leads, is much less explored. The main reasons for this gap of knowledge have been experimental difficulties to perform controlled time resolved measurements at the single junction level. The anticipation is that once the ultrafast quantum dynamics in molecular junctions could be experimentally probed, new and useful phenomena can be discovered. Indeed, while the prime interest in theoretical understanding of these systems is derived from their challenging complexity (being open many body quantum systems, out of equilibrium), an essential motivation to understand the transient behavior of molecular junctions is technological. Standard fast electronic components are operating in the gigahertz frequency regime. Molecular components with their characteristic electronic resonances and vibronic coupling time scales, are expected to be switchable on a much shorter (sub-picoseconds) time scale. Hence molecular based devices could in principle operate in the terahertz regime, with an expected improvement of three orders of magnitude in the speed of operation.

Phenomena associated with irradiation of tunneling junctions have been studied experimentally for several decades, primarily for continuous wave (ac) and in the microwave regime. Photo-assisted transport [26] was identified and analyzed in superconductors [27, 28] semiconductors [29], STM junctions [30], and carbon nanotube quantum dots [31]. In recent years, experimental efforts were focused on laser excitations of nano-junctions in the optical frequencies regime [32–42], and the role of field enhancement by surface plasmons excitations at the metallic nano-contacts was elucidated [35–40]. However, in single molecule junctions the signatures of specific molecular electronic and vibronic excitations, as well as specific local excitations at the lead near the molecule-lead contacts were not yet characterized experimentally. Indeed, such studies require an uncommon combination of molecular electronics setups along with ultra-fast laser optical tools, and the need to overcome or filter out effects of local heating [43–48] which may obscure transient dynamical effects. Yet, it is very likely that attempts to realize such experiments would appear [49], motivated by the scientific and technological interest in detecting features of the combined molecule-leads dynamical response to transient excitations with femtosecond laser pulses.

Numerous theoretical accounts for field-driven transport through nano-scale junctions have been introduced in the past decade and are detailed in comprehensive review articles [50, 51]. As early as in the 1960s, Tien and Gordon used a simple picture of an adiabatic excitation of the leads to explain the effect of continuous microwave irradiation on currents through superconductor junctions, and in particular to account for the observed enhancement and suppression of the current in the off-resonant and resonant transport regimes, respectively [52]. This simple adiabatic picture which essentially limits the effect of the driving field to a time-dependent bias voltage [52–57], does not account for direct field induced transitions between eigenstates, neither inside the molecule nor at the leads, and appears to be relevant only for sufficiently low driving frequencies. Nevertheless, it is found insightful and consistent also for interpretations of state-of-the-art experiments on nano-scale junctions at optical frequencies [37–39]. In parallel to studies of the radiation effects on the leads, the possibility to excite directly the nano-scale conductor within a junction was considered by many, introducing new ideas for light-induced rectifiers, switches and electron pumps [49, 58–68], which account for the quantum mechanical nature of the conductors, including transient coherences and interference effects.

Much work was naturally attributed to ac (monochromatic) driving fields, utilizing the Floquet theory for time-periodic systems [59–62, 68–78]. The combination of Floquet theory with quantum scattering theory [68–74] enables to account for the effect of oscillating potentials within the junctions, the leads, or both, with the benefit of analytic asymptotic solutions. The Floquet theory for ac-fields was integrated also in density matrix [59–62] or Green's function based [75–78] treatments of monochromatically driven transport. Other theoretical approaches were developed for both monochromatic and non-monochromatic fields (including short pulses) which interact directly with the nano-scale conductor (or the molecule) [79–89], using a reduced density matrix approach [49, 79–86] or non-equilibrium Green's functions [67, 87–89] techniques. While most studies did not consider non-adiabatic excitations in the leads, some works have demonstrated significant effects of specific lead excitations (e.g., phonons [90] or e-h pairs creation [91–94]) on the transport. More recent works took into account the feedback of the molecular excitation on the leads plasmons [95–97]. Rigorous attempts to account for the full leads-molecule-field system, based on semi classical [98] or quantum fields [99–102], so far addressed only the steady state problem and did not account for transient observables.

It was recently proposed [49] that steady state measurements could provide valuable information on pulse induced transient dynamics in molecular junctions. The time resolution can be obtained using a sequence of ultra-fast pulse pair excitations with a controlled delay time between the two pulses in each pair. The dependence of the steady state current on the delay time should reveal the periods of molecular dynamics on the sub-picosecond time scale, using electronic measurements on a much longer time scale. Theoretical approaches for the calculation of steady state currents induced by periodic ultra-fast pulse sequences are therefore of great interest.

Below we outline the quantum scattering approach to calculations of steady state currents for molecular junctions driven by pulse sequences. As we go along a fairly rigorous derivation, some of the fundamental problems associated with field-driven leads are considered. The approximations and limitations associated with Landauer type formulas, and the Tien–Gordon formula in particular, are discussed, and generalized formulas are introduced beyond these limitations. In Section 7.2 we formulate the generic model Hamiltonian for a driven junction, and focus on the problem of long-range time-dependent interactions in the leads. In Section 7.3 the time-independent scattering theory for time-dependent junctions Hamiltonians and the generalized Landauer transport formula for the steady state current are discussed. Simplified transport equations are derived for limiting cases in Section IV and the Tien–Gordon formula is reproduced in Section 7.4. Finally, applications for analytically solvable model systems are given in Section 7.5, where the physics underlying photo-assisted, photo-suppressed, and photo-inert currents is discussed. Conclusions are given in Section 7.6.

7.2 Dressing Long-Range Field Interactions with the Leads

We consider the following generic junction model Hamiltonian:

$$H_0 = \sum_m \varepsilon_m d_m^\dagger d_m + \sum_l \varepsilon_l b_l^\dagger b_l + \sum_r \varepsilon_r b_r^\dagger b_r + \left[\sum_{l,m} \nu_m^{(L)} \xi_l^{(L)} b_l^\dagger d_m + \sum_{r,m} \nu_m^{(R)} \xi_r^{(R)} b_r^\dagger d_m + h.c. \right], \tag{7.1}$$

where $\{\varepsilon_m\}$, $\{\varepsilon_l\}$, and $\{\varepsilon_r\}$ are single particle electronic energies at the molecule, left lead and right leads, respectively, and $d_m^\dagger(d_m)$,

$b_l^\dagger\,(b_l)$ and $b_r^\dagger\,(b_r)$ are the corresponding creation (annihilation) operators for an electron in the respective orbitals. The molecule-lead interactions $\left(\sum_{l,m} v_m^{(L)}\xi_l^{(L)} b_l^\dagger d_m + \sum_{r,m} v_m^{(R)}\xi_r^{(R)} b_r^\dagger d_m + \text{h.c.}\right)$ depend on microscopic coupling parameters, where $\{\xi_l^{(L)}\}$ and $\{\xi_r^{(R)}\}$ are the coupling parameters to particular states at the left and right leads, respectively, and $\{V_m^L\}$, $\{V_m^R\}$, define the respective couplings to the specific m-th molecular state.

Our focus below is on interactions between an external driving field and the leads. First, we limit the discussion to cases in which the semi-classical theory for the field is valid, i.e., when the interaction between the molecular junction and the radiation field can be well described in terms of a time-dependent field within the junction Hamiltonian. Second, we focus on the long-range nature of the time-dependent interaction, which implies that the entire leads Hamiltonian $\left(\sum_l \varepsilon_l b_l^\dagger b_l + \sum_r \varepsilon_r b_r^\dagger b_r\right)$ becomes time-dependent. In the most general case, the field induces transient transitions of populations and coherences between the eigenstates of the field-free leads Hamiltonian. Another effect can be the induction of uncorrelated fluctuations (pure-dephasing) of the leads eigenstates, which implies that the corresponding eigenvalues become time-dependent, $\varepsilon_r, \varepsilon_l \to \varepsilon_r(t), \varepsilon_l(t)$. In the simplest case, which is considered in detail below, the sole effect of the field is to drive the entire leads adiabatically, inducing transient changes in their chemical potentials but not affecting the relative distribution of particles over the eigenstates of the field-free system [52–57]. The time-dependent junction Hamiltonian becomes in this case,

$$H(t) = H_0 + V(t)$$

$$V(t) = f_L(t)\sum_l b_l^\dagger b_l + f_R(t)\sum_r b_r^\dagger b_r, \tag{7.2}$$

where $f_R(t)$ and $f_L(t)$ can be any time-dependent functions (ac-field, pulses, etc.). The spread of such time-dependent interactions over the entire leads prohibits the definition of stationary asymptotic lead states for scattering calculations (or quasi-equilibrium leads densities for reduce density matrix calculations).

In order to circumvent this problem, the time-dependent Schroedinger equation, $i\hbar\frac{\partial}{\partial t}|\psi(t)\rangle = H(t)|\psi(t)\rangle$ can be transformed to $i\hbar\frac{\partial}{\partial t}|\tilde{\psi}(t)\rangle = \tilde{H}(t)|\tilde{\psi}(t)\rangle$, where, $|\tilde{\psi}(t)\rangle = S(t)|\tilde{\psi}(t)\rangle$ and $\tilde{H}(t) = \left[i\hbar\frac{\partial}{\partial t}S(t)\right] S^{\dagger}(t) + S(t)H(t)S^{\dagger}(t)$. In the simple case of adiabatic driving field (Eq. 7.2), setting $S(t) = \exp\left[\frac{i}{\hbar}\int_0^t dt'V(t')\right]$, leads to

$$\tilde{H}(t) = \sum_m \varepsilon_m d_m^{\dagger} d_m + \sum_l \varepsilon_l b_l^{\dagger} b_l + \sum_r \varepsilon_r b_r^{\dagger} b_r$$
$$+ \left[g_L(t) \sum_{l,m} v_m^L \xi_l d_m^{\dagger} b_l + g_R(t) \sum_{r,m} v_m^R \xi_r b_r^{\dagger} d_m + \text{h.c.} \right], \tag{7.3}$$

where $g_{R(L)}(t) \equiv e^{\frac{i}{\hbar}\int_0^t f_{R(L)}(t')dt'}$. The time-dependent interactions in the transformed ("dressed") Hamiltonian are restricted to the molecule-leads coupling, and thus become short-range (localized) in space, and suitable for a scattering theory treatment.

We emphasize that the Hamiltonian (Eqs. 7.1–7.3) does not account for direct excitations of the molecule by the field, as well as for direct energy exchange between the molecule and the leads (phonons, excitons, e-h pairs), or for the presence of intra-molecular electronic and vibronic interactions. These terms are localized either on the molecule or at the vicinity of the molecule-leads contacts and are not directly related to the problem of long range excitation of the leads which is our focus here. Moreover, treatments of these effects for non-equilibrium transport is beyond the limits of the scattering theory as introduced below, and are beyond the scope of the present derivation.

7.3 Time-Independent Scattering Theory for Transient Excitations of the Leads

7.3.1 The Hamiltonian

Given the transformed Hamiltonian (Eq. 7.3) with time-independent lead operators, charge transport can be rigorously formulated as a single particle scattering process through the finite-range interaction region [17]. The time-dependence of the transformed Hamiltonian still needs to be accounted for, and particularly the associated time-ordering problem [81–84]. Here we follow the formulation of a time-independent scattering theory for time-dependent Hamiltonians, by mapping the problem of scattering in the presence of a finite range time-dependent interaction onto a stationary scattering problem in an extended Hilbert space [103–108]. For convenience we rewrite the Hamiltonian (Eq. 7.3) in a single particle Hilbert space,

$$\hat{H}(t)=\hat{H}_M+\hat{H}_R+\hat{H}_L+\hat{H}_{LM}(t)+\hat{H}_{ML}(t)+\hat{H}_{RM}(t)+\hat{H}_{MR}(t). \quad (7.4)$$

The molecular and the left (or right) lead Hamiltonians are $\hat{H}_M=\sum_m \varepsilon_m |m\rangle\langle m|$ and $\hat{H}_L=\sum_l \varepsilon_l |l\rangle\langle l|$ (or $\hat{H}_R=\sum_r \varepsilon_r |r\rangle\langle r|$), respectively, and the dressed time-dependent molecule-leads coupling terms read, $\hat{H}_{ML}(t)=\sum_{l,m} g_L(t)\nu_m^{(L)}\xi_l^{(L)}|m\rangle\langle l|$, $\hat{H}_{RM}(t)=\sum_{r,m} g_R(t)\nu_m^{(R)}\xi_r^{(R)}|r\rangle\langle m|$, where $\hat{H}_{LM}(t)=\hat{H}_{ML}^{\dagger}(t)$ and $\hat{H}_{MR}(t)=\hat{H}_{RM}^{\dagger}(t)$.

7.3.2 The Scattering Amplitude

Let us denote the asymptotic states corresponding to the left and right leads as $|\varphi_{\varepsilon_l}\rangle$ and $|\varphi_{\varepsilon_r}\rangle$, respectively, according to their asymptotic energies, $\hat{H}_L|\varphi_{\varepsilon_l}\rangle=\varepsilon_l|\varphi_{\varepsilon_l}\rangle$ and $\hat{H}_R|\varphi_{\varepsilon_r}\rangle=\varepsilon_r|\varphi_{\varepsilon_r}\rangle$. The probability amplitude for an incoming electron from the left lead to be transmitted through the conductor to the right lead is given by the overlap between two scattering states, $S(\varepsilon_L,\varepsilon_R)=\langle\psi_r^-|\psi_l^+\rangle$. The exact dynamics with the full time-dependent Hamiltonian maps $|\psi_l^+\rangle$ and $|\psi_r^-\rangle$ onto the asymptotic states of the respective leads, ($|\varphi_{\varepsilon_l}\rangle$ and $|\varphi_{\varepsilon_r}\rangle$), according to the asymptotic condition of scattering theory [109]. Specifically, if an electron is associated with $|\varphi_{\varepsilon_l}\rangle$ as $t\to-\infty$, then $\lim_{t\to-\infty}U(t,0)|\psi_l^+\rangle=\lim_{t\to-\infty}U_L(t,0)|\varphi_l\rangle$, where $U(t, 0)$ and $U_L(t, 0)$ are the time evolution operators associated with the full Hamiltonian and lead Hamiltonian, respectively. Similarly, if an electron is described by $|\varphi_{\varepsilon_r}\rangle$ as $t\to\infty$, then, $\lim_{t\to\infty}U(t,0)|\psi_r^-\rangle=\lim_{t\to\infty}U_R(t,0)|\varphi_r\rangle$.

A useful expression for the scattering amplitude can be obtained by applying the asymptotic condition directly on the transmission amplitude [104]:

$$S(\varepsilon_l,\varepsilon_r)=\lim_{t\to\infty}\left\langle \varphi_{\varepsilon_r}\left| e^{i\hat{H}_R t/\hbar}U(t,0)U(0,-t)e^{i\hat{H}_L t/\hbar}\right|\varphi_{\varepsilon_l}\right\rangle$$

$$=\lim_{\eta\to 0}\frac{-i}{\hbar}\int_0^{\infty}dt\Big[\left\langle \varphi_{\varepsilon_r}\left| e^{i\hat{H}_R t/\hbar}\hat{H}_{RM}(t)U(t,-t)e^{i\hat{H}_L t/\hbar}e^{-\eta t}\right|\varphi_{\varepsilon_l}\right\rangle$$

$$+\left\langle \varphi_{\varepsilon_r}\left| e^{i\hat{H}_R t/\hbar}U(t,-t)\hat{H}_{ML}(-t)e^{i\hat{H}_L t/\hbar}e^{-\eta t}\right|\varphi_{\varepsilon_l}\right\rangle\Big] \tag{7.5}$$

When the driving field is time-periodic or periodically continued, as in the case of sequences of finite pulses, the time-dependent fields read (see Eq. 7.3),

$$g_L(t)=\sum_{m=-\infty}^{\infty} g_{L,m}e^{im\omega t};\; g_R(t)=\sum_{m=-\infty}^{m=\infty} g_{R,m}e^{im\omega t}. \tag{7.6}$$

In such a case, the time evolution operator obtains the special form [104, 110],

$$U(t,-t)=\sum_{m,\lambda} e^{im\omega t}e^{-i2\lambda t/\hbar}\left\langle m|\lambda\right\rangle\rangle\langle\left\langle \lambda|0\right\rangle. \tag{7.7}$$

Here $\{|\lambda\rangle\rangle\}$ are the "quasi-energy" eigenstates of the Floquet operator [111], defined as, $\hat{H}^{(F)}=\hat{H}(t)-i\hbar\frac{\partial}{\partial t}$, i.e., $\hat{H}^{(F)}|\lambda\rangle\rangle=\lambda|\lambda\rangle\rangle$. The double bracket notation refers to the inner product in an extended Hilbert space with an additional time-coordinate, $\langle\langle f,n|g,m\rangle\rangle\equiv\frac{1}{T}\int_0^T dte^{-i(n-m)\omega t}\langle f|g\rangle\equiv\langle n|m\rangle\langle f|g\rangle$, where $\langle f|g\rangle$ is the standard inner product in the usual Hilbert space, and $T = 2\pi/\omega$ is the time period. According to the Floquet theorem, the quasi-energy states can be expressed as, $|\lambda\rangle\rangle\equiv\sum_{m=-\infty}^{\infty}e^{i\omega m}\langle m|\lambda\rangle\rangle$, and the stationary asymptotic states are mapped onto the extended space as follows, $|\varphi_\varepsilon\rangle = |\varphi_\varepsilon\rangle \otimes |0\rangle \equiv |\varphi_\varepsilon, 0\rangle\rangle$. Substitution of Eqs. (7.6 and 7.7) in Eq. (7.5) yields, after some algebra, the following expression for the transmission amplitude [104]:

$$S(\varepsilon_l,\varepsilon_r)=\lim_{\eta\to 0}\frac{1}{2}\sum_{n=-\infty}^{\infty}\left\langle\left\langle \varphi_{\varepsilon_r},n\right|\hat{H}_{RM}\frac{1}{[(\varepsilon_r+\varepsilon_l+i\eta+\hbar\omega n)/2-\hat{H}_F]}\right.$$

$$\left.+\frac{1}{[(\varepsilon_r+\varepsilon_l+i\eta+\hbar\omega n)/2-\hat{H}_F]}\hat{H}_{ML}\left|\varphi_{\varepsilon_l},0\right\rangle\right\rangle \tag{7.8}$$

Defining extended space Green's operator, $\hat{G}^{(F)}(z) \equiv [z - \hat{H}^{(F)}]^{-1}$, molecule-lead coupling operators, $\hat{H}^{(F)}_{JM} \Leftrightarrow \hat{H}_{MJ}(t)$, and a transmission operator, $\hat{T}^{(F)}(z) \equiv \hat{H}^{(F)}_{RM}\hat{G}^{(F)}(z)\hat{H}^{(F)}_{ML}$, the transmission amplitude can be written as follows:

$$S(\varepsilon_l,\varepsilon_r) = \lim_{\eta\to 0} \sum_{n=-\infty}^{\infty} \frac{\left\langle\left\langle \varphi_{\varepsilon_r}, n \right| \hat{T}^{(F)}(\varepsilon_r + \varepsilon_l + i\eta + n\hbar\omega) \left| \varphi_{\varepsilon_l}, 0 \right\rangle\right\rangle}{\varepsilon_r + \hbar\omega n - \varepsilon_l + i\eta}$$
$$+ \frac{\left\langle\left\langle \varphi_{\varepsilon_r}, n \right| \hat{T}^{(F)}(\varepsilon_r + \varepsilon_l + i\eta + n\hbar\omega) \left| \varphi_{\varepsilon_l}, 0 \right\rangle\right\rangle}{\varepsilon_l - (\varepsilon_r + \hbar\omega n) + i\eta}$$
$$= -2\pi i \sum_{n=-\infty}^{\infty} \delta(\varepsilon_r + \hbar\omega n - \varepsilon_l)$$
$$\left\langle\left\langle \varphi_{\varepsilon_r}, n \right| \hat{T}^{(F)}[(\varepsilon_r + \varepsilon_l + i\eta + \hbar\omega n)/2] \left| \varphi_{\varepsilon_l}, 0 \right\rangle\right\rangle \quad (7.9)$$

The delta function in Eq. (7.9) implies that an initial "left" asymptotic state, $|\varphi_{\varepsilon_l},0\rangle\rangle$, can scatter to the right electrode only when the electron energy satisfies $\varepsilon_r = \varepsilon_l - \hbar\omega n$. The probability amplitude for such a transition is given by the following state-to-state scattering matrix element in the extended space,

$$S(\varepsilon_l,n) \equiv -2\pi i \left\langle\left\langle \varphi_{\varepsilon_l - \hbar\omega n}, n \right| \hat{T}^{(F)}(\varepsilon_l + i\eta) \left| \varphi_{\varepsilon_l}, 0 \right\rangle\right\rangle$$
$$= -2\pi i \left\langle\left\langle \varphi_{\varepsilon_l - \hbar\omega n}, n \right| \hat{H}^{(F)}_{RM}\hat{G}^{(F)}(\varepsilon_l + i\eta)\hat{H}^{(F)}_{ML} \left| \varphi_{\varepsilon_l}, 0 \right\rangle\right\rangle, \quad (7.10)$$

where the respective transmission function is defined as $t_n^{L\to R}(\varepsilon_l) \equiv |S(\varepsilon_l,n)|^2$.

7.3.3 The Current Formula

The left-to-right current through the field-driven junction can be formulated as an integral over Fermi-weighted transmission functions, leading to a generalized Landauer transport equation. Considering the transformed Hamiltonian (Eq. 7.3), the leads Hamiltonians are time-independent and can be associated with quasi-equilibrium (Fermi) distributions of electrons and holes, $f_{\mu_J,e}(\varepsilon) = 1/(1 + e^{(\varepsilon-\mu_J)/(K_B T)})$, and $f_{\mu_J,h}(\varepsilon) = 1 - f_{\mu_J,e}(\varepsilon)$, respectively, for the Jth electrode where μ_J is the electrons' chemical potential. Integrating over the appropriate Fermi weights at the leads [17], the forward left to right current reads

$$I_{L\to R} = \frac{2e}{h}\sum_{n=-\infty}^{\infty}\int d\varepsilon\, t_n^{L\to R}(\varepsilon) f_{\mu_L,e}(\varepsilon) f_{\mu_R,h}(\varepsilon - \hbar\omega n), \tag{7.11}$$

where $t_n^{L\to R}(\varepsilon)$ is the transmission function as defined above.

Subtraction of the backward current, $I_{R\to L}$, is required in order to obtain the net current. Indeed, the current due to scattering events beginning with an electron on the left lead at energy ε_l and ending with an electron on the right lead at energy $\varepsilon_r = \varepsilon_l - \hbar\omega n$, may be compensated by opposite events, beginning with an electron on the right lead at energy $\varepsilon_r = \varepsilon_l - \hbar\omega n$, and ending with an electron on the left lead at energy ε_l. The probability amplitude for such backwards scattering events is denoted accordingly as $t_{-n}^{R\to L}(\varepsilon - \hbar\omega n)$. Accounting for the respective Fermi weights, the generalized Landauer formula for the net current reads,

$$\begin{aligned} I &= I_{L\to R} - I_{R\to L} \\ &= \frac{2e}{h}\sum_{n=-\infty}^{\infty}\Bigg[\int d\varepsilon t_n^{L\to R}(\varepsilon) f_{\mu_L,e}(\varepsilon) f_{\mu_R,h}(\varepsilon - \hbar\omega n) \\ &\quad - \int d\varepsilon t_{-n}^{R\to L}(\varepsilon - \hbar\omega n) f_{\mu_R,e}(\varepsilon - \hbar\omega n) f_{\mu_L,h}(\varepsilon)\Bigg] \\ &= \frac{2e}{h}\sum_{n=-\infty}^{\infty}\int d\varepsilon \{t_n^{L\to R}(\varepsilon) f_{\mu_L,e}(\varepsilon) - t_{-n}^{R\to L}(\varepsilon - \hbar\omega n) f_{\mu_R,e}(\varepsilon - \hbar\omega n) \\ &\quad + [t_{-n}^{R\to L}(\varepsilon - \hbar\omega n) - t_n^{L\to R}(\varepsilon)] f_{\mu_R,e}(\varepsilon - \hbar\omega n) f_{\mu_L,e}(\varepsilon)\}, \end{aligned} \tag{7.12}$$

where the energy integrals are over the respective conductance bands. (Notice that the formula for the backwards current can also be written as $I_{R\to L} = \frac{2e}{h}\sum_{n=-\infty}^{\infty}\int d\varepsilon\, t_n^{R\to L}(\varepsilon) f_{\mu_R,e}(\varepsilon) f_{\mu_L,h}(\varepsilon - \hbar\omega n)$, which yields the same result by changing integration and summation variables).

7.4 The Transmission Functions and Generalized Tien–Gordon Formulas

7.4.1 The Trace Formula

Using Eq. (7.10) and generalizing a standard derivation (see, e.g., Ref. 17), one can express the transmission function as a trace in the extended Hilbert space,

$$t_n^{L\to R}(\varepsilon) = tr[\hat{G}^{(F)}(\varepsilon)\hat{\Gamma}_{L,0}^{(F)}(\varepsilon)\hat{G}^{(F)\dagger}(\varepsilon)\hat{\Gamma}_{R,n}^{(F)}(\varepsilon - \hbar\omega n)]. \quad (7.13)$$

Similarly, the transmission function in the backwards direction reads

$$t_n^{R\to L}(\varepsilon) = tr[\hat{G}^{(F)}(\varepsilon)\hat{\Gamma}_{R,0}^{(F)}(\varepsilon)\hat{G}^{(F)\dagger}(\varepsilon)\hat{\Gamma}_{L,n}^{(F)}(\varepsilon - \hbar\omega n)]. \quad (7.14)$$

Here $\hat{G}^{(F)}(\varepsilon)$ is the extended space Green operator as defined above, where this shorter notation stands for $\hat{G}^{(F)}(\varepsilon + i\eta)$ under the limit $\eta \to +0$ for the trace. The coupling width operators in Eqs. (7.13 and 7.14) are defined as follows (for $J \in R, L$),

$$\hat{\Gamma}_{J,k}^{(F)}(z) \equiv \hat{H}_{MJ}^{(F)}|k\rangle\delta(z - \hbar\omega k - \hat{H}_J)\langle k|\hat{H}_{JM}^{(F)}$$
$$= \hat{H}_{MJ}(t)e^{i\omega kt}\delta(z - \hbar\omega k - \hat{H}_J)\frac{1}{T}\int_{-T/2}^{-T/2} dt'e^{-i\omega kt'}\hat{H}_{JM}(t'). \quad (7.15)$$

A more explicit expression for the width operators in the extended space can be obtained by using the Fourier expansions of the driving fields, Eq. (7.6). Introducing the Fourier basis set, $\left(\langle t|n\rangle = \frac{1}{\sqrt{T}}e^{in\omega t}\right)$, one obtains

$$\langle n|\hat{\Gamma}_{L,k}^{(F)}(z)|n'\rangle = g_{L,n-k}g_{L,n'-k}^{*}\hat{\Gamma}_L(z)$$
$$\langle n|\hat{\Gamma}_{R,k}^{(F)}(z)|n'\rangle = g_{R,k-n}^{*}g_{R,k-n'}\hat{\Gamma}_R(z), \quad (7.16)$$

where $\hat{\Gamma}_J(z) = \hat{H}_{MJ}(0)\delta(z - \hat{H}_J)\hat{H}_{JM}(0)$, is the coupling width operator of the field-free junction. The width operator is shown to operate exclusively within the subspace of molecular states $\{|m\rangle\}$ (see the definition of $\hat{H}_{MJ}(t)$ and $\hat{H}_{JM}(t)$ above). It follows that the trace in Eqs. (7.13 and 7.14) and therefore the extended Green operators, $(\hat{G}^{(F)}, \hat{G}^{(F)\dagger})$, can also be limited to this subspace.

7.4.2 The Molecular Green Operator

In order to obtain an explicit expression for the projection of the extended Green operator on the molecular subspace, we start from the extended space Floquet Hamiltonian, $\hat{H}^{(F)} = \hat{H}_M^{(F)} + \hat{H}_R^{(F)} + \hat{H}_L^{(F)} + \hat{H}_{LM}^{(F)} + \hat{H}_{ML}^{(F)} + \hat{H}_{RM}^{(F)} + \hat{H}_{MR}^{(F)}$, where the usual Hilbert space operators are mapped as follows ($J \in M, R, L$):

$\hat{H}^{(F)}_{JM} \Leftrightarrow \hat{H}_{JM}(t)$, $\hat{H}^{(F)}_{MJ} \Leftrightarrow \hat{H}_{MJ}(t)$ and $\hat{H}^{(F)}_{J} \Leftrightarrow \hat{H}_{J} - i\hbar\frac{\partial}{\partial t}$. Next, we introduce orthogonal projection operators to the subspaces of the right lead $\left(\hat{R} = \sum_r |r\rangle\langle r|\right)$, the left lead $\left(\hat{L} = \sum_l |l\rangle\langle l|\right)$, and the molecule $\left(\hat{M} = \sum_m |m\rangle\langle m|\right)$. Generalizing the standard derivation for time-independent junctions [17], one obtains the following equation for the extended molecular space Green operator:

$$\hat{G}^{(F)}_{M}(z) = \hat{M}\hat{G}^{(F)}(z)\hat{M} = [z - \hat{H}^{(F)}_{M,\mathrm{eff}}(z)]^{-1}. \tag{7.17}$$

The effective Floquet Hamiltonian in the molecular space is defined as

$$\hat{H}^{(F)}_{M,\mathrm{eff}}(z) = \hat{H}^{(F)}_{M} + \hat{\Sigma}^{(F)}_{L}(z) + \hat{\Sigma}^{(F)}_{R}(z), \tag{7.18}$$

where the self-energy operators representing the leads were introduced:

$$\hat{\Sigma}^{(F)}_{J}(z) = \hat{H}^{(F)}_{MJ}[z - \hat{H}^{(F)}_{J}]^{-1}\hat{H}^{(F)}_{JM}. \tag{7.19}$$

Using the Fourier basis expansions in Eq. (7.6), $\left(\langle t|n\rangle = \frac{1}{\sqrt{T}}e^{in\omega t}\right)$, one obtains $\langle n|\hat{H}^{(F)}_{ML}|k\rangle = \hat{H}_{ML}g_{L,n-k}$, $\langle n|\hat{H}^{(F)}_{RM}|k\rangle = \hat{H}_{RM}g^{*}_{R,n-k}$, and $\langle n|[z - \hat{H}^{(F)}_{J}]^{-1}|k\rangle = \frac{\delta_{k,n}}{z - \hbar\omega k - \hat{H}_{J}}$, such that the self-energy operators read

$$\begin{aligned}\langle n|\hat{\Sigma}^{(F)}_{L}(z)|n'\rangle &= \sum_{k=-\infty}^{\infty} g_{L,n-k}g^{*}_{L,k-n'}\hat{\Sigma}_{L}(z - \hbar\omega k)\\ \langle n|\hat{\Sigma}^{(F)}_{R}(z)|n'\rangle &= \sum_{k=-\infty}^{\infty} g^{*}_{R,n-k}g_{R,k-n'}\hat{\Sigma}_{R}(z - \hbar\omega k).\end{aligned} \tag{7.20}$$

Here, $\hat{\Sigma}_{J}(z) = \hat{H}_{MJ}[z - \hat{H}_{J}]^{-1}\hat{H}_{JM}$ is the self-energy of the field-free molecular junction.

7.4.3 The Wide Band Limit

We now invoke an important and useful approximation. If the density of states of the conductance bands at the leads is a sufficiently smooth function of the energy (the wide band limit), one can neglect the energy dependence of the self-energy operators, and in particular, approximate $\hat{\Sigma}_J(z-\hbar\omega k)\approx\hat{\Sigma}_J(z)$ [112]. This approximation introduces a significant simplification to Eq. (7.20), since $\hat{\Sigma}_J(z)$ is factorized out, and the remaining summations over k give $\sum_{k=-\infty}^{\infty} g_{L,n-k}g^*_{L,k-n'}=|g_L|^2_{n-n'}$, and $\sum_{k=-\infty}^{\infty} g^*_{R,n-k}g_{R,k-n'}=|g_R|^2_{n-n'}$. Considering the generic form of the time-dependent fields in the transformed junction Hamiltonian, $g_L(t)\equiv e^{\frac{i}{\hbar}\int_0^t f_L(t')dt'}$, $g_R(t)\equiv e^{\frac{i}{\hbar}\int_0^t f_R(t')dt'}$, it follows that $|g_L(t)|^2 = |g_R(t)|^2 = 1$, and therefore, $|g_L|^2_{n-n'} = |g_R|^2_{n-n'} = \delta_{n,n'}$, with the remarkable consequence that the self-energy operators become diagonal in the Fourier basis representation, i.e., $\langle n|\hat{\Sigma}^{(F)}_J(z)|n'\rangle\cong\delta_{n,n'}\hat{\Sigma}_J(z)$, for $(J\in R,L)$.

A farther simplification is inherent to cases in which the radiation field does not interact directly with the molecule, i.e., only the leads are driven by the field, and $\hat{H}_M$ is time-independent (see Eqs. (7.2–7.4)). In this case, the molecular Floquet Hamiltonian is diagonal in the Fourier basis, $\langle n|\hat{H}^{(F)}_M|n'\rangle=\hat{H}_M\delta_{n,n'}$. This, and the wide band approximation for the leads imply that the effective molecular Hamiltonian (see Eq. 7.18) is also diagonal in the Fourier basis representation, and therefore the molecular Green operator obtains the following simple form:

$$\langle n|\hat{G}^{(F)}_M(z)|n'\rangle=\delta_{n,n'}\hat{G}_M(z-\hbar\omega n), \tag{7.21}$$

where $\hat{G}_M(z)=[\hat{H}_M+\hat{\Sigma}_L(z)+\hat{\Sigma}_R(z)]^{-1}$ is the molecular Green operator of the field-free junction.

Finally, using Eqs. (7.16) and (7.21), one can rewrite the transmission function in Eqs. (7.13) and (7.14) as a sum of traces over the molecular space of the field free junction, with coefficients that depend on the particular form of the field,

$$\begin{aligned} t_n^{L\to R}(\varepsilon) &= tr[\hat{G}^{(F)}_M(\varepsilon)\hat{\Gamma}^{(F)}_{L,0}(\varepsilon)\hat{G}^{(F)\dagger}_M(\varepsilon)\hat{\Gamma}^{(F)}_{R,n}(\varepsilon-\hbar\omega n)] \\ &= \sum_{k,k'=-\infty}^{\infty} g_{L,k}g^*_{L,k'}g^*_{R,n-k'}g_{R,n-k} \\ &\quad\cdot tr_M[\hat{G}_M(\varepsilon-\hbar\omega k)\hat{\Gamma}_L(\varepsilon)\hat{G}^\dagger_M(\varepsilon-\hbar\omega k')\hat{\Gamma}_R(\varepsilon-\hbar\omega n)]. \end{aligned} \tag{7.22}$$

Similarly,

$$t_{-n}^{R\to L}(\varepsilon-\hbar\omega n)$$
$$=tr[\hat{G}_M^{(F)}(\varepsilon-\hbar\omega n)\hat{\Gamma}_{R,0}^{(F)}(\varepsilon-\hbar\omega n)\hat{G}_M^{(F)\dagger}(\varepsilon-\hbar\omega n)\hat{\Gamma}_{L,-n}^{(F)}(\varepsilon)]$$
$$=\sum_{k,k'=-\infty}^{\infty} g_{R,n-k}^{*}g_{R,n-k'}g_{L,k'}g_{L,k}^{*}$$
$$\cdot tr_M[\hat{G}_M^{\dagger}(\varepsilon-\hbar\omega k')\hat{\Gamma}_L(\varepsilon)\hat{G}_M(\varepsilon-\hbar\omega k)\hat{\Gamma}_R(\varepsilon-\hbar\omega n)]. \quad (7.23)$$

7.4.4 Generalized Tien–Gordon Formulas

In many cases of interest, the time-dependent field does not have equally strong effect on the two electrodes [52]. This is the case, e.g., when the two leads are associated with different materials and/or with different plasmon frequencies, or when the experimental design exposes only one lead to the radiation. Without loss of generality, we consider below a situation in which only the right lead is driven by the field, i.e., $g_L(t) = 1$, and $g_{L,k} = \delta_{k,0}$. Introducing this condition into Eqs. (7.22 and 7.23), one obtains

$$t_n^{L\to R}(\varepsilon)=|g_{R,n}|^2\ tr_M[\hat{G}_M(\varepsilon)\hat{\Gamma}_L(\varepsilon)\hat{G}_M^{\dagger}(\varepsilon)\hat{\Gamma}_R(\varepsilon-\hbar\omega n)] \quad (7.24)$$

$$t_{-n}^{R\to L}(\varepsilon-\hbar\omega n)=|g_{R,n}|^2\ tr_M[\hat{G}_M^{\dagger}(\varepsilon)\hat{\Gamma}_L(\varepsilon)\hat{G}_M(\varepsilon)\hat{\Gamma}_R(\varepsilon-\hbar\omega n)]. \quad (7.25)$$

In the common case, the field-free effective junction Hamiltonian satisfies time-reversal symmetry [17], and the traces over the molecular subspace are identical in these two equations. It follows that the forward and backwards transmission functions are identical, i.e.,

$$t_n^{L\to R}(\varepsilon)=t_{-n}^{R\to L}(\varepsilon-\hbar\omega n)=t_n(\varepsilon). \quad (7.26)$$

Using this in the formula for the net current through the driven junction (Eq. 7.12), one obtains a simple and intuitive result:

$$I=\sum_{n=-\infty}^{\infty}\frac{2e}{h}\int d\varepsilon t_n(\varepsilon)[f_{\mu_L,e}(\varepsilon)-f_{\mu_R+\hbar\omega n,e}(\varepsilon)]. \quad (7.27)$$

According to this formula, the total current is composed of additive contributions. Each (n-th) contribution is associated with a unique transmission function, $t_n(\varepsilon)$, and a displaced chemical potential at the right (field-driven) lead, $\mu_{R,n}=\mu_R+\hbar\omega n$.

An even simpler expression for the current is obtain by replacing, $\hat{\Gamma}_R(\varepsilon-\hbar\omega n)\approx\hat{\Gamma}_R(\varepsilon)$, in the transmission expression (Eq. 7.24), in consistency with the wide band approximation as discussed above for the self-energy terms. The result is a factorized transmission expression, in which the field determines only a pre-factor and otherwise the transmission expression corresponds to the field-free junction,

$$t_n(\varepsilon)\cong a_n t_0(\varepsilon)$$
$$t_0(\varepsilon)\equiv tr_M[\hat{G}_M(\varepsilon)\hat{\Gamma}_L(\varepsilon)\hat{G}_M^\dagger(\varepsilon)\hat{\Gamma}_R(\varepsilon)]. \tag{7.28}$$

Here, the n-dependent term is the square amplitude of the Fourier component of the transformed driving field, $a_n = |g_{R,n}|^2$. Substitution Eq. (7.28) in Eq. (7.27) yields

$$I\cong\sum_{n=-\infty}^{\infty}a_n\frac{2e}{h}\int d\varepsilon t^{L\to R}(\varepsilon)[f_{\mu_L,e}(\varepsilon)-f_{\mu_R+\hbar\omega n,e}(\varepsilon)]$$
$$=\sum_{n=-\infty}^{\infty}a_n I(eV-\hbar\omega n). \tag{7.29}$$

In the last step we introduced the notation of the static potential on the junction, $eV = \mu_L - \mu_R$. Notice that in the special case of a monochromatic driving field, i.e., $f_R(t) = \alpha\cos(\omega t)$, the transformed field reads, $g_R(t)=e^{i\frac{\alpha}{\hbar\omega}\sin(\omega t)}$, and the corresponding coefficients are the squared Bessel functions, $a_n=\left|\frac{\omega}{2\pi}\int_{-\pi/\omega}^{\pi/\omega}e^{-in\omega t}e^{i\frac{\alpha}{\hbar\omega}\sin(\omega t)}dt\right|^2=\left|J_n\left(\frac{\alpha}{\hbar\omega}\right)\right|^2$, as introduced by Tien and Gordon in Ref. 52.

Equations (7.27) and (7.29) are generalized Tien–Gordon formulas for the current. It is emphasized that these results are based on approximations beyond the particular form of the single particle Hamiltonian (Eq. 7.4) and beyond the framework of the Landauer transport equation. In particular, the field-free junction should satisfy time-reversal symmetry, as well as the wide-band limit, where the driving field should be restricted to one of the two leads. Yet, these generalizations apply to fields of essentially any shape, and in particular can account for the effect of different pulse sequences on the steady state currents through the junctions.

7.5 Analytic Model Applications

7.5.1 A Single Level Conductor

Let us consider the simplest model of a conductance junction, where the molecular conductor is associated with a single active level, denoted by the state $|0\rangle$. The molecule is coupled to two leads associated with chemical potentials, μ_L and μ_R, for the left and right leads, respectively. The right lead is adiabatically driven by a field, $f(t)$ of general shape and intensity. The model Hamiltonian for the driven junction reads,

$$\hat{H}(t) = \varepsilon_0 |0\rangle\langle 0| + \sum_l \varepsilon_l |l\rangle\langle l| + \sum_r [\varepsilon_r + f(t)] |r\rangle\langle r| \\ + \left[\sum_l \xi_l^{(L)} |l\rangle\langle 0| + \sum_r \xi_r^{(R)} |r\rangle\langle 0| + \text{h.c.} \right], \tag{7.30}$$

The transformed Hamiltonian that corresponds to this model reads (see Eq. 7.4),

$$\hat{H}(t) = \varepsilon_0 |0\rangle\langle 0| + \sum_l \varepsilon_l |l\rangle\langle l| + \sum_r \varepsilon_r |r\rangle\langle r| \\ + \left[\sum_l \xi_l^{(L)} |l\rangle\langle 0| + g_R(t) \sum_r \xi_r^{(R)} |r\rangle\langle 0| + \text{h.c.} \right], \tag{7.31}$$

where, $g_R(t) = e^{\frac{i}{\hbar}\int_0^t f_R(t')dt'} = \sum_{m=-\infty}^{\infty} g_{R,m} e^{im\omega t}$.

In the absence of a driving field, the molecular width operators in the wide-band limit is given as [17]

$$\hat{\Gamma}_J(\varepsilon) = \Gamma_J |0\rangle\langle 0|; \quad J \in L, R, \tag{7.32}$$

where Γ_J denotes the respective molecule-lead coupling width parameter. The corresponding field-free molecular Green operator reads in this case

$$\hat{G}_M(\varepsilon) = \left[\varepsilon - \varepsilon_0 + i\frac{\Gamma_L}{2} + i\frac{\Gamma_R}{2} \right]^{-1} |0\rangle\langle 0|, \tag{7.33}$$

and the state-to-state transmission (Eq. 7.28) becomes

$$t_0(\varepsilon)=\frac{\Gamma_L\Gamma_R}{(\varepsilon-\varepsilon_0)^2+(\Gamma_L+\Gamma_R)^2/4}. \tag{7.34}$$

Finally, we calculate the net left-to-right current according to Eq. (7.29):

$$I=\sum_{n=-\infty}^{\infty}|g_{R,n}|^2\frac{2e}{h}\int d\varepsilon\frac{\Gamma_L\Gamma_R[f_{\mu_L e}(\varepsilon)-f_{\mu_R+\hbar\omega n,e}(\varepsilon)]}{(\varepsilon-\varepsilon_0)^2+(\Gamma_L+\Gamma_R)^2/4}. \tag{7.35}$$

At zero temperature, the Fermi distribution functions limit the energy integration range to finite values, and the zero-temperature current reads

$$I=\sum_{n=-\infty}^{\infty}|g_{R,n}|^2\frac{2e}{h}\frac{2\Gamma_L\Gamma_R}{\Gamma_L+\Gamma_R}\cdot\left\{arctg\left[\frac{2(\mu_L-\varepsilon_0)}{\Gamma_L+\Gamma_R}\right]-arctg\left[\frac{2(\mu_R+\hbar\omega n-\varepsilon_0)}{\Gamma_L+\Gamma_R}\right]\right\}. \tag{7.36}$$

Equation (7.36) is a convenient framework for analysis of the effect of time-dependent fields on the current in limiting cases.

7.5.2 A Narrow Transmission Resonance

First we consider the case of a narrow resonance, typically associated with the limit of weak molecule-lead coupling. For this purpose we assume that Γ_L, Γ_R are sufficiently small such that electron hopping rates are the slowest of all possible physical processes in the junction. Formally, we assume $\Gamma_L, \Gamma_R << |\varepsilon_0-\mu_L|, |\mu_R+\hbar\omega n-\varepsilon_0|$, and therefore the *arctg* functions can be approximated by their asymptotic values, $arctg(x)\xrightarrow[x\pm\infty]{}\pm\pi/2$. It follows that

$$\left\{arctg\left[\frac{2(\mu_L-\varepsilon_0)}{\Gamma_L+\Gamma_R}\right]-arctg\left[\frac{2(\mu_R+\hbar\omega n-\varepsilon_0)}{\Gamma_L+\Gamma_R}\right]\right\}\cong\begin{cases}0\,; & \mu_L>\varepsilon_0 \text{ and } \mu_R+\hbar\omega n>\varepsilon_0\\ \pi\,; & \mu_L>\varepsilon_0 \text{ and } \mu_R+\hbar\omega n<\varepsilon_0\\ -\pi; & \mu_L<\varepsilon_0 \text{ and } \mu_R+\hbar\omega n>\varepsilon_0\\ 0\,; & \mu_L<\varepsilon_0 \text{ and } \mu_R+\hbar\omega n<\varepsilon_0\end{cases} \tag{7.37}$$

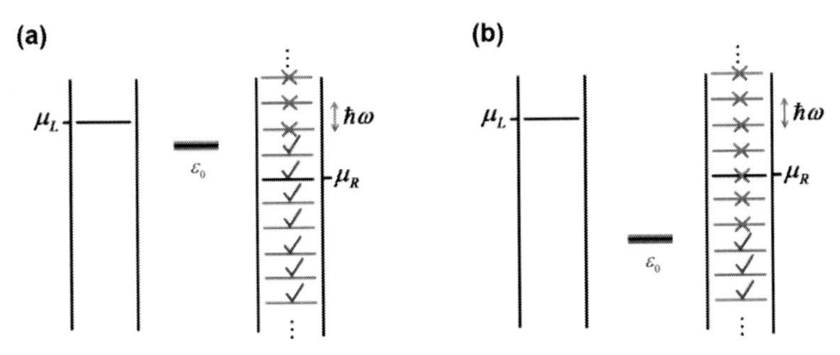

Figure 7.1 (a) Left: Illustration of current suppression by the field in resonant tunneling. When the molecular level is positioned between the chemical potentials of the field free leads, the overall transmission is distributed over different scattering channels associated with a displaced chemical potential, $\mu_{R,n} = \mu_R + \hbar\omega n$. Only channels for which $\varepsilon_0 > \mu_{R,n}$ (denoted by √) can contribute to the current, while others are Pauli blocked. (b) Right: Illustration of current enhancement in off-resonant tunneling. When the molecular level is below the two chemical potentials of the field free leads, the distribution of the transmission over scattering channels associated with a displaced chemical potential, $\mu_{R,n} = \mu_R + \hbar\omega n$, facilitates transport when $\varepsilon_0 > \mu_{R,n}$ (as denoted by √) and contributes to the current.

Let us consider first situations in which the left chemical potential is above the conducting level (see Fig. 7.1), i.e., $\mu_L > \varepsilon_0$. In this case, Eqs. (7.36) and (7.37) yield the current,

$$I \cong \sum_{n=-\infty}^{(\varepsilon_0-\mu_R)/(\hbar\omega)} |g_{R,n}|^2 \frac{2e}{\hbar}\frac{\Gamma_L\Gamma_R}{\Gamma_L+\Gamma_R}. \tag{7.38}$$

Two different scenarios are illustrated in Fig. 7.1. The first scenario corresponds to “resonant tunneling” in the absence of the time-dependent field, i.e., $\mu_L > \varepsilon_0 > \mu_R$, where the molecular level lies between the chemical potentials of the two leads. The corresponding resonant tunneling current in the field-free junction equals $I_0 = \frac{2e}{\hbar}\frac{\Gamma_L\Gamma_R}{\Gamma_L+\Gamma_R}$. Substitution in Eq. (7.38) gives $I \cong I_0 \sum_{n=-\infty}^{(\varepsilon_0-\mu_R)/(\hbar\omega)} |g_{R,n}|^2$, which implies that the field-free current is multiplied by a sum over contributions from the different harmonics of the transformed field. This sum is limited, however, by the condition, $\hbar\omega n < \varepsilon_0 - \mu_R$. Considering the normalization of the driving field, $\sum_{n=-\infty}^{\infty} |g_{R,n}|^2 = 1$, one obtains $I < I_0$, i.e., the field suppresses the current through the junction in the case of resonant tunneling. The physical reasoning for this is that the field distributes the overall transmission probability among different scattering processes, each one associated with a different chemical potential at the right lead ($\mu_{R,n} = \mu_R + \hbar\omega n$). Since some of these scattering events are Pauli blocked due to the respective Fermi distributions, the overall current decreases.

The second scenario corresponds to the case of off-resonant tunneling, where the molecular level is outside the Fermi conductance window of the field-free junction, i.e., $\varepsilon_0 < \mu_R, \mu_L$. In this case, the tunneling current in the absence of the field is negligible. Turning on the field induces additional scattering processes in which the right lead chemical potential is displaced ($\mu_{R,n} = \mu_R + \hbar\omega n$). Although only a part of these scattering events is Pauli allowed, they contribute to a finite transmission probability (see Eq. 7.38) such that $I > 0$, and therefore the field promotes off-resonant tunneling. This well-studied phenomenon is often termed photon-assisted tunneling [26].

Similar considerations for the case $\mu_L < \varepsilon_0$, using Eq. (7.38) in Eq. (7.37) show that in the case of resonant tunneling ($\mu_L < \varepsilon_0 < \mu_R$) and off-resonant tunneling ($\mu_R, \mu_L < \varepsilon_0$), the driving field suppresses and increases the current, respectively, where in these cases the sign of the current is reversed, $I \cong -\sum_{n=(\varepsilon_0-\mu_R)/(\hbar\omega)}^{\infty} |g_{R,n}|^2 \frac{2e}{\hbar}\frac{\Gamma_L\Gamma_R}{\Gamma_L+\Gamma_R}$.

Notice that both current suppression in the resonant tunneling regime and current enhancement in the off-resonant regime were observed experimentally in early experiments of photo-induced transport through tunnel junctions [27]. In complex junctions, however, the detailed energy dependence of the transmission function (which reflects the shape of the orbitals at the conductor and the density of states at the leads) determines whether the current at a given voltage is suppressed or enhanced [26].

So far, our qualitative considerations did not account for the particular form of the periodic field. Figure 7.2 demonstrates enhancement and suppression of the current according to Eq. (7.36) in the case of an ac field, $f_R(t) = \alpha\cos(\omega t)$. For positive bias voltage ($\mu_L > \varepsilon_0$), the off-resonant current ($\mu_R > \varepsilon_0$) is shown to increase in the presence of the time-dependent field. Each step in the current is associated with $-\frac{eV}{2} = \mu_{R,n} = \mu_R + \hbar\omega n$, where the step height equals $|g_{R,n}|^2\ (2e/\hbar)\Gamma_L\Gamma_R/(\Gamma_L+\Gamma_R)$. As the right chemical potential drops below the conductance energy level ($\mu_R < \varepsilon_0$) and the resonant transport regime is reached, the current in the presence of the field is shown to drop below its value in the absence of the field, as discussed above (see Fig. 7.1).

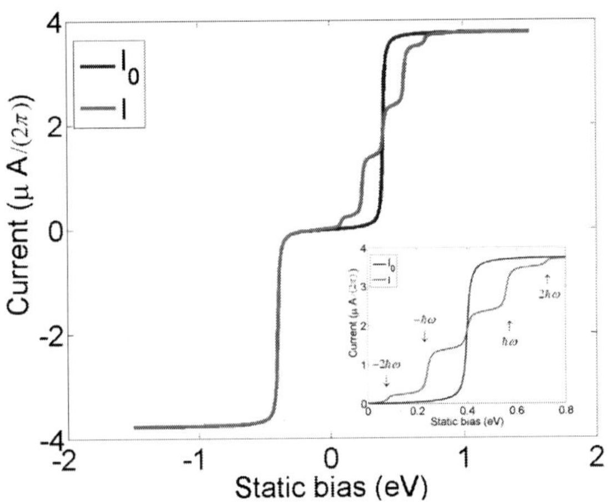

Figure 7.2 $I(V)$ curves for field-free (I_0, blue) and field-driven (I, red) junctions, with a driving field, $f_R(t) = \alpha\cos(\omega t)$. The model parameters are: $\Gamma_L = \Gamma_R = 0.005$, $\varepsilon_0 = -0.2$, $\hbar\omega = 0.08$, $\alpha = 0.12$ (all in eV). The potential is distributed evenly on the two contacts, $\mu_L = eV/2 = -\mu_R$, where V is the bias potential. Inset: A zoom into the transition from current enhancement in the off-resonant transport regime to current suppression in the resonant transport regime. The step heights correspond to $|J_n[\alpha/(\hbar\omega)]|^2$.

In Fig. 7.3 different fields are considered. In particular a Gaussian pulse, $f_R(t) = \alpha e^{-t^2/2\tau^2}$ with a pulse width parameter, $\tau = 200$ f/sec, is periodically continued, with a period $T = 1.8$ psec. The current is suppressed with respect to the field-free result for $\alpha > 0$, where for $\alpha < 0$ the current is enhanced. Indeed, this pulse has a non-zero dc (zero frequency) component that either increases transiently the effective right lead potential (for $\alpha > 0$) or decreases it ($\alpha < 0$), with the respective effects on the current.

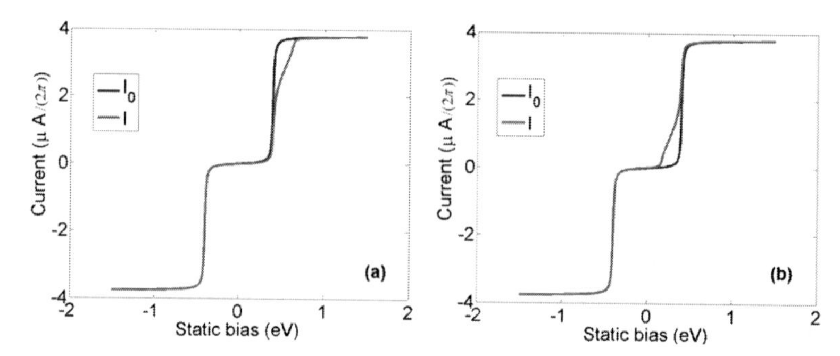

Figure 7.3 $I(V)$ curves for field-free (I_0, blue) and field-driven (I, red) junctions, with a periodically continued Gaussian pulse, $f_L(t) = \alpha e^{-t^2/2\tau^2}$. The model parameters are: $\Gamma_L = \Gamma_R = 0.005$, $\varepsilon_0 = -0.2$, $\alpha = 0.12$ (all in eV). The bias is distributed evenly on the two contacts, $\mu_L = eV/2 = -\mu_R$, where V is the bias potential. (a) and (b) correspond to positive and negative α, respectively.

7.5.3 A Perfect Conductor

We now turn to analysis of the field effect of a perfect one-dimensional conductor, (a G_0 contact). For this purpose the couplings between the conductor and the leads in the wide band limit should satisfy [17], $\Gamma_L = \Gamma_R \equiv \Gamma$, and $|E - \mu_R|, |E - \mu_L| << \Gamma$. It follows that the *arctg* functions in Eq. (7.36) can be approximated by their first order Taylor expansion which yields, $I = \sum_{n=-\infty}^{\infty} |g_{R,n}|^2 \frac{2e}{h}(\mu_L - \mu_R - \hbar\omega n)$. Denoting the potential bias, $\mu_L = \mu_R \equiv eV$, and the quantum of conductance, $G_0 = \frac{2e^2}{h}$, we obtain

$$I \cong G_0 V - \sum_{n=-\infty}^{\infty} \frac{2e}{h} |g_{R,n}|^2 \hbar\omega n. \tag{7.39}$$

For any field that satisfies the condition, $|g_{R,n}|^2 = |g_{R,-n}|^2$ (the ac field, $f_R(t) = \alpha \cos(\omega t)$, for example), the current equals $G_0 V$. This is also the current in the absence of a field, which implies that the perfect conductor is inert to the field in this case. Other fields (e.g., any driving field with a non-vanishing dc component) would enhance/suppress the current through the perfect wide band conductor according to Eq. (7.39).

7.6 Conclusions

In conclusion, this work reviewed some of the challenges associated with observing and utilizing transient sub picosecond scale dynamics in molecular junctions. A theory for calculating the effect of periodic sequences of pulses of general shape and intensity on the steady state current was outlined and applied for simple test cases revealing characteristics of the physics of transport through molecular junctions with periodically driven leads. General transport formulas for field-driven leads were derived and the approximations leading from these formulas to simpler generalized Tien–Gordon formulas were discussed. This chapter focused on the scattering approach, and therefore some physical phenomena associated with molecular junctions out of equilibrium, such as direct excitations of the molecule, inelastic transport processes and the effect of electronic and vibronic coupling were not considered. Accounting for these phenomena requires more elaborate theories of non-equilibrium transport. Many of the limitations and approximations considered in this article should find their analogs in different theoretical considerations and should guide their development.

Acknowledgments

Yoram Selzer is acknowledged for stimulating discussions. This research was supported by the Israel Science Foundation.

References

1. Van der Molen, S. J., Naaman R., Scheer, E., Neaton, J. B., Nitzan, A., Natelson, D., Tao, N. J., van der Zant, H., Mayor, M., Ruben, M., Reed, M., and Calame, M. (2013). Visions for a molecular future, *Nat. Nanotechnol.,* **8**, 385–389.
2. Aradhya, S. V., and Venkataraman, L. (2013). Single-molecule junctions beyond electronic transport, *Nat. Nanotechnol.,* **8**, 399.
3. Cuniberti, G., Fagas, G., and Richter, K. (2005). *Introducing Molecular Electronics,* Springer.
4. Chen, F., Hihath, J., Huang, Z. F., Li, X. L., and Tao, N. J. (2007). Measurement of single-molecule conductance, *Ann. Rev. Phys. Chem.,* **58**, 535.

5. Roth, S., and Joachim, C. (1997). *Atomic and Molecular Wires,* Kluwer, Dordrecht.
6. Joachim, C., Gimzewski, J. K., and Aviram, A. (2000). Electronics using hybrid-molecular and mono-molecular devices, *Nature,* **408**, 541.
7. Nitzan, A., and Ratner, M. A. (2003). Electron transport in molecular wire junctions, *Science,* **300**, 1384.
8. Datta, S. (2004). Electrical resistance: An atomistic view, *Nanotechnology,* **15**, 433.
9. Read, M. A., Zhou, C., Muller, C. J., Burgin, T. P., and Tour, J. M. (1997). Conductance of a molecular junction, *Science,* **278**, 252.
10. Yazdani, D., Eigler, M., and Lang, N. D. (1997). *Atomic and Molecular Wires,* NATO ASI Series E, vol. 341, Kluwer, Dordrecht.
11. Xiao, X., Xu, B., and Tao, N. J. (2004). Measurement of single molecule conductance: Benzenedithiol and benzenedimethanethiol, *Nano. Lett.,* **4**, 267.
12. Smit, R. H. M., Noat, Y., Unteidt, C., Lang, N. D., van Hemert, M. C., and van Ruitenbeek, J. M. (2002). Measurement of the conductance of a hydrogen molecule, *Nature,* **419**, 906.
13. Reicher, J., Ochs, R., Weber, H. B., Mayor, M., and von Lohneysen, H. (2002). Driving current through single organic molecules, *Phys. Rev. Lett.,* **88**, 176804.
14. Elbing, M., Ochs, R., Koentopp, M., Fischer, M., von Hanisch, C., Weigend, F., Evers, F., Weber, H. B., and Mayor, M. (2005). A single-molecule diode, *PNAS,* **102**, 8815.
15. Galperin, M., Ratner, M. A., and Nitzan, A. (2007). Molecular transport junctions: Vibrational effects, *J. Phys. Condens. Matter,* **19**, 103201.
16. van der Molen, S. J., and Liljeroth, P., Charge transport through molecular switches, *J. Phys. Condens. Matter,* **22**, 133001.
17. Peskin, U. (2010). An introduction to the formulation of steady-state transport through molecular junctions, *J. Phys. B,* **43**, 153001.
18. Ballmann, S., Härtle, R. Coto, P. B., Elbing, M., Mayor, M., Bryce, M. R., Thoss, M., and Weber, H. B. (2012). Experimental Evidence for quantum interference and vibrationally induced decoherence in single-molecule junctions, *Phys. Rev. Lett.,* **109**, 056801.
19. Elbing, M., Ochs, R., Koentopp, M., Fischer, M., von Hanisch, C., Weigend, F., Evers, F., Weber, H. B., and Mayor, M. (2005). A single molecule diode, *PNAS,* **102**, 8815.

20. Ioffe., Z., Shamai, T., Ophir, A., Noy, G., Yutsis, I. Cheshnovsky, O., and Selzer, Y. (2008), Detection of heating in current carrying molecular junctions *Nat. Nano,* **3**, 727.

21. Cheng, Z. L., Skouta, R., Vazquez, H., Widawsky, J. R., Schneebeli, S., Chen, W., Hybersten, M. S., Breslow, R., and Venkataraman, L. (2011). In-situ formation of highly conducting covalent Au-C contacts for single molecule junctions, *Nat. Nano,* **6**, 353.

22. Schmaus, S. Bagrets, A., Nahas, Y., Yamada, T. K., Bork, A., Bowen, M., Beaurepaire, E., Evers, F., and Wulfhekel, W. (2011). Giant magnetoresistance through a single molecule *Nat. Nano,* **6**, 185.

23. Reddy, P., Jang, S. Y., Segalman, R. A., and Majumdar, A. (2007). Thermoelectricity in molecular junctions, *Science,* **315**, 1568.

24. Ismael, D. P., Zhihai, L., Mullen, K., and Tao, N. J. (2010). Gate controlled electron transport in coronenes as a bottom-up approach towards graphene transistors *Nat. Comm.*, **1**, 31.

25. Wu, S. M., Gonzalez, M. T., Huber, R., Grunder, S., Mayor, M., Schonenberger, C., and Calame, M. (2008). Molecular junctions based on aromatic coupling *Nat. Nano,* **3**, 569.

26. Cuevas, J. C., and Scheer, E. (2010). *Molecular Electronics: An Introduction to Theory and Experiment,* World Scientific, Singapore.

27. Dayem, A. H., and Martin, R. J. (1962). Quantum interaction of microwave radiation with tunneling between superconductors, *Phys. Rev. Lett.*, **8**, 246.

28. Chauvin, M., vom Stein, P., Pothier, H., Joyez, P., Huber, M. E., Esteve, D., and Urbina, C. (2006). Superconducting atomic contacts under microwave irradiation, *Phys. Rev. Lett.*, **97**, 067006.

29. Platero, G., and Aguado, R. (2004). Photon-assisted transport in semiconductor nanostructures, *Phys. Rep.*, **395**, 1.

30. Tu, X. W., Lee, J. H., and Ho, W. (2006). Atomic-scale rectification at microwave frequency, *J. Chem. Phys.*, **124**, 021105.

31. Meyer, C., Elzerman, J. M., and Kouwenhoven, L. P. (2007). Photon-assisted tunneling in a carbon nanotube quantum dot, *Nano Lett.,* **7**, 295.

32. Wu, S. W., Ogawa, N., and Ho, W. (2006). Atomic-scale coupling of photons to single-molecule junctions, *Science,* **312**, 1362.

33. Guhr, D. C., Rettinger, D., Boneberg, J., Erbe, A., Leiderer, P., and Scheer, E. (2007). Influence of laser light on electronic transport through atomic-size contacts, *Phys. Rev. Lett.*, **99**, 086801.

34. Ward, D. R., Scott, G. D., Keane, Z. K., Halas, N. J., and Natelson, D. (2008). Electronic and optical properties of electromigrated molecular junctions, *J. Phys. Condens. Matter,* **20**, 374118.

35. Banerjee, P., Conklin, D., Nanayakkara, S., Park, T.-Hong, Therien, M. J., and Bonnell, D. A. (2010). Plasmon-induced electrical conduction in molecular devices, *ACS Nano,* **4**, 1019.

36. Noy, G., Ophir, A., and Selzer, Y. (2010). Response of molecular junctions to surface plasmons polaritons, *Angew. Chem. Int. Ed.*, **49**, 5734.

37. Ittah, N., Noy, G., Yutsis, I., and Selzer, I. (2009). Measurement of electronic transport through $1G_0$ gold contacts under laser irradiation. *Nano Lett.*, **9**, 1615.

38. Arielly, R., Ofarim, A., Noy, G., and Selzer, Y. (2011). Accurate determination of plasmonic fields in molecular junctions by current rectification at optical frequencies, *Nano Lett.*, **11**, 2968.

39. Ittah, N., and Selzer, Y. (2011). Electrical detection of surface plasmon polaritons by $1G_0$ gold quantum point contacts, *Nano Lett.*, **11**, 529.

40. Savage, K. J., Hawkeye, M. M., Esteban, R., Borisov, A. G., Aizpurua, J., and Baumberg, J. J. (2012). Revealing the quantum regime in tunnelling plasmonics, *Nature,* **491**, 574.

41. Battacharyya, S., Kibel, A., Kodis, G., Liddell, P. A., Gervaldo, M., Gust, D., and Lindsay, S., Optical modulation of molecular conductance. *Nano Lett.,* **11**, 2709.

42. Lara-Avila, S. et al., (2011). Light-triggered conductance switching in single-molecule dihydroazulene/vinylheptafulvene junctions, *J. Phys. Chem. C,* **115**, 18372–18377.

43. Grafstrom, S., Kowalski, J., Neumann, R., Probst, O., and Wortge, M. J. (1991). Analysis and compensation of thermal effects in laser-assisted scanning tunneling microscopy, *J. Vac. Sci. Tech. B.,* **9**, 568.

44. Grafstrom, S., Schuller, P., Kowalski, J., and Neumann, R. (1998). Thermal expansion of scanning tunneling microscopy tips under laser illumination, *J. Appl. Phys.*, **83**, 3453.

45. Gerstner, V., Thon, A., and Pfeiffer, W. (2000). Thermal effects in pulsed laser assisted scanning tunneling spectroscopy, *J. Appl. Phys.*, **87**, 2574.

46. Ukraintsev, V. A., and Yates, J. T. (1996). Nanosecond laser induced single atom deposition with nanometer spatial resolution using a STM, *J. Appl. Phys.*, **80**, 2561.

47. Lyubinetsky, I., Dohnalek, Z., Ukraintsev, V. A., and Yates, J. T. (1997). Transient tunneling current in laser assisted scanning tunneling spectroscopy, *J. Appl. Phys.*, **82**, 4115.

48. Gerstner, V., Knoll, A., Pfeiffer, W., Thon, A., and Gerber, G. (2000). Femtosecond laser assisted scanning tunneling spectroscopy, *J. Appl. Phys.*, **88**, 4851.

49. Selzer, Y., and Peskin, U. (2013). Transient dynamics in molecular junctions: Picosecond resolution from dc measurements by a laser pulse pair sequence excitation, *J. Phys. Chem. C*, **117**, 22369.

50. Kohler, S., Lehmann, J., and Hänggi, P. (2005). Driven quantum transport, *Phys. Rep.*, **406**, 379.

51. Galperin, M., and Nitzan, A. (2012). Molecular optoelectronics: The interaction of molecular conduction junctions with light, *Phys. Chem. Chem. Phys.*, **14**, 9421.

52. Tien, P. K., and Gordon, J. P. (1963). Multiphoton process observed in the interaction of microwave fields with the tunneling between superconductor films, *Phys. Rev.*, **129**, 647.

53. Zhu, Y., Maciejko, J., Ji, T., Guo, H., and Wang, J. (2005). Time-dependent quantum transport: Direct analysis in the time domain, *Phys. Rev. B*, **71**, 075317.

54. Maciejko, J., Wang, J., and Guo, H. (2006). Time-dependent quantum transport far from equilibrium: An exact nonlinear response theory, *Phys. Rev. B*, **74**, 085324.

55. Ovchinnikov, I. V., and Neuhauser, D. (2005). A Liouville equation for systems which exchange particles with reservoirs: Transport through a nanodevice, *J. Chem. Phys.*, **122**, 024707.

56. Viljas, J. K., and Cuevas, J. C. (2007). Role of electronic structure in photoassisted transport through atomic-sized contacts, *Phys. Rev. B.*, **75**, 075406.

57. Viljas, J. K., Pauly, F., and Cuevas, J. C. (2007). Photoconductance of organic single-molecule contacts. *Phys. Rev. B*, **76**, 033403.

58. Stafford, C. A., and Wingreen, N. S. (1996). Resonant photon-assisted tunneling through a double quantum dot: An electron pump from spatial Rabi oscillations, *Phys. Rev. Lett.*, **76** 1916.

59. Lehmann, J., Kohler, S., Hänggi, P., and Nitzan, A. (2002). Molecular wires acting as coherent quantum ratchets, *Phys. Rev. Lett.*, **88**, 228305.

60. Lehmann, J., Kohler, S., Hänggi, P., and Nitzan, A. (2003). Rectification of laser-induced electronic transport through molecules, *J. Chem. Phys.*, **118**, 3283.

61. Kohler, S., Lehmann, J., Camalet, S., and Hänggi, P. (2002). Resonant laser excitation of molecular wires, *Israel J. Chem.*, **42**, 135.
62. Lehmann, J., Camalet, S., Kohler, S., and Hanggi, P. (2003). Laser controlled molecular switches and transistors, *Chem. Phys. Lett.*, **368**, 282.
63. Galperin, M., and Nitzan, A. (2005). Current-induced light emission and light-induced current in molecular-tunneling junctions, *Phys. Rev. Lett.*, **95**, 206802.
64. Thanopulos, I., and Paspalakis, E. (2007). Laser-operated porphyrin-based molecular current router, *Phys. Rev. B*, **76**, 035317.
65. Volkovich, R., and Peskin, U. (2011). Transient dynamics in molecular junctions: Coherent bichromophoric molecular electron pumps, *Phys. Rev. B.*, **83**, 033403.
66. White, A. J., Peskin, U., and Galperin, M. (2013). Coherence in charge and energy transfer in molecular junctions, *Phys. Rev. B*, **88**, 205424.
67. Prociuk, A., and Dunietz, B. D. (2010). Photoinduced absolute negative current in a symmetric molecular electronic bridge, *Phys. Rev. B.*, **82**, 125449.
68. Moskalets, M., and Buttiker, M. (2004). Adiabatic quantum pump in the presence of external ac voltages, *Phys. Rev. B.*, **69**, 205316.
69. Moskalets, M., and Buttiker, M. (2004). Floquet scattering theory for current and heat noise in large amplitude adiabatic pumps, *Phys. Rev. B.*, **70**, 245305.
70. Tikhonov, A., Coalson, R. D., and Dahnovsky, Y. (2002). Calculating electron transport in a tight binding model of a field-driven molecular wire: Floquet theory approach, *J. Chem. Phys.*, **116**, 10909.
71. Tikhonov, A., Coalson, R. D., and Dahnovsky, Y. (2002). Calculating electron current in a tight-binding model of a field-driven molecular wire: Application to xylyl-dithiol, *J. Chem. Phys.*, **117**, 567.
72. Keller, A., Atabek, O., Ratner, M., and Mujica, V. (2002). Laser-assisted conductance of molecular wires, *J. Phys. B*, **35**, 4981.
73. Urdaneta, I., Keller, A., Atabek, O., and Mujica, V. (2005). A simple model for laser-electrode interaction and its role in photo-assisted electron transport processes in molecular interfaces, *J. Phys. B: At. Mol. Opt. Phys.*, **38**, 3779.
74. Urdaneta, I., Keller, A., Atabek, O., and Mujica, V. (2007). Laser-induced nonlinear response in photoassisted resonant electronic transport, *J. Chem. Phys.*, **127**, 154110.

75. Camalet, S., Lehmann, J., Kohler, S., and Hänggi, P. (2003). Current, *Phys. Rev. Lett.*, **90**, 210602.

76. Kohler, S., Lehmann, J., Strass, M., and Hänggi, P. (2004). Molecular wires in electromagnetic fields, *Adv. Solid State Phys.*, **44**, 157.

77. Hänggi, P., Kohler, S., Lehmann, J., and Strass, M. (2005). AC-driven transport, in: *Introducing Molecular Electronics* (Cuniberti, G., Fagas, G., and Richter, K., eds.), *Lect. Notes Phys.*, **680**, pp. 55–75, Springer-Verlag, Berlin, Heidelberg, New York.

78. Kienle, D., Vaidyanathan, M., and Leónard, F. (2010). Self-consistent ac quantum transport using nonequilibrium Green functions, *Phys. Rev. B*, **81**, 115455.

79. Franco, I., Shapiro, M., and Brumer, P., Laser-induced currents along molecular wire junctions, *J. Chem. Phys.*, **128**, 244906.

80. Welack, S., Schreiber, M., and Kleinekathofer, U. (2006). The influence of ultrafast laser pulses on electron transfer in molecular wires studied by a non-Markovian density-matrix approach, *J. Chem. Phys.*, **124**, 044712.

81. Kleinekathöfer, U., Li, G.-Q., Welack, S., and Schreiber, M. (2006). Switching the current through model molecular wires with Gaussian laser pulses, *Europhys. Lett.*, **75**, 139.

82. Kleinekathöfer, U., Li, G.-Q., Welack, S., and Schreiber, M. (2006). Coherent destruction of the current through molecular wires using short laser pulses, *Phys. Stat. Sol. (B)*, **243**, 3775.

83. Li, G.-Q., Schreiber, M., and Kleinekathöfer, U. (2007). Coherent laser control of the current through molecular junctions, *Europhys. Lett.*, **79**, 27006.

84. Li, G.-Q., Welack, S., Schreiber, M., and Kleinekathöfer, U. (2008). Tailoring current flow patterns through molecular wires using shaped optical pulses, *Phys. Rev. B*, **77**, 075 321.

85. May, V., and Kühn, O. (2008). Optical field control of charge transmission through a molecular wire. I. Generalized master equation description, *Phys. Rev. B*, **77**, 115439.

86. May, V., and Kühn, O. (2008). Photoinduced removal of the Franck-Condon blockade in single-electron inelastic charge transmission, *Nano Lett.*, **8**, 1095.

87. Stefanucci, G., Kurth, S., Rubio, A., and Gross, E. K. U. (2008). Time-dependent approach to electron pumping in open quantum systems, *Phys. Rev. B*, **77**, 075339.

88. Peskin, U., and Galperin, M. (2012). Coherently controlled molecular junctions, *J. Chem. Phys.*, **136**, 044107.
89. Galperin, M., and Tretiak, S. (2008). Linear optical response of current-carrying molecular junction: A nonequilibrium Green's function–time-dependent density functional theory approach, *J. Chem. Phys.*, **128**, 124705.
90. Dong, B., Cui, H. L., and Lei, X. L. (2004). Photon-phonon-assisted tunneling through a single-molecule quantum dot, *Phys. Rev. B*, **69**, 205315.
91. Li, G. Q., Fainberg, B., Nitzan, A., Kohler, S., and Hänggi, P. (2010). Coherent charge transport, *Phys. Rev. B*, **81**, 165310.
92. Wang, L. X., and May, V. (2010). Optical switching of charge transmission through a single molecule: Effects of contact excitations and molecule heating, *J. Phys. Chem. C*, **114**, 4179.
93. Haertle, R., Peskin, U., and Thoss, M. (2013). Vibrationally coupled electron transport in single-molecule junctions: The importance of electron–hole pair creation processes, *Phys. Status Solidi B*, **250**, 2635–2647.
94. Kornbluth, M., Nitzan, A., and Seideman, T., Light induced electronic non-equilibrium in plasmonic particles, *J. Chem. Phys.*, **138**, 174707.
95. Fainberg, B., Jouravlev, M., and Nitzan, A. (2007). Theory of light-induced current in molecular-tunneling junctions excited with intense shaped pulses, *Phys. Rev. B*, **76**, 245329.
96. Sukharev, M., and Galperin, M. (2010). Transport and optical response of molecular junctions driven by surface plasmon polaritons, *Phys. Rev. B*, **81**, 165307.
97. Fainberg, B. D., Sukharev, M., Park, T. H., and Galperin, M. (2011). Light-induced current in molecular junctions: Local field and non-Markov effects, *Phys. Rev. B*, **83**, 205425.
98. Jauho, A. P., Wingreen, N. S., and Meir, Y. (1994). Time-dependent transport in interacting and noninteracting resonant-tunneling systems, *Phys. Rev. B*, **50**, 5528.
99. Galperin, M., and Nitzan, A. (2006), Optical properties of current carrying molecular wires, *J. Chem. Phys.*, **124**, 234709.
100. Fainberg, B. D., Hanggi, P., Kohler, S., and Nitzan, A. (2009). Exciton- and Light-induced Current in Molecular Nanojunctions, *AIP Conf. Proc.*, **1147**, 78.

101. Zelinskyy, Y., and May, V. (2011). Photoinduced switching of the current through a single molecule: Effects of surface plasmon excitations of the leads, *Nano Lett.*, **12**, 446.

102. White, A. J., Fainberg, B. D., and Galperin, M. (2012). Collective plasmon-molecule excitations in nanojunctions: Quantum consideration, *J. Phys. Chem. Lett.*, **3**, 2738.

103. Peskin, U., and Moiseyev, N. (1993). The solution of the time-dependent Schroedinger equation by the (t, t') method: Theory, computational algorithm and applications, *J. Chem. Phys.*, **99**, 4590–4596.

104. Peskin, U., and Moiseyev, N. (1994). Time-independent scattering theory for time-periodic Hamiltonians: Formulation and complex scaling calculations of above threshold ionization spectra, *Phys. Rev. A.*, **49**, 3712.

105. Peskin, U., Kosloff, R., and Moiseyev, N. (1994). The solution of the time-dependent Schroedinger equation by the (t, t') method: The use of global polynomial propagators for time-dependent Hamiltonians, *J. Chem. Phys.*, **100**, 8849.

106. Peskin, U., Alon, O. E., and Moiseyev, N. (1994). The solution of the time-dependent Schroedinger equation by the (t, t') method: Multiphoton ionization/dissociation probabilities in different gauges of the electromagnetic potentials, *J. Chem. Phys.*, **100**, 7310.

107. Peskin, U., and Miller, W. H. (1995). Reactive scattering theory for molecular transitions in time-dependent fields, *J. Chem. Phys.*, **102**, 4084.

108. Peskin, U., Edlund, A., and Miller, W. H. (1995). Quantum time-evolution in time-dependent fields and time-independent reactive scattering calculations via an efficient Fourier grid preconditioner, *J. Chem. Phys.*, **103**, 10030.

109. Taylor, J. R. (1972). *Scattering Theory*, Wiley, New York.

110. Shirley, J. H. (1965). Solution of the Schroedinger equation with a Hamiltonian periodic in time, *Phys. Rev.*, **138**, B979.

111. Floquet, G. (1883). Sur les équations différentielles linéaires à coefficients périodiques, *Annales de l'École Normale Supérieure* **12**, p. 47.

112. We emphasize that the assumption regarding the smooth energy-dependence of the leads density of states can not hold in its own right as $\hbar\omega k \to \pm\infty$, since the change in energy becomes infinite.

However, the contribution of different k values in Eq. (7.20) is scaled by the frequency components of the transformed driving field $\{g_{J,k}\}$. Therefore, a strict condition for the validity of the wide band approximation involves additional restrictions on the field, such that the contribution of large $|k|$ values to the summation in Eq. (7.20) drops to zero. This condition is typically satisfied but needs to be verified.

Chapter 8

Biomolecular Electronics

Juan Manuel Artés Vivancos,[a] Joshua Hihath,[a] and Ismael Díez-Pérez[b]

[a]*Electrical and Computer Engineering, University of California Davis, One Shields Ave., Davis, California 95616, USA*
[b]*Physical Chemistry Department, University of Barcelona, Marti I Franquès, 1 Barcelona, 08028 Barcelona, Spain*

juanmanuel.artes@gmail.com, jhihath@ucdavis.edu, isma_diez@ub.edu

This chapter provides an introduction to the charge transport (CT) characteristics of molecular devices built using complex biological backbones. This field has emerged as a subset of the molecular electronics discipline, which has brought a large variety of new tools to study CT in molecular electronic devices down to the single-molecule level. We try to show the importance of this biomolecular electronics field from both the fundamental and technological perspectives.

The appearance of the first examples of single-molecule three-terminal devices within molecular electronics evolved into electrochemically gated single-molecule devices, allowing the first fundamental studies of CT in individual biomolecules bridged between two macroscopic electrodes. This chapter describes the latest examples of such novel biomolecular wires that are contributing to the definition of the emerging discipline of biomolecular electronics. Along with these detailed descriptions,

Molecular Electronics: An Experimental and Theoretical Approach
Edited by Ioan Bâldea

ISBN 978-981-4613-90-3 (Hardcover), 978-981-4613-91-0 (eBook)
www.panstanford.com

a survey of the latest advances on interfacing proteins and nucleic acids to electrodes is presented.

8.1 Introduction

Building up efficient electronic nanodevices by profiting from nature optimized biomolecular machinery is a clear challenge for today's scientific and engineering communities [1]. Thousands of years of natural evolution have resulted in highly sophisticated and specialized molecular architectures that are able to transport charge and/or energy with astonishing efficiency [2–4]. The dream of exploiting such properties has motivated researchers for decades in fields such us bioelectronics, and with the advent of nanoscience, the field has experienced a new boom [5–7]. The ultimate goal of this field is the integration of a biological motif into an electronic platform that allows the transduction of the target signal and its easy implementation into a device. Taking advantage of biomolecular properties to design functional electronic devices poses two clear challenges: First, we need to fill the knowledge gap of the fundamental mechanisms behind biological energy/charge transfer, and second, we need to efficiently engineer the biomolecule/electrode interface to achieve a stable hybrid bio-electronic platform.

Generally speaking, the ongoing research within the field of bioelectronics is still at a very phenomenological level. A number of fundamental questions need to be answered in order to push this field to the next level. These include: *What are the main charge/energy pathways located in a complex biomolecule? How can they be effectively hybridized with an electrode surface? Where should "the electrical plugs" be allocated in such biological molecular structures?* And *what anchoring chemistry should then be developed?* To this aim, a huge amount of information has yet to be experimentally addressed such as the exact role that different molecular motifs have on biological charge/energy transport processes, e.g., α-helix, β-sheet, random coil, etc., the role of intramolecular interactions like for instance molecular orbital coupling, H-bonding, etc., and the role of the molecular dynamics, e.g., fluctuations between different conformers, interactions with partner molecules, etc. Answering these and other fundamental questions will place us in position to address a systematic strategy to fully integrate active biomolecular components onto metal/semiconductor platforms.

The complexity of this particular topic calls for the use of novel interdisciplinary approaches in order to shed new light onto the fundamental mechanisms behind biological charge transport and fill the gap between the well-known crystallographic structure of a biomolecular complex and the final functional bioelectronic device.

We are convinced that the momentum in the emerging field of molecular electronics, and its progressive immersion into the study of more complex biomolecular moieties will at some point provide important breakthroughs in our knowledge about how to engineer biomolecule/electrode interfaces. Such knowhow will surely show us new directions toward the design of the next generation of bioelectronic devices.

8.1.1 The Birth of Molecular Electronics

It is difficult to set an exact kickoff date for the field of molecular electronics. This field is often depicted as being conceived from Aviram and Ratner's paper [8]. However, beyond this seminal event, the field progressively evolved at the interface between disciplines such as nanotechnology and chemistry, with the ideal motivation of developing new approaches to build nanoscale circuits with atomic control. This idea was largely boosted by the increasing pressure of the contemporary microelectronics industry to find solutions to miniaturize circuits and achieve smaller electrical components with better performance. The miniaturization needs in electronics were clear shortly after the first integrated circuits were developed, when many scientists and engineers used neologisms such as "atomic electronics" or "angstronics" [9]. Reducing the size of already microfabricated silicon-based components has an inherent spatial limitation due to the intrinsic heterogeneity of such structures, i.e., heterogeneous distribution of dopants, defects in the crystalline structure, poorly developed interfaces between materials, etc. Indeed, in the 1950s, the concept of molecular electronics appeared as an alternative to the classical top-bottom approaches, suggesting the possibility to build nanoscale electrical devices from their atomic/molecular components. The futuristic vision was an appeal to synthesize new materials with predetermined electrical properties that could be used as active components in devices with specific functions.

What may be considered one of the first experimental attempts to understand charge transport in molecular systems with reduced dimensionality took place at the beginning of the 1960s with the development of Langmuir–Blodgett films. Scientists such as H. Kuhn at the University of Göttingen pioneered this area presenting consistent experimental measurements on charge transport through a molecular monolayer sandwiched between two metal contacts (Fig. 8.1). Even though the lateral cross-section of such molecular contacts were in the micro-scale range, the current is measured through a one-molecule thick contact whereby the transport is dominated by quantum effects, as deduced from the exponential decay of the conductance as a function of the molecular length [10].

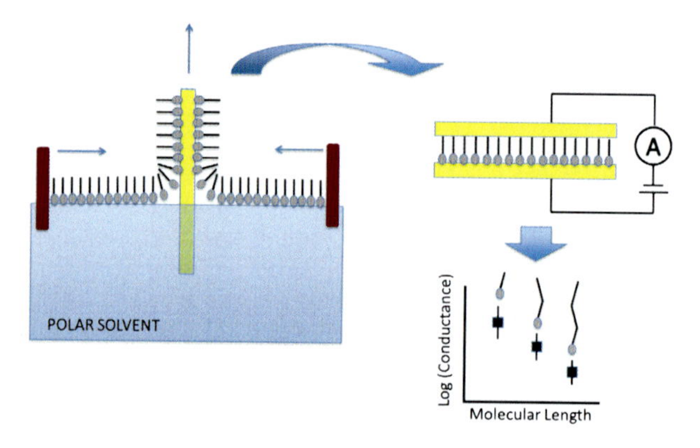

Figure 8.1 Schematic concept of the use of the Langmuir–Blodgett method to study transport in a unimolecular organic layer.

Contemporary to these first experimental examples, the idea of integrating organic conductors into well-developed integrated circuits was gestating at companies such as IBM. Ari Aviram presented a project on electron diffusion through tetrathiafulvalene-tetracyanoquinodimethane (TTF-TCNQ) moieties to the theorist Professor Mark Ratner at New York University, who saw the potential interest in such studies. From this collaboration, in 1974 both researchers published the now-famous paper on "molecular rectifiers" where they proposed a single-molecule electrical contact that could operate as a diode in a circuit [8]. The molecular structure was inspired on a TTF-TCNQ structure with one unit rich in electrons (TTF) and the other unit deficient in electrons (TCNQ), which resembles the structure of traditional semiconductor diodes with electrically adjacent electron-poor and electron-rich regions. When a voltage difference is applied to such interface, electrons will easily pass from the electron-rich to the electron-poor regions generating substantial current, while in the opposite voltage polarity electrons will poorly pass through the electron-poor region, overall resulting in electrical rectification behavior (Fig. 8.2). Although Aviram and Ratner's molecular rectifier is now considered the founding statement of modern molecular electronics, at that time it was seen as a mere theoretical curiosity difficult to be experimentally addressed, and it went virtually unnoticed for more than a decade.

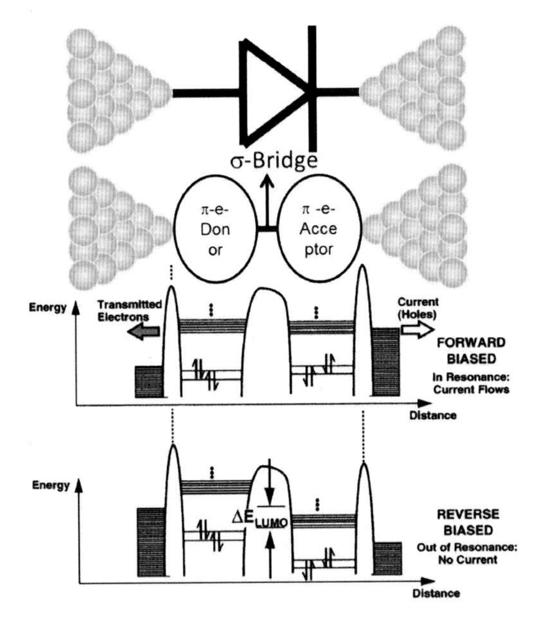

Figure 8.2 General representation of the single-molecule diode concept showing the asymmetric energy levels distribution across the single-molecule electrical contact. Adapted from [8] with permission, copyright Elsevier 1974.

By the late 1980s, the invention of the scanning tunneling microscope (STM) [11], followed by other experimental tools for nanoscience, gave new momentum to molecular electronics. Two key experiments were presented during the 1990s, demonstrating for the first time the feasibility of building and characterizing charge transport in an individual molecule connected to two metallic beads. In 1995, Joachim and Gimzewski used an STM tip to contact a C_{60} molecule absorbed on a metal substrate [12], and two years later, Reed and Tour used a mechanical controllable break-junction (MCBJ) device to bridge a small benzene backbone between two metal electrodes [13]. In the latter, the benzene molecule was functionalized with two axial thiol groups that enabled specific covalent binding between the molecule and the metal electrodes. Such experiments constituted a direct demonstration of the single-molecule circuit concept. Following this achievement, the turn of the millennium was filled with a number of related experimental approaches that allowed for improved reliability of the formation of single-molecule junctions and more robust transport characterization. This advancement was largely enabled by improving the measurements statistics. To name some, electromigrated nanogaps [14], MCBJs [13, 15, 16], and the STM-based break-junction [17] methods (Fig. 8.3) have spread around the world, and are now established as the central experimental methodologies for studying fundamental charge transport properties in molecular systems.

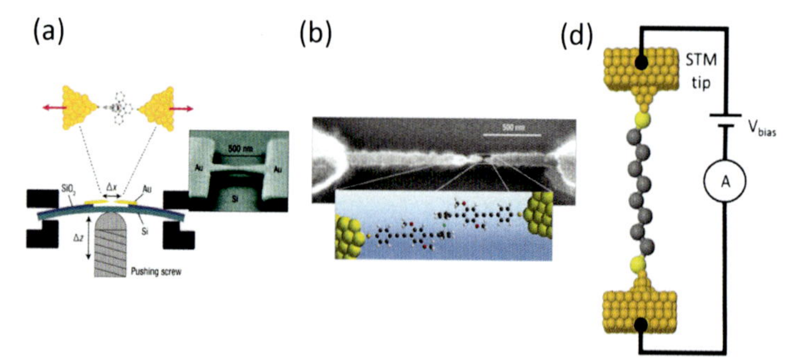

Figure 8.3 Representation of the latest methods to measure single-molecule conductance: (a) Mechanical controllable break-junction (MCBJ). Reproduced with permission from [18]. Copyright 2010, American Association for the Advancement of Science. (b) Electromigrated Nanoscale Junctions. Reproduced with permission from [19]. Copyright 2004 Nature Publishing Group. (c) Scanning tunneling break-junction (STM-BJ) scheme [17].

These single-molecule junction (SMJ) approaches have allowed key fundamental insights into the relationships between chemical structure and transport, for instance, how selected chemical substitutions in a molecular backbone modify its charge transport behavior [20] or how different chemical backbones (e.g., conjugation, presence of hetero-atoms, etc.) give rise to a wide variety of electrical behaviors [21, 22]. Precise interfacial information such as the different contact geometries that can be established between the molecule and the electrode have been also elucidated by means of SMJs [23, 24]. Relevant clues for the fundamental comprehension of charge transport in biological molecules have been also brought about from SMJ approaches. Some examples are the measured molecular conductance as a function of both the dihedral angle between two conjugated rings [25, 26], and between a conjugated wire and an electrode surface [27, 28]. The charge transport in both cases appears to be enhanced by the π-orbital coupling, a mechanism that has been extensively discussed as a key ingredient in biological long-range electron transport [29].

Although many technical challenges have to be addressed before the field of molecular electronics is ready to deliver real functional devices, today it is a true scientific field that is constantly revealing new fundamental effects in chemistry-related charge transport processes. It is likely that such a rich scientific field will eventually hit a new concept in molecular circuitry that could result in a real breakthrough for the future electronics technology.

8.1.2 Introduction to CT Mechanisms: From Tunneling to Hopping

When considering the pool of theoretical models used to describe charge transport through molecules one can find two extreme situations which are relevant for a conceptual understanding of the processes: (i) *coherent electron tunneling* describes the charge transport through an organic moiety without a significant molecule-electron/hole interaction, thus preserving the coherence of the electron wave. (ii) *Thermally activated hopping transport* mechanisms refers to fully incoherent charge transport process where the itinerant electron/hole resides for a significant amount of time (typically in the order of the molecular reorganization time) at specific low energy orbitals located along the molecular wire. Between these two well-distinguished cases, one can find intermediate situations such as the *two-step sequential tunneling* mechanism that describes a tunneling process with partial molecular relaxation. The latter has been extensively applied to charge transport studies in various complex redox metalloproteins [30–33].

8.1.2.1 Coherent Tunneling

The coherent tunneling (Fig. 8.4a) or super-exchange mechanism is a quantum mechanical effect that describes the probability of an electron passing through a classically forbidden energy barrier [34]. When studying molecular transport, the frontier molecular orbitals define the energy barrier height for the *tunneling* process and thus the transmission probability. Because this is a tunneling process, transmission probability strongly depends on the barrier width, which is equivalent to the molecular length. In this scenario, the conductance decays exponentially as the molecular length is increased:

$$G = G_0 \exp(-\beta L), \tag{8.1}$$

where L is the gap distance between the two electrodes and is equivalent to the molecular length in a single-molecule device (Fig. 8.4), β is the decay constant (typically given in nm^{-1}) and G_0 is the conductance quantum $2e^2/h$ = 77.4 μS, e being the elementary charge and h the Planck constant. Here, the electron is directly transferred through the molecular structure in a single coherent step. The current decay rate is described by the constant β, which, in turn, depends on the chemical nature and folding structure of the molecular backbone. This model has been satisfactorily used to describe charge transport in molecules displaying large energy gaps (several eV) between the highest occupied (HOMO) and the lowest unoccupied (LUMO) molecular orbitals. One of the most prominent examples are simple alkane chains, such as octanedithiol, which present a clear coherent tunneling CT mechanism with large decay constant values $\beta = 8–10\ nm^{-1}$ [17].

8.1.1.2 Hopping Transport

In sharp contrast, the hopping mechanism is an incoherent process. In this case the electron travels through the molecular backbone by *jumping* (hopping) between different low-energy molecular orbital sites present along the molecular backbone (Fig. 8.4b). As can be intuitively deduced, this regime presents a much shallower dependence on the molecular distance L than tunneling does. Moreover, the electron/hole *hopping* process between sites is typically thermally activated following an Arrhenius dependence [35, 36]:

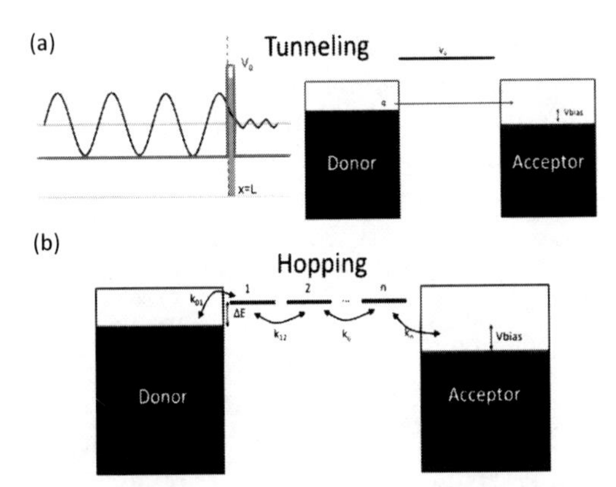

Figure 8.4 Charge transport mechanisms. (a) Superexchange tunneling: (left) representation of the wavefunction in a tunneling junction, and (right) energy diagram for a donor acceptor system where the electron crosses an energy barrier V_0. (b) Hopping: a charge undergoes a thermally activated CT through different states (1, 2, ... N) in the donor acceptor gap. Adapted with permission from ref [37]. Copyright 2014 Elsevier Ltd.

$$g = \left(\frac{q^2}{k_B T}\right) \frac{k e^{-(E_B/k_B T)}}{\dfrac{k}{k_{R,N}} + \dfrac{k}{k_{L,1}} + (N-1)}, \tag{8.2}$$

where k_B is Boltzmann's constant, q is the electron charge, E_B is the average bridge energy level, T is the temperature, k is the transfer rate between bridge sites, $k_{R,N}$ and $k_{L,1}$ are the rates from the right and left molecule onto or off of the bridge, respectively, and N is the number of bridge sites in the system. Interestingly the conductance of a system displaying thermally activated hopping is inversely proportional to the length (N) of the molecule. Also important to note is that the conductance is temperature independent in the tunneling case but has an Arrhenius dependence in a hopping CT.

8.1.3 Two-Terminal versus Three-Terminal Devices: Towards the Molecular Transistor

The main block of information concerning single-molecule transport mechanisms has been brought about by two-terminal experimental configurations, meaning the simplest device version with a molecule bridging between two biased (V_{bias}) electrodes. Although such single-molecule, two-terminal platforms have become the primary testbed for the field, three-terminal devices arranged in a field-effect transistor (FET) configuration are required when it comes to the implementation of logic functionalities [38]. In such an ideal platform, one should have a single molecular backbone connected to both a source and drain electrode, thus acting as the channel for the charge flow. A third, *gate electrode* should then be built very close to the molecular backbone allowing the opening and closing of the channel by applying an electric field (see Fig. 8.5a). In order to effectively shift the molecular energy levels in a single-molecule bridge by an external gate electrode, such an electrode must be placed in close proximity to the molecular channel so that the electric field at the molecule/gate interface surpasses that at the molecule/source or drain interfaces. In long molecular channels such as a nanotube or graphene sheet connecting two metallic pads, this is not a significant problem [39, 40]. However, when the channel is scaled down to the size of a small molecule the technical challenge of fabricating a gate electrode close enough to the molecular junction becomes obvious. In this vein, pioneering works performed at cryogenic temperatures by Park [41] and Bjørnholm [42] demonstrated the experimental feasibility of adding a third electrode into a single-molecule electrical device and measuring gate-dependent transport. Despite the experimental advance in designing solid-state, three-terminal, single-molecule devices during the last decade [43–45], the microfabrication of these systems still involves complex clean room processing and often electron beam lithographic processing. These steps significantly diminish the success rate of fabrication and make it rather difficult to integrate them into real functional devices. On top of fabrication difficulties, one important limitation on such devices is their low gate efficiency (α), being α defined as the ratio between the molecular orbital energy shift and the applied gate voltage (V_g). By measuring α, one can evaluate the amount of electric field from the gate electrode that is actually felt by the confined molecule. α can be evaluated from the slope of the transition voltage (V_{trans}) as a function of V_g (Fig. 8.5b), where V_{trans} is extracted from the minimum in the transition voltage spectroscopy (TVS) results [46].

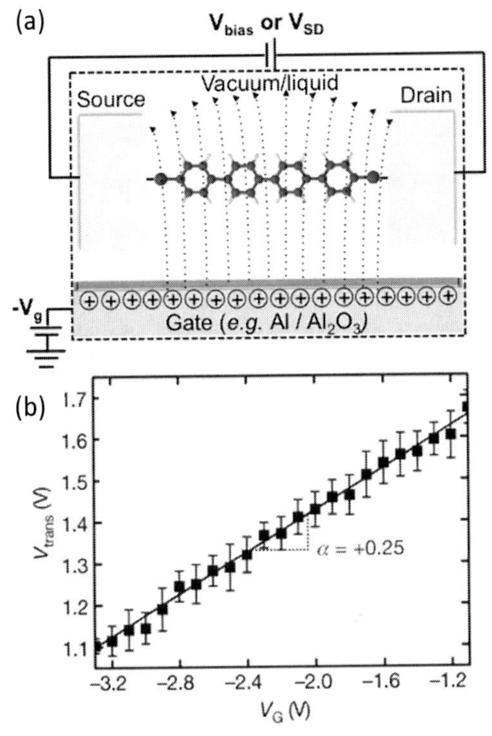

Figure 8.5 (a) Schematic of a solid-state, single-molecule Field-effect transistor and (b) the corresponding transition voltage versus gate voltage plot to evaluate the gate efficiency α. (a) Reprinted from [47] with permission, copyright 2013 Elsevier Ltd., and (b) reprinted from [44, 46] with permission, copyright 2009 Nature Publishing Group.

An alternative method to overcome the technical difficulties in the fabrication and operation of three-terminal single-molecule devices is to use an electrochemical gate. In this approach, the molecular junction is immersed in an electrolyte solution and the gate voltage is applied through the electrochemical double layer developed in the vicinity of the molecular channel. Figure 8.6 shows a schematic representation of a single-molecule electrical contact coupled to an electrochemical gate. The electrolyte can be either an aqueous environment or a polar organic solvent, e.g., acetonitrile containing millimolar concentrations of a salt that increases the dielectric constant of the liquid medium. A third electrode (reference) immersed within the solution acts as the gating electrode to modulate the current that flows through the molecule. The way the electrochemical gate is implemented in the actual experimental setup is by using an electrochemical bipotentiostat so that independent electrochemical potentials can be applied to both junction electrodes (Fig. 8.6: W1 and W2 or source and drain following the FET notation). The V_{bias} will be now defined as the electrochemical potential difference between electrodes W1 and W2. When an electrochemical potential is applied to the single-molecule junction, an electrochemical double layer (EDL) capacitor builds up at every electrified interface, namely the source/electrolyte, the drain/electrolyte, and the reference/electrolyte interfaces. The two plates of these interfacial capacitors are formed by the charged electrode interface on one side, and the solvated ions in solution counteracting this charge on the other (Fig. 8.6). The gate voltage applied through the reference electrode will drop primarily across the two main EDLs: the one formed at the gate electrode/electrolyte interface and the one at the molecular junction/electrolyte interface. This configuration corresponds to an effective gate thickness on the order of a few solvated ions (encircled +/– signs in Fig. 8.6), thus resulting in a large highly localized gate field. It is easier to picture the electrochemical gate principle as a direct tuning of the Fermi energy of the source and drain electrodes while the frontier molecular orbitals are pinned. The final result is a net shift of the molecular energy levels with respect to the electrodes Fermi energies. The electrochemical gate has been already proven very effective for tuning the device current flow in solid-state EDL transistors displaying gating efficiencies α close to 1 [48, 49].

The electrochemically gated single-molecule approach has allowed the first single-biomolecule transport studies performed in a physiological environment at nearly neutral pHs [30–33, 50, 51], and has provided insights into CT mechanisms in these systems. Key biological motifs such as double-stranded DNA, short peptide sequences, and redox metalloproteins have been studied, and the results set the starting point of the biomolecular electronics discipline. The main advances of such systems will be treated in more detail within the next sections.

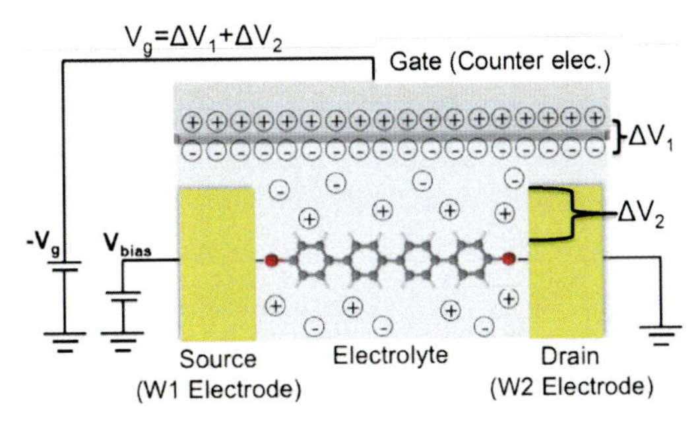

Figure 8.6 Schematics of an electrochemically gated single-molecule field-effect transistor. W1 and W2 are the two working electrodes (source and drain, respectively) that can be gated by the gate electrode (the electrochemical potential). Reprinted from [47] with permission, copyright 2013 Elsevier Ltd.

8.2 Peptides and Proteins

Proteins are fascinating molecules not only from the biological point of view (they are involved in most of the biological processes) but also from a nanotechnology perspective. Proteins are optimized molecular machinery capable of performing almost any imaginable function. In that sense, understanding the CT mechanisms in proteins could open new perspectives in molecular electronics. Charge is required to move over long distances through proteins in many biological energy transduction pathways [52], and developing a complete understanding of the CT processes in proteins will have a huge impact on several fields. Recently, several theoretical and experimental approaches have addressed how proteins support CT, and in the following sections we review some of them with a special emphasis on approaches using molecular electronic techniques to study CT and/or fundamental biological studies that opened new perspectives on the molecular electronics field.

The basis of CT in proteins has been studied at different levels of structure (primary, secondary and tertiary), and in the next section we will review the first CT studies going from simpler peptide chains to more complex protein structures.

8.2.1 Peptides in Molecular Electronics

Theoretical efforts have established models for the electronic coupling and CT pathways along biomolecular structures [53, 54], and they predict that the distance decay factors of biomolecules depend on their structure [52, 53]. In the case of proteins, β-values of 10 nm^{-1} were expected for a beta sheet secondary structure [55, 56] and 12.6 nm^{-1} for an alpha-helix [56, 57]. Other theoretical reports have focused on fluctuations and interferences among different CT pathways that account for the coherence of the CT mechanism [54, 58, 59]. It is now clear that the secondary structure of the protein and its fluctuations can deeply impact the CT characteristics.

Some experimental CT studies on alpha-helical peptides have been reported, and they demonstrate the feasibility of synthesizing α-helical peptides with different amino acid sequences, preparing SAMs with peptides of different lengths, and measuring CT with STM-based techniques [60, 61]. Tao and co-workers presented one of the first reports on CT in peptides [62, 63]. Although these examples represented the starting point of single-molecule CT studies in peptidic architectures, the important mechanistic points are still pending. The discussion about CT in helical peptides is extremely controversial. Reports excluding the suitability for helical peptides as charge intermediary have been published [64], even though several groups have stated that the helical backbone is an excellent CT intermediate [65]. The mechanistic discussion behind it involves transitions between a coherent electron tunneling process and an incoherent hopping mechanism.

A good example of the previous mechanistic discussion is the work of Mandal et al. [66], showing the electrochemical response of SAMs of redox-active ferrocene-labeled helical peptides of increasing length. A clear transition in the electron transfer rate versus peptide length is observed and ascribed to a tunneling-to-hopping transition (Fig. 8.7). Changes in the β-value should not be directly ascribed to a change in the CT mechanism. The transition in electrochemical rates can be also interpreted as structural changes as the length of the peptides increases [66–70].

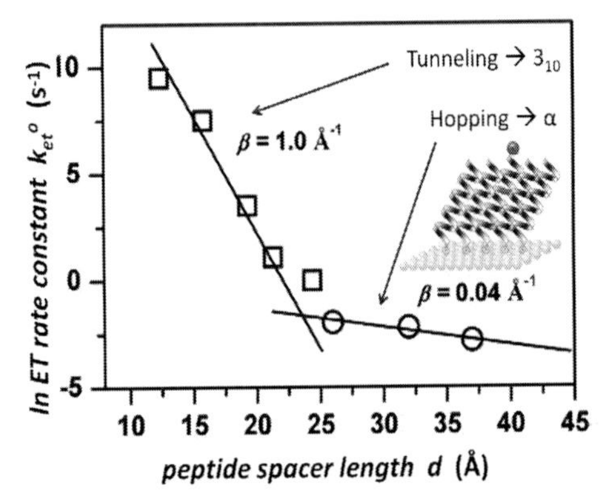

Figure 8.7 Standard electrochemical rate constants versus peptide length (D–A distance). Adapted from [66] with permission. Copyright 2012 American Chemical Society.

Another fact to take into account when interpreting CT in peptides is that the internal H-bond networking is well known to assist the tunneling process. The presence of intramolecular H-bonds in the helix could assist in defining "electron pathways," where tunneling is facilitated by better electronic coupling and resulting in lower β-values [66, 70]. These works clearly demonstrate the importance of the secondary structure on the CT process [66].

Besides offering valuable insights into the fundamental CT processes in proteins at the secondary structure level, experimental molecular electronics approaches using peptides could exploit some unique biomolecular characteristics for device applications. In particular, the concept of chiral CT, described in detail by Mujica et al. [71], could be used to create novel spintronic devices. The concept states that electrons transmitted through a chiral molecular structure will favor one particular electron-spin polarization over another depending on its handedness of the chirality. The work describes how spin-orbit contributions are responsible for spin-dependent scattering. This chiral CT effect has been also experimentally demonstrated in DNA by Naaman et al. [72].

8.2.2 Proteins

Besides primary and secondary structure, the higher tertiary levels of organization are the ones that determine the final activity of the protein and its final CT characteristics. In order to address how CT takes places in a protein with tertiary structure, in the next sections, we will focus on the azurin metalloprotein as a benchmark for molecular electronics studies in proteins.

8.2.2.1 Azurin: Photochemical and electrochemical approaches to CT

Pseudomonas aeruginosa Azurin (Fig. 8.8a) is a 14.6 KDa globular protein that has an unimetallic copper center. The Cu center is located 8 Å below the protein surface and the protein has a solvent-exposed disulfide bridge between *Cys* 3 and *Cys* 26 [73], which was used in pulse radiolysis experiments to investigate intraprotein electron transfer pathways [55]. These native *Cys* residues have been also used to immobilize the protein on gold surfaces to study the CT process using electrochemical [74] and scanning probe microscopy techniques [75–81] (see following sections).

One of the most extensively used approaches to study CT through molecular bridges is the use of redox probes that can be attached to molecules and excited optically. The CT rate can then be studied by monitoring the fluorescence following a flash-quenching photolysis measurement (scheme in Fig. 8.8b). This approach has revealed to be extremely useful to study bulk CT in proteins [82] and in nucleic acids [83]. Using this photochemical approach, a pioneering work by Gray et al. explored the intraprotein distance decay CT behavior for azurin [56], finding $\beta \approx 10$ nm^{-1} (Fig. 8.8b), in agreement with previous calculations [53]. This result suggested coherent tunneling as the dominant intraprotein CT mechanism. Figure 8.8c shows the experimental results for different redox proteins in the context of tunneling transport in different media for comparison [84]. However, in order to achieve efficient intermolecular CT in proteins, maximum donor-acceptor tunneling distance of about 2 nm are needed, which is far below the actual distances found between many redox enzyme centers [52]. CT between distant redox centers could occur by a multistep tunneling mechanism, which displays smoother distance dependence (blue and red lines, respectively, in Fig. 8.8d), assisted by protein residues with low ionization potentials like tryptophan and tyrosine [85].

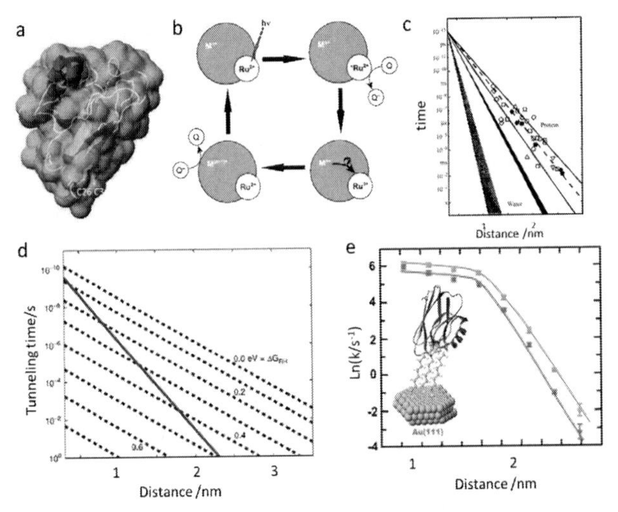

Figure 8.8 (a) Azurin 3-D representation (PDB accession code: 1AZU [73]). Reproduced from [37] with permission, copyright 2014 Elsevier. (b) The flash-quench scheme for measuring CT rates. (c) Tunneling time versus donor–acceptor separation in Ru-modified proteins: Azurin (solid circle); Cytochrome C (*cyt* c, empty circle); Myoglobin (empty triangle); *cyt* b562 (empty square). The solid lines illustrate the tunneling pathway predictions for coupling along beta-strands and alpha-helices, respectively; the dashed line illustrates a 1.1 $Å^{-1}$ distance decay factor. (b) and (c) Reproduced from [3] with permission. Copyright Cambridge University Press 2011. (d) Distance dependences of the tunneling time of single-step and two-step electron tunneling reactions. Red (continued) solid line indicates theoretical distance dependence for a single-step, ergoneutral tunneling process. Blue dashed lines indicate distance dependence calculated for two-step ergoneutral tunneling with the indicated standard free-energy changes for the first step. Reprinted with permission from [52]. Copyright 2010 Elsevier. (e) Distance-dependent electron transfer kinetics shown by a plot of the rate constant (in natural logarithm) versus the distance for experiments performed in H_2O (green) and in D_2O (purple). *Inset*: Schematic illustration of azurin molecules wired through an octanethiol monolayer self-assembled on a Au(111) surface. Adapted with permission from [30], copyright National Academy of Sciences 2005.

Alternatively, bulk electrochemical studies can be performed by immobilizing the redox protein onto a well-defined electrode surface. Generally, a biomolecular backbone can be modified by introducing a redox reporter on one end of the molecule and attaching the other side to an electrode in order to measure an electrochemical current. This procedure can then be related to the electrochemical rate constant. In the case of redox proteins, the redox center(s) can mediate the process, avoiding the use of redox reporters. In order to perform electrochemical experiments, azurin can be covalently bound to a gold electrode through either the native or other introduced *Cys* residues. Another strategy is to use non-covalent interactions with a linking self-assembled monolayer (SAM) covering the working electrode. This has been used to study the distance dependence of the CT reaction involving azurin using variable-length alkanethiol SAMs [86]. By these means, the previously reported distance decay factor (β = 9–10 nm^{-1}) for intraprotein electron transfer was confirmed [30, 87]. In addition, a combination of cyclic voltammetry (CV) with fluorescence detection allowed the observation of thermodynamic and kinetic heterogeneity of azurin populations on an electrode surface [88, 89].

8.2.2.2 Full azurin electronic characterization using ECSTM

As introduced in the previous section, CT in simple azurin models has been characterized at the bulk level and has provided important insights into the single-protein CT process. Electrochemical STM (ECSTM)-based approaches offer the possibility of studying the CT process at the single-protein level while still maintaining a physiological aqueous environment (Fig. 8.9b) [75, 90]. In addition, ECSTM offers the possibility of performing spectroscopic experiments at fixed points on the surface by introducing perturbations on the probe to sample distance, electrochemical potentials, etc., while monitoring the tunneling current. This is known as electrochemical tunneling spectroscopy (ECTS) [91] and, in the following sections, we present its applications to understand CT in the azurin model.

8.2.2.2.1 ECTS current–distance characteristics

Figure 8.9a shows ECSTM images of azurin molecules bound to a gold surface. The apparent height reported for individual azurin molecules studied by ECSTM was one order of magnitude lower than expected from the protein X-ray structure [75, 90]. This effect was assumed to be due to the constant-current STM imaging conditions and the different conductance of the protein and the metallic substrate, but a detailed understanding required deeper study using current-distance ECTS. As introduced previously, an important parameter characterizing the CT process is the distance decay factor β, as noted in Section 8.1.2 above. Current-distance ECTS provides a direct experimental determination of β by measuring the current flowing between the electrodes while varying the distance between them [92]. In systems described by a metal|insulator|metal model, β is directly related to the effective tunneling barrier height in the tunneling process (Eq. 8.1) [92].

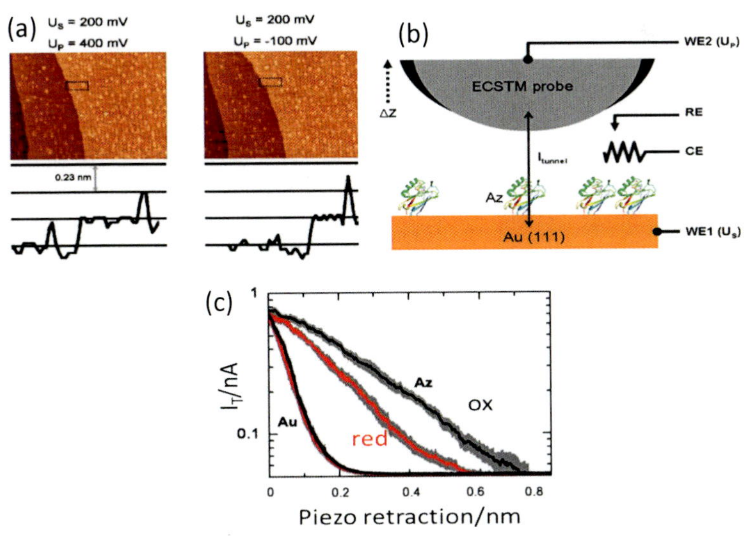

Figure 8.9 (a) Azurin proteins on an atomically flat gold surface imaged by ECSTM in 50 mM ammonium acetate solution. Representative images (100 × 100 nm and 1 nm colored height scale). (b) Experimental setup for ECSTM imaging and current-distance spectroscopy studies of azurin under bipotentiostatic control. RE and CE denote the reference and counter electrodes in solution, respectively. (c) Averaged current-distance plots on bare gold at U_S = 200 mV (black traces) and U_S = −300 mV (red traces) at constant bias of 200 mV. Current-distance plots recorded on azurin-coated gold (labeled "Az"). Reprinted with permission from [78]. Copyright 2011 American Chemical Society.

In the case of azurin, the intramolecular β-value has been extensively studied using electrochemical techniques. However, the β corresponding to the intermolecular process, which defines the physiological activity of the protein, remained elusive with only a few theoretical predictions [3]. ECSTM studies at variable electrochemical potentials have helped to clarify this point suggesting a resonant-like mechanism for the CT mechanism [75], such as resonant tunneling or a sequential two-step electron transfer with partial vibrational relaxation [93]. Figure 8.9c shows experimental current versus distance curves ($I(z)$) obtained by retracting the tip from an initial position on top of an azurin-modified electrode while recording the current. The data was used to quantify the distance decay factor β for the CT process involving azurin [78]. The current as the distance is increased decays exponentially suggesting a tunneling process (Fig. 8.9c). $I(z)$ resulted in β = 11 nm^{-1} for a bare Au electrode surface, characteristic of a coherent tunneling process. In the case of azurin samples, β = 5 nm^{-1} was obtained, matching the predicted value for a multistep tunneling process (slope of the continuous red line in Fig. 8.8c) [3].

It is particularly interesting to note that different decay constants were observed for both the reduced-Cu(I) and oxidized-Cu(II) azurin cases (red and black in Fig. 8.9c). These differences in β directly correlate to the redox activity of the protein, suggesting different energy barriers for hole and electron transfer process along the protein backbone [85].

8.2.2.2.2 Single-protein conductance and the Azurin transistor

The ECSTM can also be used to perform conductance measurements on a single molecule electrically connected to two electrodes [17, 94].

Figure 8.10 describes experimental conductance measurements performed on an individual azurin molecule using the STM-BJ approach [17, 79]. The results show current steps comparable to those found in studies performed on organic molecules having a simpler structure, indicating that stable junctions can be established between a wild-type azurin and two Au electrodes, and the data was used to construct conductance histograms that showed peaks not found in clean Au control experiments, corresponding to the single-protein signature. This procedure allowed the conductance of the single-protein wire to be determined (Fig. 8.10b).

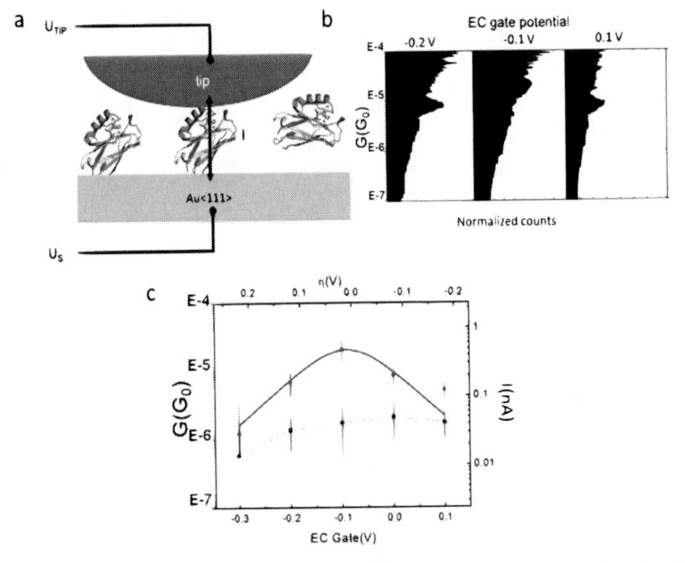

Figure 8.10 (a) Scheme of a single-azurin junction formation by the STM-BJ approach. (b) Semilogarithmic conductance histograms. (c) Conductance values obtained from histograms as a function of electrochemical gate potential at constant −0.3 V bias for azurin (red circles) and nonredox Zn–azurin junctions (black squares). The red plot shows a fit of the numerical version corresponding to the formalism for a two-step electron transfer process. Reprinted with permission from [33]. Copyright 2012 American Chemical Society.

Figure 8.10c shows that the single-azurin conductance varies between 10^{-6} and $10^{-5}G_0$ depending upon the applied electrochemical potentials, demonstrating that the single-azurin wire conductance can be modulated by the redox behavior of the protein. The dependence of the single-azurin conductance on electrochemical potentials was found to be consistent with a two-step electron transfer scenario [93], and a numerical version of the model [95] was used to fit the data and obtain an approximation to different parameters such as a 0.3 eV reorganization energy for the transient oxidation/reduction cycle during the electron passage. The conductance in the junction is controlled in this case by the alignment of the Fermi levels of the electrodes with the energy level of the redox center (using an electrochemical gate approach as introduced in previous sections) [47]. These results constituted a proof of concept of a single-protein electrochemically gated field-effect transistor, with on/off ratios typically around 20 at very low operation voltages of ~100 mV [79].

Other proteins have been imaged using ECSTM- and STM-based molecular junction techniques [32, 96, 97]. The interfacial electron transfer process involving cytochrome was studied and led to similar conclusions as in the case of azurin [98]. In that study, cytochrome conductance was estimated to be in the order of 1 nS ($10^{-5}G_0$) and in a later report, the STM-BJ approach was used in air to find conductance characteristics of cytochrome mutants, finding an orientation dependent conductance ranging from 1×10^{-5} to $3.6 \times 10^{-5}G_0$, depending on the CT pathway through the protein [99]. Finally, cytochrome conductance was studied in electrochemical conditions [32], modulating the conductance of the redox center by electrochemical potentials, in analogy to the previous studies on azurin.

8.2.2.2.3 Current–voltage characteristics

The ECTS approach enables the collection of current-voltage characteristics from individual proteins. The potential difference between the electrodes is ramped with the tip held at a fixed distance from the sample while the current is monitored (Fig. 8.11a–b). Variable bias potential versus current curves can be then recorded for two distinct situations: (i) when azurin is probed by the STM tip (meaning no physical contact) and (ii) when azurin is wired between the electrodes [80].

(i) Azurin I(V)s by ECTS: Redox probing

Current-potential curves obtained from experiments in tunneling regime, i.e., with the STM probe is parked at a tunneling distance from the azurin surface, are shown in Fig. 8.11c and they display a marked current-rectifying behavior as opposed to the linear behavior obtained when probing the bare Au substrate. The rectifying behavior depends on the sample potential (the redox state of azurin) and is in agreement with the metalloprotein electrochemical behavior. Transition voltage spectroscopy [46] can be then used to find a transition voltage (TV) for the CT process involving azurin (minima in

Fig. 8.11d), which is directly related to the effective tunneling barrier. The TV was found to be 0.4 V and, in order to interpret this low TV we examined the numerical version of the Kuznetsov–Ulstrup formalism for adiabatic two-step electron transfer with partial vibrational relaxation [95]. The expression gave a minimum in the TVS representation as function of the electrochemical parameters. We used this expression to fit our experimental data and obtain values for the various parameters describing the CT process. Numerical simulations of the TVS representation of the two-step electron transfer formalism showed that the primary modulator of the TV is the electronic coupling between the ECSTM tip electrode and the azurin molecule. The TV diminishes when the probe coupling increases, indicating that the TV is related to the limiting energy barrier in the CT process. TV values are in agreement with the distance decay constants obtained by current distance ECTS (see Section 8.2.2.2.1).

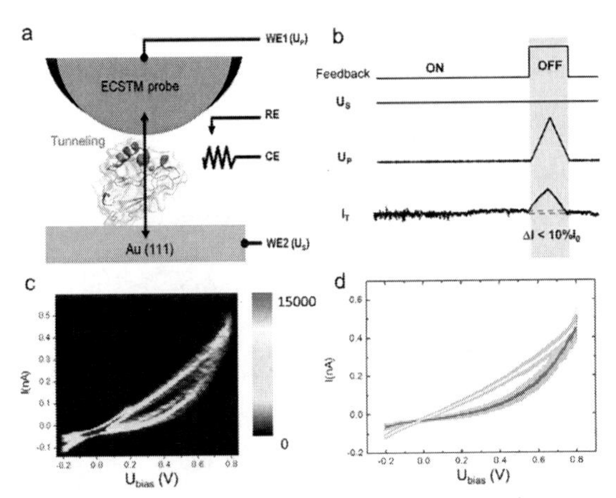

Figure 8.11 Current/voltage measurements on azurin in a physiological environment under electrochemical control. (a) Experimental setup. (b) Protocol for *I–V* recording: a triangular ramp is applied to the probe under feedback loop off. The current signal is recorded at constant sample potential. (c) 2D *I–V* histogram showing two populations of curves in azurin-modified Au(111) electrode: a nearly linear behavior (from Au) and a rectifying (from azurin). (d) Typical TVS curves showing transition voltage minima. Bottom histogram show the most probable TV values. Reprinted with permission from [80]. Copyright 2012 American Chemical Society.

(ii) Wired Azurin I(V)s

TVS was also used to study azurin wired between electrodes [80]. In some of the recordings a lower TV value was found, as predicted by the numerical simulations of a strong electronic coupling between ECSTM tip and protein. The TV indicates the transition between two conductance regimes in high and low electronic coupling with a redox partner, in agreement with the biological activity of azurin: exchange and transport charges. These results suggest that biological CT can be fine-tuned by the electronic coupling between redox partners.

More rigorous theoretical approaches have recently been presented to interpret TVS measurements on redox active metalloproteins [100]. This model, based on an extended Newns–Anderson framework, fits experimental results of azurin from different groups and it is promising for future TVS investigations of biomolecules, as it provides a reliable value for parameters such as the reorganization energy, a magnitude difficult to address experimentally and often overestimated in previous experimental and theoretical approaches. In view of these recent findings, previous interpretations and discussions on the CT mechanism should be reconsidered.

Together, these results demonstrated the feasibility of using a single biomolecule in a device in a three terminal configuration by using an electrochemical gate and the intrinsic natural design of a protein optimized to transfer charges.

8.2.2.3 Other proteins

Scanning probe microscopy (SPM) techniques are beginning to be used to study more complex systems and ECSTM was used to study laccase (a multicopper oxidase) immobilized on gold electrodes in electrochemical environment [97]. Figure 8.12a shows an ECSTM image of hydrogenase molecules immobilized on a gold electrode through a SAM linker [96]. ECSTM imaging combined with a complete electrochemical study were used to estimate the enzyme catalytic turnover rate at the single molecule level.

Using scanning near-field optical microscopy, the photocurrent generated in junctions comprising photosynthetic junctions was studied (see Fig. 8.12e–f) [101], providing the first photocurrent measurement of a single photosynthetic protein.

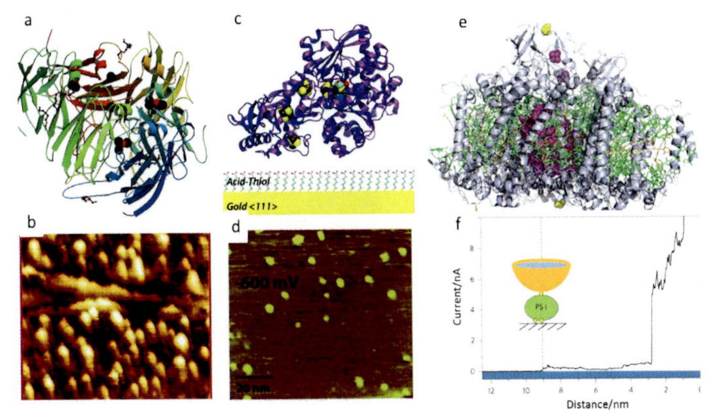

Figure 8.12 Other proteins studied using SPM techniques. (a) Laccase from Streptomyces coelicolor (pdb accession code 3CG8 [102]). (b) In situ STM image of ScL on a butanethiol-modified Au(111). Pure oxygen atmosphere. Image at 0.44 V RHE. V_{Bias}: −0.30 V. Adapted with permission from [97]. Copyright 2011 American Chemical Society. (c) Homology model of CaHydA [FeFe]-hydrogenase on a 6-mercaptohexanoic acid-modified gold electrode. (d) EC-STM image of the protein on a 3-mercaptopropionic acid/ethanethiol SAM (1:1) on a Au (111); V_{bias} = 200 mV, I_t = 400 pA, (c)-(d) are adapted with permission from [96]. Copyright 2011 American Chemical Society. (e) Molecular structure image of PS I based on crystallographic data. The PS I can covalently bind to the tip and substrate through cysteine mutations. (f) Current distance profile for a submonolayer of bipolar PS I measured in ultrahigh vacuum at V_{bias} = 0.25 V, (e)-(f) are adapted with permission from [101]. Copyright 2012 Nature Publishing Group.

Besides the approaches based on SPM, a recent work shows another possible strategy to address the conductance of a single biomolecule, taking profit of the use of antibodies. Figure 8.13 shows that approach, antibodies (immunoglobulin G) were attached to CdSe quantum dots and a protein transistor modulated by an optical field was demonstrated [103]. This approach offers the advantage of using antibodies that could be engineered to bind any biomolecule in the molecular junction, but the conductance properties of protein complexes (including antibody-antigen interactions) are not characterized yet and further research is needed in that sense in order to study the feasibility of this approach.

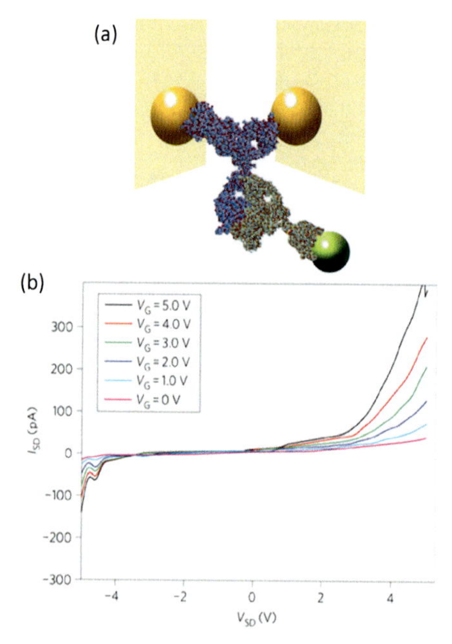

Figure 8.13 (a) Schematic illustration the antibody-based single-protein transistor. (b) I_{SD} versus V_{SD} plot for different V_G at 298 K. Adapted from [103] with permission. Copyright 2012 Nature Publishing Group.

Overall, these fundamental advances may be technologically relevant because the use of proteins as functional components in nanoscale devices is beginning to be considered and different approaches are proposed to this end [103]. The use of biotechnological approaches combined with nanoscience is likely to provide new fundamental results regarding the CT process in biomolecules, as well as new devices that could benefit from the exceptional properties of proteins as nanomachines.

8.3 Conductance of DNA Duplexes

Just a decade after the discovery of the DNA structure [104], it was suggested that hybridization between the orbitals in the bases of double stranded DNA (dsDNA) could lead to conductive behavior [105]. And because of this many theoretical and experimental approaches have been reported addressing the question of the electronic conductance in DNA over the last several decades. The problem of how the charges move through this biomolecule has been addressed using different techniques including bulk and single molecule studies, in environments ranging from ultra high vacuum to solid state configurations in air as well as in buffer and in electrochemical environments [50, 51, 83, 106].

Despite not having a clear unified picture of the CT process in DNA, it has been suggested that electron transfer plays an important role in a variety of DNA biological processes. In particular, it is thought to be instrumental to the detection and repair of oxidative damage in DNA sequences, which is of course related to many wide spread diseases [107].

DNA has emerged as one of the most important materials in nanoscience and technology today. Its 4 bases guanine, cytosine, adenine, and thymine provide a potential basis for programming and computation [108, 109]. Its self-recognition and self-assembly processes enable it to be engineered into structures and assemblies with unparalleled precision at the nanometer scale [110]. Its extended π-stack suggests that it should be a capable electrical conductor similar to graphite [106, 111]. However, despite this promise the CT properties of DNA have remained a contentious field [112]. A wide variety of results for the conductance of DNA have been reported in the literature [113], but in recent years the advent of statistically driven single-molecule conductance measurements using techniques like the break-junction system described in Section 8.1.1 have begun to provide insights into the transport properties of this unique material.

In this section, we will explore the electronic properties of DNA from the standpoint of an electrical conductor attached to two electrodes. Although this viewpoint is somewhat different from the photochemical and electrochemical measurements on charge transfer [114–117], the results from these techniques will be leveraged as necessary to describe the observed conductance phenomena. This section will proceed by discussing the history of DNA conductance measurements by focusing on long molecules (>10 nm in length), and then switch to shorter sequences and statistically driven measurements that have allowed a variety of nanoscopic transport behaviors to be observed, and will conclude with a brief discussion of some of the more recent DNA conductance experiments that go beyond understanding conductance mechanisms.

8.3.1 Early DNA Conductance Experiments

Despite the desire to use electrodes as a probe the electronic properties of DNA, direct contact measurements have proven to be difficult to carry out experimentally [118]. Because of this difficulty, the reported electrical properties of DNA from direct contact measurements vary over a wide range, including insulating [119, 120], semiconducting [121, 122], and conducting [123, 124]. A brief summary of some of the initial direct contact measurements is shown in Table 8.1. However, such a table is somewhat misleading because as one carefully examines the literature a couple of trends become apparent. The first point is that certainly the specific experimental details of the direct contact measurements seem to affect and perhaps even dominate the measured electrical properties and conductance in some experiments [118]. In some cases bundles or networks are measured [125], often in a dry environment or even in vacuum [124]. Sometimes, covalent bonding is used [50], and other times, nonspecific binding is the only contact made to the DNA [126]. All of these details make it difficult to determine exactly what effects are dominating the measurement and causing the variety of experimental results.

It is also important to note that in these cases, the emphasis tends to relate to the conductance of a specific molecule or duplex rather than the conductivity, or the conductivity of base-pair additions. This terminology is used because depending on the specific environment the actual length and cross-sectional area vary with time. This point lends itself to the notion that it remains difficult to keep DNA in a biological conformation and access enough appropriate experimental variables not just to determine if DNA is a capable conductor, but what type of conductor it is. Recently, the conductance measurements for single molecules have advanced to the point where it is possible to change various parameters such as temperature and electrochemical potential, and see the resultant changes in the conductance of a single molecule through a statistical analysis [20, 127, 128]. Given the large dispersion in the long-distance transport measurements, and the emphasis on controlling the environment, the remainder of this section will focus on short DNA sequences in controlled environments.

Table 8.1 Conductance properties of DNA in different experimental environments

Year	Author	Properties	Sample type	No. of strands	Method	Ref.
1998	Braun et al.	Insulator	λ-DNA	Many	Thiolated contacts	[129]
1999	Fink & Shönen-berger	Metallic	λ-DNA	Bundles	LEEPS–Tungsten tip TEM grid	[130]
2000	Dekker et al.	Semiconductor	30 base pair Poly(GC)	One	Field trapping	[131]
2000	Cai et al.	Metallic	Poly(GC) Poly(AT)	Networks/ Bundles	AFM on Mica	[125]
2000	de Pablo et al.	Insulator	λ-DNA	Several	Gold deposited on Top–AFM	[132]
2001	Dekker et al.	Insulator	50–200 nm mixed and poly(G:C)	Few	Thiol-bonded to EBL defined electrodes	[119]
2001	Rakitin et al.	Metallic	M-DNA	Bundles	Vacuum across electrodes	[133]
2001	Yoo et al.	Semiconductor	Poly(AT)–μm	One	FET Strands	[121]
2001	Yoo et al.	Semiconductor	Poly(GC)–μm	One	FET Strands	[121]
2001	Kasumov et al.	Metallic	λ-DNA non-specific contact	Few	Re/C contacted	[134]
2003	Shigematsu et al.	Non-crystalline hopping mech.	λ-DNA with CNT contacts	One	Triple probe conducting AFM	[135]
2004	Xu et al.	Hopping/Tunneling sequence dependence	8–14 base pairs strands	One	STM-BJ	[50]

8.3.2 Conductance of Short DNA Duplexes

The second thing one notices when looking beyond the dispersion that is so apparent in Table 8.1 and concentrating more specifically on small scale measurements, less than or equal to 20 nm in length, is that a representative conductance is typically measured on this scale regardless of the specific types of contacts, or the solution environment. This fact suggests that at least at the mesoscopic scale there are some inherent conductance properties in DNA [50, 122, 135–137]. And, although the mechanisms in these cases still depend on the experimental conditions, this simply demonstrates the electronic flexibility of DNA, and the necessity of a controlled environment for reliable and repeatable DNA conductance measurements.

Despite these difficulties, in recent years photochemical measurements have begun to demonstrate that the coupling of the donor and the acceptor units to the energy levels of the bases in the DNA stack play one of the key roles in the ability of DNA to transfer charge [138].

These systems have shown that in some cases thermally activated hopping is the most likely transport mechanism [139], while in other cases it is apparent that superexchange is dominant [140]. In some cases, it has been shown that there is a transition between hopping and superexchange tunneling [141]. It is vital that we understand the parameters that control these transitions and the interplay between them if DNA is to be used as a nanoscale material system for either electrical or mechanical purposes.

In Section 8.1.2, we introduced the two more important transport mechanisms that must be understood in order to discuss the conductance studies of short DNA duplexes, tunneling, and hopping [115, 142, 143]. As mentioned above, photochemical measurements initially indicated that both mechanisms were possible for short DNA sequences.

8.3.2.1 Statistical DNA conductance measurements

With the apparent sensitivity of DNA conductance to environmental or experimental conditions it was necessary to begin perform large numbers of experiments in fixed environments to demonstrate that reproducible conductance values could be obtained in short DNA sequences and demonstrate that the results of the photochemical measurements could also be obtained for direct contact measurements. Although considerable strides have been made in the development of lithographically defined systems [144–147], we will focus primarily on measurements performed using the STM-break junction system described above. This system lends itself directly to these types of measurements because it is well situated for short molecules, it depends on acquiring a large number of conductance measurements to obtain statistically significant results, and the measurements can be performed in aqueous solution to ensure that the molecules maintain a certain conformation.

One of the first experiments along these lines was performed by Xu et al. (Fig. 8.14) [50]. In these experiments, an alternating GC sequence ranging from 8 to 14 base pairs (bp) in length was measured, and as can be seen in Fig. 8.14a, the conductance of this family was inversely proportional to the length. Alternatively, when AT bp were inserted into the stack, the conductance decreased exponentially with increasing length, with a β-value of 4.3 nm^{-1} (Fig. 8.14b). These initial experiments suggest that the conductance of GC-rich sequences may be thermally activated hopping, and that inserting AT pairs into the stack creates a barrier that is tunneled through for small numbers of AT pairs. These results match well with expectations from the oxidation potentials of the various bases which show that guanine has the lowest oxidation potential and so hole-hopping along this base may be the dominant mechanism, and other bases may represent a small barrier that is tunneled through [148].

However, to truly prove the supposition that the dominant conduction mechanism in the GC sequences is hopping, temperature dependent measurements must also be performed. Two break-junction based experiments have been performed on short GC-rich sequences [149, 151] and interestingly, neither of these experiments showed a temperature dependence of the conductance (Fig. 8.14c). There are a variety of potential reasons for this outcome. First, short DNA molecules have a low melting temperature, T_m, where the two strands dissociate, meaning that the upper limit on the temperature is restricted. Moreover, since DNA has to be hydrated to maintain its physiological conformation, the lower limit on the temperature range is ~0°C. Thus, neither of the experiments were able to measure the conductance over a temperature range larger than ~50°C. So, one potential reason why the thermally activated conductance was not confirmed may be that the activation energy, E_B in Eq. (8.2) above, is small, and the change in conductance is not observable in such a small temperature window. Alternatively, other transport mechanisms may also be possible. It could be that the system tunnels over long distances with a very small barrier, or that some sequential tunneling process, such as the two-step sequential tunneling described in previous section for redox proteins, is occurring. Nevertheless, understanding the inherent charge transport mechanisms in DNA conductance experiments still remains an open question at this time.

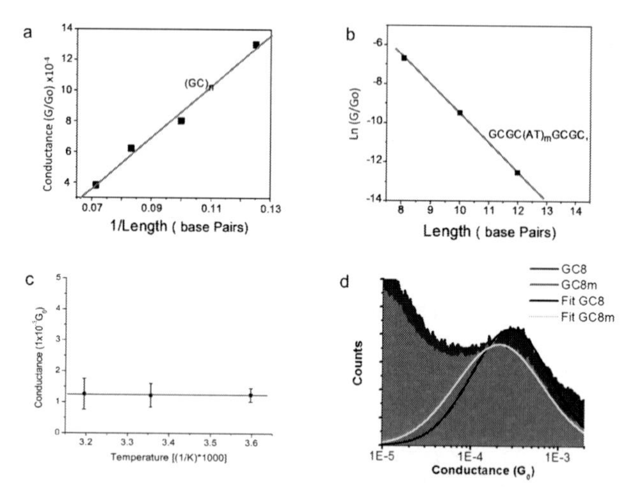

Figure 8.14 Charge transport properties of DNA. (a) The conductance is inversely proportional to the length of the molecule for $(GC)_n$ sequences. (b) The conductance displays an exponential dependence with length as AT bp are added to the system. (c) For a $(GC)_4$ sequence the conductance no measureable change in conductance was observed between 5 and 40°C. (d) Conductance histograms for a DNA sequence with and without methylated cytosines showing a change in conductance due to this modification. (a), (b) Reproduced with permission from ref. [50], copyright 2004 American Chemical Society. (c) Reproduced with permission from ref. [149], copyright 2007 IOP Publishing Ltd. (d) Reproduced from ref. [150], copyright 2012 IOP Publishing Ltd.

Despite the fact that the inherent transport mechanisms are not fully understood in this molecular system, it may have relevant technological impacts. Since DNA is such an important molecule biologically, the ability to contact and electrically interrogate a single DNA duplex may allow relevant biological information be read out electrically. Several initial studies have shown the feasibility of this. Hihath et al. demonstrated that a single base mismatch in a DNA sequence can change the conductance by as much as 1 order of magnitude in some cases [136]. It has also been demonstrated that the binding of a methyltransferase enzyme can be detected in a DNA-carbon nanotube device [152]. And, that epigenomic changes such as the methylation of cytosine bases can induce a change in the conductance (see Fig. 8.14d) [150]. These series of experiments demonstrate that measuring the electrical properties of DNA is important beyond electronics and may open up new avenues for biological diagnostics.

8.3.3 Outlook for DNA Charge Transport

Considerable progress has been made in the direct contact conductance measurements of DNA over the last decade. However, there is still uncertainty about the mechanisms that dominate the charge transport properties, and additional measurements, and perhaps modified transport models, are needed to fully address these issues [112]. However, despite these exciting opportunities for advanced studies, it is rapidly becoming apparent that measuring the conductance of individual DNA molecules may be relevant beyond using it as a material for nanoscale electronics, and it may yield a novel way of extracting biologically relevant information electrically. These capabilities may yield new sensor platforms or advanced diagnostic capabilities for biology or health-care. Moreover, only limited studies have been performed beyond DNA looking at peptide-nucleic acids, or RNA, which would yield even more insights into nanoscale transport phenomena [153, 154]. Clearly, the study of the conductance of nucleic acids is still in its infancy, and will continue to be an important field for years to come.

8.4 Conclusion and Future Perspectives

This chapter has discussed the state-of-the-art in the study of CT in biomolecules and the current understanding of the fundamentals behind it. Proteins have become an attractive candidate for molecular electronic devices, and proof-of-concept examples of single-protein devices are starting to appear in the literature. In the coming years, it is likely that CT pathways in proteins will be unraveled through SPM and other molecular junction approaches in combination with biotechnology (engineering the sequence to study the relative importance of the residues in the CT pathway). Once the CT pathways and the factors affecting them are completely understood, the ability of modulating them could open new possibilities for the use of proteins as active components in molecular devices with new functionalities (i.e., benefiting from quantum interference or constituting spin valves). On the other hand, single-protein conductance results also pave the way to new studies about molecular recognition and the possibility of detecting biomolecular interactions by monitoring electrical signals. This latest aspect could revolutionize applications in the study of fundamental biological processes at unprecedented levels and help in the design of biomedical sensors based on single molecule conductance.

The study of CT in DNA has been a very active field of research for decades. Different experimental approaches have allowed establishing the CT mechanisms in short DNA molecules, namely tunneling and hopping as function of sequence and length. Other possible factors affecting the CT process are beginning to be considered, such as structural fluctuations. Single molecule studies on oligonucleotides systematically addressing these factors could help to this end, providing results useful to derive a general theory for CT in oligonucleotides. Besides, a complete electronic characterization of oligonucleotides has potential technological applications. Biomolecular devices based on oligonucleotides could offer many advantages thanks to their high specificity and self-assemble capabilities and the fine-tuning of their electronic properties. Once the conductance of DNA and related oligonucleotides is understood, this knowhow could be used in conjunction with technologies such as DNA origami. Being able to design DNA sequences capable of folding in nanoscale patterns and with a predictable electronic behavior could establish DNA as a unique platform for self-assembled molecular devices with novel functionalities.

References

1. Lee, B. Y., Zhang, J., Zueger, C., Chung, W.-J., Yoo, S. Y., Wang, E., Meyer, J., Ramesh, R., Lee, S.-W., *Nat Nano*, 2012, **7**(6), 351–356. DOI http://www.nature.com/nnano/journal/v7/n6/abs/nnano.2012.69.html#supplementary-information.
2. Blankenship, R. E., *Molecular Mechanisms of Photosynthesis*. John Wiley & Sons: 2014.
3. Gray, H. B., Winkler, J. R., *Q. Rev. Biophys.*, 2003, **36**(3), 341–372. DOI 10.1017/s0033583503003913.
4. Giese, B., Graber, M., Cordes, M., *Curr. Opin. Chem. Biol.*, 2008, **12**(6), 755–759. DOI http://dx.doi.org/10.1016/j.cbpa.2008.08.026.
5. Offenhäusser, A., Rinaldi, R., *Nanobioelectronics: For Electronics, Biology, and Medicine*. Springer: 2009.
6. Palazzo, G., Magliulo, M., Mallardi, A., Angione, M. D., Gobeljic, D., Scamarcio, G., Fratini, E., Ridi, F., Torsi, L., *ACS Nano*, 2014, **8**(8), 7834–7845. DOI: 10.1021/nn503135y.
7. Fu, T.-M., Duan, X., Jiang, Z., Dai, X., Xie, P., Cheng, Z., Lieber, C. M., *Proc. Natl. Acad. Sci.*, 2014, **111**(4), 1259–1264. DOI 10.1073/pnas.1323389111.
8. Aviram, A., Ratner, M. A., *Chem. Phys. Lett.*, 1974, **29**(2), 277–283.
9. Gartner, W. W., *Semiconductor Prod.*, 1959, **3**, 41–46.
10. Mann, B., Kuhn, H., *J. Appl. Phys.*, 1971, **42**(11), 4398–4405. DOI doi: http://dx.doi.org/10.1063/1.1659785.
11. Binnig, G., Rohrer, H., *Helvetica Phys. Acta*, 1982, **55**(6), 726–735.
12. Joachim, C., Gimzewski, J. K., Schlittler, R. R., Chavy, C., *Phys. Rev. Lett.*, 1995, **74**(11), 2102–2105.
13. Reed, M. A., Zhou, C., Muller, C. J., Burgin, T. P., Tour, J. M., *Science*, 1997, **278**(5336), 252–254.
14. Park, H., Lim, A. K. L., Alivisatos, A. P., Park, J., McEuen, P. L., *Appl. Phys. Lett.*, 1999, **75**(2), 301–303. DOI doi:http://dx.doi.org/10.1063/1.124354.
15. Moreland, J., Ekin, J. W., *J. Appl. Phys.*, 1985, **58**(10), 3888–3895. DOI doi:http://dx.doi.org/10.1063/1.335608.
16. Muller, C. J., van Ruitenbeek, J. M., de Jongh, L. J., *Phys. C Superconductivity* 1992, **191**(3–4), 485–504. DOI http://dx.doi.org/10.1016/0921-4534(92)90947-B.

17. Xu, B. Q., Tao, N. J. J., *Science*, 2003, **301**, 1221–1223.
18. Parks, J. J., Champagne, A. R., Costi, T. A., Shum, W. W., Pasupathy, A. N., Neuscamman, E., Flores-Torres, S., Cornaglia, P. S., Aligia, A. A., Balseiro, C. A., Chan, G. K.-L., Abruña, H. D., Ralph, D. C., *Science*, 2010, **328**(5984), 1370–1373. DOI 10.1126/science.1186874.
19. Reed, M. A., *Nat. Mater.*, 2004, **3**(5), 286–287.
20. Venkataraman, L., Park, Y. S., Whalley, A. C., Nuckolls, C., Hybertsen, M. S., Steigerwald, M. L., *Nano Lett.*, 2007, **7**(2), 502–506. DOI 10.1021/nl062923j.
21. Zhihai, L., Ilya, P., Bo, H., Thomas, W., Alfred, B., Marcel, M., *Nanotechnology*, 2007, **18**(4), 044018.
22. Diez-Perez, I., Hihath, J., Lee, Y., Yu, L. P., Adamska, L., Kozhushner, M. A., Oleynik, II, Tao, N. J., *Nat. Chem.*, 2009, **1**(8), 635–641. DOI 10.1038/nchem.392.
23. Chen, F., Li, X., Hihath, J., Huang, Z., Tao, N., *J. Am. Chem. Soc.*, 2006, **128**(49), 15874–15881. DOI 10.1021/ja065864k.
24. Ayano, N., Junichi, T., Yasuyuki, K., Shintaro, F., Masaaki, S., Masamichi, F., *Nanotechnology*, 2007, **18**(42), 424005.
25. Venkataraman, L., Klare, J. E., Nuckolls, C., Hybertsen, M. S., Steigerwald, M. L., *Nature*, 2006, **442**(7105), 904–907. DOI http://www.nature.com/nature/journal/v442/n7105/suppinfo/nature05037_S1.html.
26. Mishchenko, A., Vonlanthen, D., Meded, V., Bürkle, M., Li, C., Pobelov, I. V., Bagrets, A., Viljas, J. K., Pauly, F., Evers, F., Mayor, M., Wandlowski, T., *Nano Lett.*, 2009, **10**(1), 156–163. DOI 10.1021/nl903084b.
27. Haiss, W., Wang, C., Grace, I., Batsanov, A. S., Schiffrin, D. J., Higgins, S. J., Bryce, M. R., Lambert, C. J., Nichols, R. J., *Nat. Mater.*, 2006, **5**(12), 995–1002. DOI http://www.nature.com/nmat/journal/v5/n12/suppinfo/nmat1781_S1.html.
28. Diez-Perez, I., Hihath, J., Hines, T., Wang, Z.-S., Zhou, G., Mullen, K., Tao, N., *Nat Nano*, 2011, **6**(4), 226–231. DOI http://www.nature.com/nnano/journal/v6/n4/abs/nnano.2011.20.html#supplementary-information.
29. Gray, H. B., Winkler, J. R., *Proc. Natl. Acad. Sci. U. S. A.*, 2005, **102**, 3534–3539. DOI 10.1073/pnas.0408029102.
30. Chi, Q. J., Farver, O., Ulstrup, J., *Proc. Natl. Acad. Sci. U. S. A.*, 2005, **102**(45), 16203–16208. DOI 10.1073/pnas.0508257102.
31. Alessandrini, A., Salerno, M., Frabboni, S., Facci, P., *Appl. Phys. Lett.*, 2005, **86**(13). DOI 10.1063/1.1896087.

32. Della Pia, E. A., Chi, Q., Macdonald, J. E., Ulstrup, J., Jones, D. D., Elliott, M., *Nanoscale*, 2012, **4**(22), 7106–7113. DOI 10.1039/C2NR32131A.
33. Artes, J. M., Diez-Perez, I., Gorostiza, P., *Nano Lett.*, 2012, **12**(6), 2679–2684.
34. Lindsay, S., *Introduction to Nanoscience*. Oxford University Press: 2010.
35. Nitzan, A., *Isr. J. Chem.*, 2002, **42**(2–3), 163–166. DOI 10.1560/ylbd-yf7y-4j4e-epqe.
36. Nitzan, A., Ratner, M. A., *Science*, 2003, **300**, 1384–1389.
37. Artés, J. M., López-Martínez, M., Díez-Pérez, I., Sanz, F., Gorostiza, P., *Electrochim. Acta*, 2014, **140**(0), 83–95. DOI http://dx.doi.org/10.1016/j.electacta.2014.05.089.
38. Ami, S., Hliwa, M., Joachim, C., *Chem. Phys. Lett.*, 2003, **367**(5), 662–668.
39. Tans, S. J., Verschueren, A. R., Dekker, C., *Nature*, 1998, **393**(6680), 49–52.
40. Jiao, L., Zhang, L., Wang, X., Diankov, G., Dai, H., *Nature*, 2009, **458**(7240), 877–880.
41. Liang, W., Shores, M. P., Bockrath, M., Long, J. R., Park, H., *Nature*, 2002, **417**(6890), 725–729.
42. Kubatkin, S., Danilov, A., Hjort, M., Cornil, J., Brédas, J., Stuhr-Hansen, N., *Nature*, 2003, **425**, 698–701.
43. Kervennic, Y. V., Thijssen, J. M., Vanmaekelbergh, D., Dabirian, R., Jenneskens, L. W., van Walree, C. A., van der Zant, H. S., *Angew. Chem. Int. Ed.*, 2006, **45**(16), 2540–2542.
44. Song, H., Kim, Y., Jang, Y. H., Jeong, H., Reed, M. A., Lee, T., *Nature*, 2009, **462**(7276), 1039–1043. DOI http://www.nature.com/nature/journal/v462/n7276/suppinfo/nature08639_S1.html.
45. Christian, A. M., Jan, M. V. R., Herre, S. J. V. D. Z., *Nanotechnology*, 2010, **21**(26), 265201.
46. Beebe, J. M., Kim, B., Gadzuk, J. W., Daniel Frisbie, C., Kushmerick, J. G., *Phys. Rev. Lett.*, 2006, **97**(2), 026801.
47. Guo, S., Artés, J. M., Díez-Pérez, I., *Electrochim. Acta*, 2013, **110**, 741–753. DOI http://dx.doi.org/10.1016/j.electacta.2013.03.146.
48. Yuan, H., Liu, H., Shimotani, H., Guo, H., Chen, M., Xue, Q., Iwasa, Y., *Nano Lett.*, 2011, **11**(7), 2601–2605. DOI 10.1021/nl201561u.
49. Ye, J. T., Inoue, S., Kobayashi, K., Kasahara, Y., Yuan, H. T., Shimotani, H., Iwasa, Y., *Nat. Mater.*, 2010, **9**(2), 125–128. DOI http://www.nature.

com/nmat/journal/v9/n2/abs/nmat2587.html#supplementary-information.

50. Xu, B. Q., Zhang, P. M., Li, X. L., Tao, N. J., *Nano Lett.*, 2004, **4**(6), 1105–1108. DOI 10.1021/nl0494295.
51. Hihath, J., Xu, B. Q., Zhang, P. M., Tao, N. J., *Proc. Natl. Acad. Sci. U. S. A.*, 2005, **102**(47), 16979–16983. DOI 10.1073/pnas.0505175102.
52. Gray, H. B., Winkler, J. R., *Biochim. Biophys. Acta (BBA)–Bioenerg.*, 2010, **1797**(9), 1563–1572. DOI 10.1016/j.bbabio.2010.05.001.
53. Beratan, D. N., Betts, J. N., Onuchic, J. N., *Science*, 1991, **252**(5010), 1285–1288.
54. Beratan, D. N., Skourtis, S. S., Balabin, I. A., Balaeff, A., Keinan, S., Venkatramani, R., Xiao, D., *Acc. Chem. Res.*, 2009, **42**(10), 1669–1678. DOI 10.1021/ar900123t.
55. Farver, O., Pecht, I., *Coordination Chem. Rev.*, 2011, **255**(7–8), 757–773. DOI 10.1016/j.ccr.2010.08.005.
56. Langen, R., Chang, I. J., Germanas, J. P., Richards, J. H., Winkler, J. R., Gray, H. B., *Science*, 1995, **268**(5218), 1733–1735.
57. Winkler, J. R., Di Bilio, A. J., Farrow, N. A., Richards, J. H., Gray, H. B., *Pure Appl. Chem.*, 1999, **71**(9), 1753–1764. DOI 10.1351/pac 199971091753.
58. Page, C. C., Moser, C. C., Chen, X., Dutton, P. L., *Nature*, 1999, **402**(6757), 47–52.
59. Prytkova, T. R., Kurnikov, I. V., Beratan, D. N., *Science*, 2007, **315**(5812), 622–625.
60. Sek, S., Swiatek, K., Misicka, A., *J. Phys. Chem. B*, 2005, **109**(49), 23121–23124. DOI 10.1021/jp055709c.
61. Sek, S., Misicka, A., Swiatek, K., Maicka, E., *J. Phys. Chem. B*, 2006, **110**(39), 19671–19677. DOI 10.1021/jp063073z.
62. Xiao, X. Y., Xu, B. Q., Tao, N. J., *J. Am. Chem. Soc.*, 2004, **126**(17), 5370–5371. DOI 10.1021/ja049469a.
63. Xiao, X., Xu, B., Tao, N., *Angew. Chem. Int. Ed.*, 2004, **43**(45), 6148–6152. DOI 10.1002/anie.200460886.
64. Inai, Y., Sisido, M., Imanishi, Y., *J. Phys. Chem.*, 1991, **95**(9), 3847–3851. DOI 10.1021/j100162a074.
65. Morita, T., Kimura, S., *J. Am. Chem. Soc.*, 2003, **125**(29), 8732–8733. DOI 10.1021/ja034872n.
66. Mandal, H. S., Kraatz, H.-B., *J. Phys. Chem. Lett.*, 2012, **3**(6), 709–713. DOI 10.1021/jz300008s.

67. Scholtz, J. M., Baldwin, R. L., *Ann. Rev. Biophys. Biomol. Struct.*, 1992, **21**(1), 95–118. DOI 10.1146/annurev.bb.21.060192.000523.

68. Harrison, R. S., Shepherd, N. E., Hoang, H. N., Ruiz-Gómez, G., Hill, T. A., Driver, R. W., Desai, V. S., Young, P. R., Abbenante, G., Fairlie, D. P., *Proc. Natl. Acad. Sci. U. S. A.*, 2010, **107**(26), 11686–11691.

69. Su, J. Y., Hodges, R. S., Kay, C. M., *Biochemistry*, 1994, **33**(51), 15501–15510. DOI 10.1021/bi00255a032.

70. Hungerford, G., Martinez-Insua, M., Birch, D. J. S., Moore, B. D., *Angew. Chem. Int. Ed. Engl.*, 1996, **35**(3), 326–329. DOI 10.1002/anie.199603261.

71. Yeganeh,S.,Ratner,M.A.,Medina,E.,Mujica,V.,*J.Chem.Phys.*,2009,**131**(1), 014707–014716. DOI doi:http://dx.doi.org/10.1063/1.3167404.

72. Göhler, B., Hamelbeck, V., Markus, T. Z., Kettner, M., Hanne, G. F., Vager, Z., Naaman, R., Zacharias, H., *Science*, 2011, **331**(6019), 894–897. DOI 10.1126/science.1199339.

73. Adman, E. T., Jensen, L. H., *Isr. J. Chem.*, 1981, **21**(1), 8–12.

74. Chi, Q. J., Zhang, J. D., Friis, E. P., Andersen, J. E. T., Ulstrup, J., *Electrochem. Commun.*, 1999, **1**(3–4), 91–96.

75. Alessandrini, A., Corni, S., Facci, P., *Phys. Chem. Chem. Phys.*, 2006, **8**(38), 4383–4397. DOI 10.1039/b607021c.

76. Friis, E. P., Andersen, J. E. T., Madsen, L. L., Moller, P., Ulstrup, J., *J. Electroanal. Chem.*, 1997, **431**(1), 35–38.

77. Davis, J. J., Wang, N., Morgan, A., Tiantian, Z., Jianwei, Z., *Faraday Discuss./Faraday Discuss.*, 2005, (131), 167–179.

78. Artes, J. M., Diez-Perez, I., Sanz, F., Gorostiza, P., *Acs Nano*, 2011, **5**(3), 2060–2066. DOI 10.1021/nn103236e.

79. Artés, J. M., Díez-Pérez, I., Gorostiza, P., *Nano Lett.*, 2011. DOI 10.1021/nl2028969.

80. Artés, J. M., López-Martínez, M., Giraudet, A., Díez-Pérez, I., Sanz, F., Gorostiza, P., *J. Am. Chem. Soc.*, 2012, **134**(50), 20218–20221. DOI 10.1021/ja3080242.

81. Artés, J. M., López-Martínez, M., Díez-Pérez, I., Sanz, F., Gorostiza, P., *Small*, 2014, **10**(13), 2537–2541. DOI 10.1002/smll.201303753.

82. Winkler, J. R., Gray, H. B., *Chem. Rev.*, 2013. DOI 10.1021/cr4004715.

83. Sontz, P. A., Muren, N. B., Barton, J. K., *Acc. Chem. Res.*, 2012, **45**(10), 1792–1800. DOI 10.1021/ar3001298.

84. Wenger, O. S., Leigh, B. S., Villahermosa, R. M., Gray, H. B., Winkler, J. R., *Science*, 2005, **307**(5706), 99–102. DOI 10.1126/science.1103818.

85. Shih, C., Museth, A. K., Abrahamsson, M., Blanco-Rodriguez, A. M., Di Bilio, A. J., Sudhamsu, J., Crane, B. R., Ronayne, K. L., Towrie, M., Vlcek, A., Richards, J. H., Winkler, J. R., Gray, H. B., *Science*, 2008, **320**(5884), 1760–1762. DOI 10.1126/science.1158241.

86. Zhang, J. D., Kuznetsov, A. M., Medvedev, I. G., Chi, Q. J., Albrecht, T., Jensen, P. S., Ulstrup, J., *Chem. Rev.*, 2008, **108**(7), 2737–2791. DOI 10.1021/cr068073+.

87. Fujita, K., Nakamura, N., Ohno, H., Leigh, B. S., Niki, K., Gray, H. B., Richards, J. H., *J. Am. Chem. Soc.*, 2004, **126**(43), 13954–13961. DOI 10.1021/ja047875o.

88. Salverda, J. M., Patil, A. V., Mizzon, G., Kuznetsova, S., Zauner, G., Akkilic, N., Canters, G. W., Davis, J. J., Heering, H. A., Aartsma, T. J., *Angew. Chem.*, 2010, **122**(33), 5912–5915.

89. Patil, A. V., Davis, J. J., *J. Am. Chem. Soc.*, 2010, **132**(47), 16938–16944. DOI 10.1021/ja1065448.

90. Chi, Q., Zhang, J., Jensen, P. S., Christensen, H. E. M., Ulstrup, J., *Faraday Discuss./Faraday Discuss.*, 2005, (131), 181–195.

91. Diez-Perez, I., Guell, A. G., Sanz, F., Gorostiza, P., *Anal. Chem.*, 2006, **78**(20), 7325–7329. DOI 10.1021/ac0603330.

92. Halbritter, J., Repphun, G., Vinzelberg, S., Staikov, G., Lorenz, W. J., *Electrochim. Acta*, 1995, **40**(10), 1385–1394.

93. Kuznetsov, A. M., Ulstrup, J., *Electrochim. Acta*, 2000, **45**(15–16), 2339–2361.

94. Haiss, W., Nichols, R. J., van Zalinge, H., Higgins, S. J., Bethell, D., Schiffrin, D. J., *Phys. Chem. Chem. Phys.*, 2004, **6**(17), 4330–4337. DOI 10.1039/b404929b.

95. Pobelov, I. V., Li, Z. H., Wandlowski, T., *J. Am. Chem. Soc.*, 2008, **130**(47), 16045–16054. DOI 10.1021/ja8054194.

96. Madden, C., Vaughn, M. D., Díez-Pérez, I., Brown, K. A., King, P. W., Gust, D., Moore, A. L., Moore, T. A., *J. Am. Chem. Soc.*, 2011. DOI 10.1021/ja207461t.

97. Climent, V., Zhang, J., Friis, E. P., Østergaard, L. H., Ulstrup, J., *J. Phys. Chem. C*, 2011. DOI 10.1021/jp2086285.

98. Della Pia, E. A., Chi, Q. J., Jones, D. D., Macdonald, J. E., Ulstrup, J., Elliott, M., *Nano Lett.*, 2011, **11**(1), 176–182. DOI 10.1021/nl103334q.

99. Della Pia, E. A., Elliott, M., Jones, D. D., Macdonald, J. E., *ACS Nano*, 2011. DOI 10.1021/nn2036818.

100. Bâldea, I., *J. Phys. Chem. C*, 2013, **117**(48), 25798–25804. DOI 10.1021/jp408873c.

101. Gerster, D., Reichert, J., Bi, H., Barth, J. V., Kaniber, S. M., Holleitner, A. W., Visoly-Fisher, I., Sergani, S., Carmeli, I., *Nat. Nano*, 2012, **7**(10), 673–676. DOI http://www.nature.com/nnano/journal/v7/n10/abs/nnano.2012.165.html#supplementary-information.

102. Skalova, T., Dohnalek, J., Ostergaard, L. H., Ostergaard, P. R., Kolenko, P., Duskova, J., Stepankova, A., Hasek, J., *J. Mol. Biol.*, 2009, **385**(4), 1165–1178. DOI 10.1016/j.jmb.2008.11.024.

103. Chen, Y.-S., Hong, M.-Y., Huang, G. S., *Nat. Nanotechnol.*, 2012, **7**(4). DOI 10.1038/nnano.2012.41.

104. Watson, J. D., Crick, F. H. C., *Nature*, 1953, **171**(4356), 737–738.

105. Eley, D. D., Spivey, D. I., *Trans. Faraday Soc.*, 1962, **58**(0), 411–415. DOI 10.1039/TF9625800411.

106. Endres, R. G., Cox, D. L., Singh, R. P., *Rev. Modern Phys.*, 2004, **76**, 195–214.

107. Barton, J. K., Olmon, E. D., Sontz, P. A., *Coordination Chem. Rev.*, 2011, **255**(7–8), 619–634. DOI 10.1016/j.ccr.2010.09.002.

108. Lund, K., Williams, B., Ke, Y., Liu, Y., Yan, H., *Curr. Nanosci.*, 2006, **2**(2), 113–122.

109. Adleman, L., *Science*, 1994, **266**(5187), 1021–1024. DOI 10.1126/science.7973651.

110. Torring, T., Voigt, N. V., Nangreave, J., Yan, H., Gothelf, K. V., *Chem. Soc. Rev.*, 2011, **40**(12), 5636–5646. DOI 10.1039/c1cs15057j.

111. Eley, D. D., and D. I. Spivey, *Trans. Faraday Soc.*, 1962, **58**, 411.

112. Venkatramani, R., Keinan, S., Balaeff, A., Beratan, D. N., *Coordination Chem. Rev.*, 2011, **255**(7–8), 635–648. DOI http://dx.doi.org/10.1016/j.ccr.2010.12.010.

113. Porath, D., Cuniberti, G., Di Felice, R., *Long-Range Charge Transfer In Dna Ii*, 2004, **237**, 183–227.

114. Lewis, F., Wu, T., Zhang, Y., Letsinger, R., Greenfield, S., Wasielewski, M., *Science*, 1997, **277**, 673–676.

115. Giese, B., Amaudrut, J., Kohler, A.-K., Spormann, M., Wessely, S., *Nature (London)*, 2001, **412**(6844), 318–320.

116. Dandliker, P. J., Holmlin, R. E., Barton, J. K., *Science*, 1997, **275**(5305), 1465–1468.

117. Genereux, J. C., Barton, J. K., *Chem. Rev. (Washington, DC, United States)*, 2010, **110**(3), 1642–1662.

118. Hipps, K. W., *Science*, 2001, **294**(5542), 536–537.

119. Storm, A. J., Noort, J. V., Vries, S. D., Dekker, C., *Appl. Phys. Lett.*, 2001, **79**(23), 3881–3883.

120. Aboul-ela, F., Koh, D., Tinoco, I., Jr, Martin, F., *Nucleic Acids Res.*, 1985, **13**(13), 4811–4824.

121. Yoo, K. H., Ha, D. H., Lee, J.-O., Park, J. W., Kim, J., Kim, J. J., Lee, H.-Y., Kawai, T., Choi, H. Y., *Phys. Rev. Lett.*, 2001, **87**(19), 198102.

122. Porath, D., Bezryadin, A., de Vries, S., Dekker, C., *Nature (London)*, 2000, **403**, 635–638.

123. Fink, H. W., Schonenberger, C., *Nature (London)*, 1999, **398**, 407–410.

124. Kasumov, A. Y., Kociak, M., Gueron, S., Reulet, B., Volkov, V. T., Klinov, D. V., Bouchiat, H., *Science*, 2001, **291**, 280–282.

125. Cai, L., Tabata, H., Kawai, T., *Appl. Phys. Lett.*, 2000, **77**(19), 3105–3106.

126. Watanabe, H., Manabe, C., Shigematsu, T., Shimotani, K., *Appl. Phys. Lett.*, 2001, **79**(15), 2462–2464.

127. Li, X., Hihath, J., Chen, F., Masuda, T., Zang, L., Tao, N., *J. Am. Chem. Soc.*, 2007, **129**(37), 11535–11542.

128. Haiss, W., Van Zalinge, H., Bethell, D., Ulstrup, J., Schiffrin, D. J., Nichols, R. J., *Faraday Discuss.*, 2006, **131**, 253–264.

129. Braun, E., Eichen, Y., Sivan, U., Ben-Yoseph, G., *Nature*, 1998, **391** (February 1998), 775–778.

130. Fink, H.-W., Schonenberger, C., *Nature*, 1999, **398**(1 April), 407–410.

131. Porath, D., Bezryadin, A., Vries, S. D., Dekker, C., *Nature*, 2000, **403**(10 February 2000), 635–639.

132. Pablo, P. J. D., Moreno-Herrero, F., Colchero, J., Herrero, J. G., Herrero, P., Baro, A. M., Ordejon, P., Soler, J. M., Artacho, E., *Phys. Rev. Lett.*, 2000, **85**(23), 4992–4995.

133. Rakitin, A., Aich, P., Papadopoulos, C., Kobzar, Y., Vedeneev, A. S., Lee, J. S., Xu, J. M., *Phys. Rev. Lett.*, 2001, **86**(16), 3670.

134. Kasumov, A. Y., Kociak, M., Gueron, S., Reulet, B., Volkov, V. T., Klinov, D. V., Bouchiat, H., *Science*, 2001, **291**(5502), 280–282.

135. Shigematsu, T., Shimotani, K., Manabe, C., Watanabe, H., Shimizu, M., *J.Chem. Phys.*, 2003, **118**(9), 4245–4252.

136. Hihath, J., Xu, B., Zhang, P., Tao, N., *Proc. Natl. Acad. Sci. U. S. A.*, 2005, **102**(47), 16979–16983.

137. Van Zalinge, H., Schiffrin, D. J., Bates, A. D., Haiss, W., Ulstrup, J., Nichols, R. J., *ChemPhysChem*, 2006, **7**(1), 94–98.

138. Delaney, S., Barton, J. K., *J. Org. Chem.*, 2003, **68**(17), 6475–6483.

139. Bixon, M., Giese, B., Wessely, S., Langenbacher, T., Michel-Beyerle, M. E., Jortner, J., *Proc. Natl. Acad. Sci. U. S. A.*, 1999, **96**(21), 11713–11716.

140. Lewis, F. D., Wu, T., Zhang, Y., Letsinger, R. L., Greenfield, S. R., Wasielewski, M. R., *Science*, 1997, **277**(5326), 673–676.

141. Giese, B., Amaudrut, J., Kohler, A.-K., Spormann, M., Wessely, S., *Nature*, 2001, **412**(6844), 318–320.

142. Bixon, M., Jortner, J., *J. Am. Chem. Soc.*, 2001, **123**(50), 12556–12567.

143. Berlin, Y. A., Burin, A. L., Ratner, M. A., *J. Am. Chem. Soc.*, 2001, **123**(2), 260–268.

144. Roy, S., Vedala, H., Roy, A. D., Kim, D.-H., Doud, M., Mathee, K., Shin, H.-K., Shimamoto, N., Prasad, V., Choi, W., *Nano Lett.*, 2007, **8**(1), 26–30. DOI 10.1021/nl0716451.

145. Guo, X., Gorodetsky, A. A., Hone, J., Barton, J. K., Nuckolls, C., *Nat. Nano*, 2008, **3**(3), 163–167. DOI http://www.nature.com/nnano/journal/v3/n3/suppinfo/nnano.2008.4_S1.html.

146. Mahapatro, A. K., Lee, G. U., Jeong, K. J., Janes, D. B., *Appl. Phys. Lett.*, 2009, **95**(8), 083106 DOI 10.1063/1.3186056.

147. Liu, S., Clever, G. H., Takezawa, Y., Kaneko, M., Tanaka, K., Guo, X., Shionoya, M., *Angew. Chem. Int. Ed.*, 2011, **50**(38), 8886–8890. DOI 10.1002/anie.201102980.

148. Treadway, C. R., Hill, M. G., Barton, J. K., *Chem. Phys.*, 2002, **281**(2–3), 409–428.

149. Hihath, J., Chen, F., Zhang, P., Tao, N., *J. Phys. Condensed Matter*, 2007, **19**(21), 215202/1–215202/9.

150. Hihath, J., Guo, S., Zhang, P., Tao, N., *J. Phys. Condensed Matter*, 2012, **24**(16). DOI 10.1088/0953-8984/24/16/164204.

151. Van Zalinge, H., Schiffrin, D. J., Bates, A. D., Starikov, E. B., Wenzel, W., Nichols, R. J., *Angew. Chem. Int. Ed. Engl.*, 2006, **45**(33), 5499–5502.

152. Wang, H., Muren, N. B., Ordinario, D., Gorodetsky, A. A., Barton, J. K., Nuckolls, C., *Chem. Sci.*, 2012, **3**(1), 62–65. DOI 10.1039/c1sc00772f.

153. Paul, A., Watson, R. M., Wierzbinski, E., Davis, K. L., Sha, A., Achim, C., Waldeck, D. H., *J. Phys. Chem. B*, 2010, **114**(45), 14140–14148. DOI 10.1021/jp906910h.

154. Venkatramani, R., Davis, K. L., Wierzbinski, E., Bezer, S., Balaeff, A., Keinan, S., Paul, A., Kocsis, L., Beratan, D. N., Achim, C., Waldeck, D. H., *J. Am. Chem. Soc.*, 2011, **133**(1), 62–72. DOI 10.1021/ja107622m.

Chapter 9

EC-STM/STS of Redox Metalloproteins and Co-Factors

Andrea Alessandrini[a,b] and Paolo Facci[c]

[a]*Department of Physics, Informatics and Mathematics, University of Modena and Reggio Emilia, Via Campi 213/A, 41125 Modena, Italy*
[b]*S3, Nanoscience Institute, CNR, Via Campi 213/A, 41125 Modena, Italy*
[c]*Institute of Biophysics, National Research Council—CNR, Via De Marini 6, Genova, 16149 Italy*

paolo.facci@cnr.it

This chapter reviews some of the main experimental and theoretical results in the field of single-molecule characterization of electron transport through redox adsorbates. Both redox metalloproteins and co-factors have been object of intense investigation by EC-STM/STS techniques. Particularly, the paradigmatic case of the redox metalloprotein azurin and of derivatized quinone molecules adsorbed on atomically flat gold provide a valuable insight into the potential of the technique and into the attainable level of understanding of the electron transfer phenomenon. The results are discussed critically in view of their impact on the field of molecular electronics.

Molecular Electronics: An Experimental and Theoretical Approach
Edited by Ioan Bâldea

ISBN 978-981-4613-90-3 (Hardcover), 978-981-4613-91-0 (eBook)
www.panstanford.com

9.1 Introduction

The appeal of measuring and controlling redox reactions at the single-(bio)molecule level stems, on the one hand, from the general motivations behind any single-molecule investigation and, on the other hand, from the technological challenges posed by the field of molecular electronics.

Indeed, it is clear that bulk measurements, providing information on the average behavior of a large number of molecules, can get a robust description of the mean behavior of a molecular population that can be even quite heterogeneous. As such, the distribution of any relevant parameter will be affected both by homogeneous (i.e., dependent on the intrinsic properties) and inhomogeneous (i.e., dependent on the differences between molecular microenvironment, orientation, etc.) broadening. Hence, its mean value will reflect both these sources of uncertainty in a tangled fashion.

The single-molecule approach allows one to get rid of the effects brought about by inhomogeneous broadening, probing directly the "intrinsic" nature of a given parameter. It is, furthermore, evident that the distribution of events giving rise to a certain average or most probable value contains much more information than the two values alone. In the case at issue, for example, electron transfer in single biomolecules, having access to a large set of possible isoenergetic pathways that an electron can follow to reduce or oxidize a redox molecule, provides a unique clue to the possible physical-chemical mechanisms ruling this important phenomenon.

There is, however, a further reason prompting single-molecule study of electron transfer that is related to the possible technological exploitation of redox proteins and co-factors. Indeed, they represent the ultimate size units for the biological electron transfer.

Given their nanometer extent, it is straightforward to think of them in terms of artificial, hybrid nanodevices whose functionality is intrinsic to them rather than emerging from the way the device is assembled (as in the case of more conventional, top-down devices such as electronic, solid-state ones). Therefore, studying and achieving control of functionality at the single-molecule level emerges in the context of nanoscience as an extremely appealing possibility, towards the implementation of single-molecule devices.

This is the case, for instance, of bioelectronics devices, i.e., devices that exploit the electron-transfer capabilities of a redox metalloprotein (or co-factor) sandwiched between a set of nanometer-spaced electrodes. It is predicted that these devices could display typical power consumption as low as 10 pW, fast commutation rates and large on/off ratios. Other charming scenarios are connected to the full exploitation of the self-assembly properties inherent to many biological systems, such as physiologically interacting partners, with the aim of achieving self-assembling (e.g., self-positioning) functional devices.

In general, for investigating such single-molecule systems, one has to refer to tools coming from the realm of nanoscience and nanotechnology.

In the case of redox metalloproteins and co-factors, those tools are primarily the electrochemical scanning tunneling microscope (ECSTM) and spectroscopy (ECSTS).

Indeed, scanning tunneling microscopy and spectroscopy with full control of substrate and tip potential in an electrochemical cell represent powerful tools for investigating the phenomenology and mechanisms of interfacial electron transport in redox molecules down to single-molecule level [1]. These techniques have been used for studying transition metal complexes [2–4], molecular wires containing viologen [5, 6], redox metalloproteins (e.g., the blue copper protein azurin [7–10], cytochrome b562 [11], and derivatives of redox co-factors (e.g., quinones with thio-alkyl chains) [12] chemisorbed on noble metal surfaces (e.g., Pt and Au). Both ECSTM and ECSTS can provide information on the role of the molecular redox moieties in mediating electron transport processes across the tunneling gap between two electrodes. In comparison to measurements performed in UHV conditions, these techniques allow operation in buffered aqueous media, especially important in case of metalloproteins, while enabling potential control of both substrate and tip with respect to solution (i.e., a reference electrode placed in it). In the following paragraphs, the main ECSTM/STS results on redox metalloproteins and co-factors will be reviewed and interpretation of the registered behavior in terms of current theoretical frameworks will be discussed.

9.2 The Electrochemical Scanning Tunneling Microscope

The introduction of scanning probe microscopy (SPM) in the early 1980s [13] represented a genuine revolution in surface science and solid-state physics, allowing probe and manipulate the atomic structure of the matter in the direct space.

Soon after the appearance of the STM, it was shown that this microscope could also be used in liquid environment [14], provided the current that enters the feedback loop arises mainly from tunneling of electrons between tip and substrate. Moreover, working in liquid on a conductive substrate enables the interesting opportunity of establishing an electrochemical cell where the role of working electrode is played by the substrate [15, 16]. Adding a reference and a counter electrode, a typical three-electrode electrochemical cell is implemented. However, when an STM is operated in solution, the conductive tip could represent another polarizable electrode. In an electrochemical scanning tunneling microscope (EC-STM), the tip can be considered as a second working electrode of a four-electrode electrochemical cell [17]. Such an experimental setup allows one to study electrode surfaces at high spatial resolution under bi-potentiostatic control, Fig. 9.1.

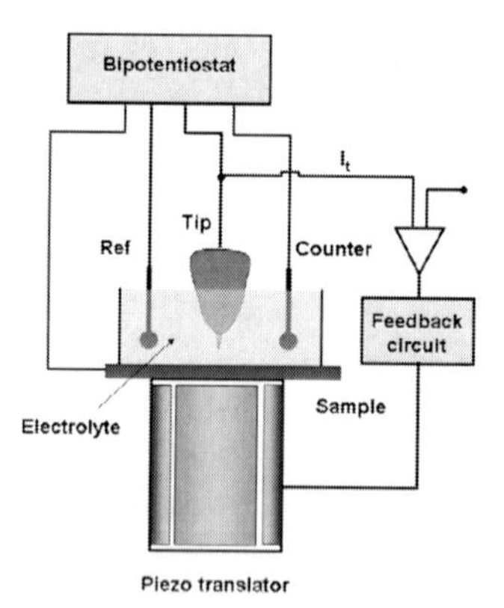

Figure 9.1 Schematic representation of an EC-STM setup with bipotentiostatic control. Both substrate and tip, connected to a bipotentiostat, play the role of working electrodes, whose potential can be driven independently. From Alessandrini, A., et al. *Phys. Chem. Chem. Phys.*, 2006, **8**, 4383–4397. Reproduced by permission of the PCCP Owner Societies.

The total measured current results from the sum of Faradaic and non-Faradaic contributions (including tunneling current). To reduce non-tunneling currents to a small percentage of the set-point current the tip surface area exposed to the electrolytic solution has to be minimized. This is usually achieved by coating the tip with an electrical insulating material such as Apiezon wax [18] or electropolymerizable paints [19]. Furthermore, Faradaic currents that might enter the feedback loop, arising from electron or ion exchange between the working electrodes and solution, can be limited by controlling the electrode potentials with a bi-potentiostat. Measuring the current that enters the feedback loop at large tip–sample separation as a function of both tip and substrate potential allows one to find out a potential window where non-tunneling currents can be kept to a very small level ($<1 \div 2$ pA), not influencing microscope operation in the tunneling regime [20].

In a typical electrochemical cell, the potential of a reference electrode is known and does not change over time. In EC-STM experiments, the reference electrode is usually a silver wire placed in contact with solution, representing a "quasi-reference" whose potential can be measured with respect to other common reference electrodes such as saturated calomel electrode (SCE–Hg/$HgCl_2$) or silver/silver chloride electrode (Ag/AgCl). The counter electrode is usually made of platinum and it is the one the current flows through, playing the role of sacrificial electrode. The bi-potentiostatic approach allows one to control independently tip and substrate potential with respect to a reference and their difference represents the tunneling bias voltage.

Similarly to conventional STM, also the substrates used in EC-STM experiments have to be conductive and atomically flat, the last requirement being critical for measurements of small objects at high spatial resolution. Usually, highly oriented pyrolytic graphite (HOPG) or well-defined crystallographic faces of a metal are used. Among the second type of substrates, Au(111) is often the best choice because of the ease of fabrication [21], the resistance to oxidation and the possibility of exploiting thiols' reactivity for immobilizing suitable molecules on it.

The presence of a bi-potentiostat in the typical experimental setup enables the important features of spectroscopic-like imaging and tunneling spectroscopy. In the first case, constant current images of the same sample area can be repeatedly acquired as a function of substrate/tip potential at constant tip bias voltage. The retrieved information will be the apparent height variation of the imaged features as a function of the working electrode potential. In the second case, one can have direct access to tunneling current as a function of the working electrode potential [22].

In the framework of classical surface science, the EC-STM has been used to study underpotential deposition of metals [23], phase transitions controlled by electrode-potential [24] and electro-chemical reactions of adsorbed molecules [25]. In the remainder of this chapter, we will show how EC-STM/STS can be exploited to study the behavior of electron transfer molecules, such as redox biomolecules and co-factors, as a function of the potential of the electrode onto which these are immobilized [26]. The reported experiments help shedding light on the mechanisms ruling electron transport across these kinds of molecules, allowing one to assess the role of redox moieties and specific environment in influencing electron flow through molecular adsorbates.

9.3 Redox Metalloproteins

The transfer of electrons in biological macromolecules represents the mean by which a relevant number of complex biological functions are partially or fully accomplished [27]. Key phenomena such as photosynthesis, respiration, catalytic reactions, and many others, involve as crucial steps the transfer of one or more electrons between molecular partners along free energy cascades. In biological electron transfer, a peculiar role is played by a special class of proteins, called redox metalloproteins. These molecules belong to the large ensemble of metalloproteins, large polypeptides containing one or more metal ions or clusters that represent a fraction between 25% and 30% of the entire proteome [28].

Generally speaking, the set of metals important for biological functions include iron, magnesium, copper, nickel, zinc, manganese, molybdenum, cobalt, tungsten and vanadium [29]. These metals and their ligands define prosthetic groups that are usually covalently bound to the polypeptide backbone (e.g., heme group in cytochromes) by endogenous ligands belonging to the amino acid side chains.

Redox or electron transfer metalloproteins have one or more metal ions in their active site, whose oxidation state can change reversibly, being the basis of their ability to exchange electrons with other biomolecular partners. The most relevant metalloproteins can be classified in three families: blue copper proteins (or cuprodoxins), cytochromes (containing heme groups) and iron-sulfur complexes; moreover, one can add the chlorophyll-based photosynthetic complexes (e.g., reaction centers, antenna complexes) that are redox metalloproteins where a photoionization event triggers an electron transfer cascade [30].

Given their peculiar functional activity and biological relevance, redox metalloproteins have been intensively studied by theoretical, computational, and experimental approaches [31]. These comprise classical and semi-classical (QM/MM) molecular dynamics, various spectroscopies (UV-Vis and IR absorption, fluorescence, Raman scattering, EXAFS, quasi-elastic and low-angle neutron scattering, NMR, EPR, XPS) diffractions (X-rays, electrons), pulsed radiolysis, electrochemistry, and, more recently, scanning probe microscopy (scanning tunneling microscopy [STM], electrochemically assisted STM, atomic force microscopy [AFM]), often assisted by protein engineering to help clarify proteins' fine structural and functional details by means of site-directed mutagenesis. The scientific effort directed towards metalloprotein characterization has provided a deep understanding of the structure and function of several of these redox molecules and, in many cases, it has allowed us to shed light on the mechanisms used by these biomolecules to exchange electrons between molecular partners.

This aspect is, of course, very important in view of the possible technological exploitation of redox metalloprotein electron transfer properties. Indeed, the advent of nanoscience has provided an unprecedented level of understanding of the electron transfer mechanisms ruling the functional activity of redox metalloproteins, that of the single molecule. Indeed, at that level, one can get rid of the statistical average typical of "bulk" investigation methods, (that smear out the fine differences in the possible various electron transfer mechanisms) enabling us to attain a quantitative, mechanistic description of the observed phenomena and a related control.

It is worthy of note that the efficiency by which redox metalloproteins perform the task of transferring electrons, along with their nanometer size and the self-assembly ability inherent to many biological molecules, make them very appealing candidates for technological applications [32, 33], namely for the implementation of bioelectronics devices.

9.3.1 The Azurin

Among electron transfer metalloproteins, an important family is that of the "blue copper proteins." It comprises redox biomolecules bearing a single Cu ion in their active site, characterized by a distorted co-ordination geometry that causes their typical intense blue color. Blue-copper proteins, or cuprodoxins, can be found in both bacteria and plants where they are usually involved in some of the early steps of respiration and photosynthesis [34]. Their family mainly comprises azurin, plastocyanin, amicyanin, rusticyanin, and caeruloplasmin. In particular, azurin has been thoroughly investigated by electrochemically assisted scanning probe techniques due to a series of peculiar features which will be highlighted in the following.

Azurin from *Pseudomonas aeruginosa* is a water-soluble, relatively small protein (molecular mass 14600) that is involved in the oxidative phosphorylation of the bacterium *P. aeruginosa*. Its functional role is believed to be that of exchanging electrons between two molecular partners, cytochrome c551 and nitrite reductase, that act as azurin primary donor and acceptor, respectively [35]. From a redox chemistry point of view, azurin functional behavior depends on the reversible oxidation of Cu^{1+} to Cu^{2+}. Peculiar electronic properties are connected to the special structural features of its redox site. Indeed, it is formed by a copper ion ligated to five amino acids (two hystidines and a cysteine, strongly bound to copper, and two more weaker axial ligands: a methionine and a main chain carbonyl oxygen) [36]. The resulting geometry provides the site with unusual spectroscopic and electrochemical characteristics. Among them are an intense optical absorption band at 628 nm (associated with a S(Cys)–Cu bonding to antibonding transition), a small hyperfine splitting in the electron paramagnetic spectrum [37], and a surprisingly large standard potential (+116 mV versus SCE) [38] in comparison to that of the Cu(II/I) aqua redox couple (–89 mV versus SCE) [39]. Molecular structure has a globular shape characterized by a marked β-barrel conformation, Fig. 9.2. Azurin active redox site is buried inside the molecule and appears located ≈ 0.8 nm underneath the globule surface; it features also an exposed S–S bridge that is formed by the thiol moieties of Cys3 and Cys26 side chains.

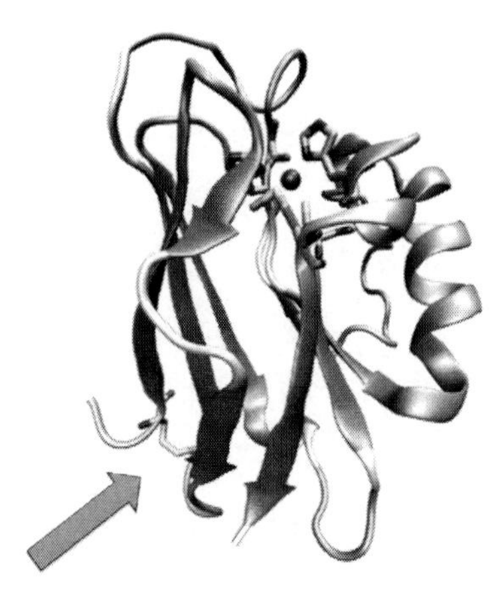

Figure 9.2 Schematic representation of the structure of azurin from *Pseudomonas aeruginosa*. Note the Cu ion represented as a dot in the upper part of the structure; the arrow points to the disulfide bridge formed by Cys3 and Cys26 thiol moieties. Structural data from PDB file 1E5Y.

Both its conformation and the presence of the mentioned disulfide bridge make azurin very appealing as a candidate for single-molecule level investigation by scanning probe techniques.

Indeed, the presence of an exposed disulfide provides a very convenient anchoring point to chemisorb azurin at the surface of electronically soft metals such as Au, Ag, Pt, and Cu, by means of the formation of one or two Me-S bonds. A relevant number of experimental techniques, including FTIR absorption spectroscopy in ATR configuration, quartz crystal microbalance, scanning force microscopy, XPS, reductive desorption and cyclic voltammetry confirm that azurin chemisorbs on gold retaining its redox activity once arranged in submonolayer [40]. Finally, its β-barrel conformation provides the protein with a remarkable resistance to the mechanical action of the scanning probe, particularly useful in case of repeated scans. This is a very desirable feature that unfortunately is not very common in biomolecules, which usually tend to be damaged by the action of a scanning tip, even in those cases where the tip–sample interaction can be kept to a minimum (e.g., alternating contact mode AFM, [41]).

9.4 Redox Co-Factors

A redox co-factor is a non-proteinaceous chemical compound that is bound to a metalloprotein and is required for protein's functional activity. Proteins involving co-factors are often, but not exclusively, metalloenzymes, and co-factors can be considered as "support molecules" that assist an enzyme in performing the biochemical transformations it is deputed to.

Co-factors can be either organic or inorganic molecules. They can also be classified according to how tightly they bind to an enzyme; loosely bound co-factors are named co-enzymes, whereas tightly bound co-factors are termed prosthetic groups (e.g., heme group). An enzyme depleted by its co-factors is called an apo-enzyme, while the complete enzyme with co-factor is the holo-enzyme [42].

Some enzymes or enzyme complexes need several co-factors to function [43].

Organic co-factors are often vitamins or are made from vitamins. Many of them contain the nucleotide adenosine monophosphate (AMP) as part of their structures, such as *ATP*, co-enzyme *A*, *FAD*, and *NAD*$^+$. This common structure possibly reflects a common evolutionary origin as part of ribozymes in an ancient RNA-based world. Also, metal ions are common co-factors. In many cases, the co-factor includes both an inorganic and an organic component. A typical example is that of heme group, which consist of a porphyrin ring coordinated to iron.

Iron-sulfur clusters are complexes of iron and sulfur atoms held within proteins by cysteine residues. They play both structural and functional roles, including electron transfer and redox sensing, as structural modules [44].

Vitamins too can serve as precursors to many organic co-factors (e.g., vitamins *B*1, *B*2, *B*6, *B*12, niacin, folic acid) or as co-enzymes themselves (e.g., vitamin *C*). Most of these co-factors are found in a huge variety of species, and some are universal to all forms of life.

The classes of co-factors that will focus on in the following paragraphs are those derived from quinones. Quinones are a class of organic compounds that are formally "derived from aromatic compounds (such as benzene or naphthalene) by conversion of an even number of –CH= groups into –C(=O)– groups with any necessary rearrangement of double bonds" [45]. The set includes some heterocyclic compounds. Among quinones, a paradigmatic one is 1,4-benzoquinone (Fig. 9.3), which will be extensively discussed in ECSTM/STS experiments as a surface-bound species (following suitable derivatization with different thiol-terminated chains, vide infra).

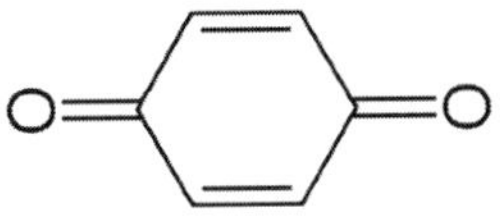

Figure 9.3 The structure of 1,4-benzoquinone.

Derivatives of quinones are common constituents of biologically relevant molecules. Some of them play the role of electron acceptors in electron transport chains such as those in photosynthesis (plastoquinone, phylloquinone), and aerobic respiration (ubiquinone) [30]. Phylloquinone, also known as Vitamin K1, is used by animals to help form certain proteins which are involved in blood coagulation, bone formation, and other processes.

Here, we will focus on two different molecules containing benzoquinone, namely, 2-(6-mercaptoalkyl)hydroquinone and 4-(2′,5′-dihydroxystyryl)benzyl thioacetate. Their choice allowed us to study different degree of electronic coupling to the substrate due to the different conductivity of their thiol-terminated chains. Indeed, the former has an alkyl chain terminated with a SH group, whereas the latter features an oligo phenylene vinylene (OPV) chain, terminated with a thioacetate. Both of them readily chemisorb on gold. The alkyl chain, however, has a much lower conductivity than the OPV counterpart. Thus, 2-(6-mercaptoalkyl) hydroquinone is predicted to have a much lower electronic coupling to substrate electronic states than the OPV-derivatized benzoquinone.

9.5 ECSTM of Azurin

An application of ECSTM spectroscopic-like imaging that is paradigmatic both of the potential of this technique and of the behavior of redox metalloproteins in a tunneling gap is the study of the blue copper protein azurin immobilized on a gold substrate.

As we have seen in Section 9.3.1, azurin chemisorbs readily on gold via its exposed disulfide bridge that, in proximity of the metal surface, reduces and forms up to two S–Au bonds.

Atomically flat (e.g., Au(111)), azurin-coated gold can be imaged by ECSTM taking advantage of the possibility of performing spectroscopic imaging as a function of substrate potential. This kind of experiment aims at studying the role of redox levels associated with the molecule, i.e., with the copper ion in the protein's active site, in the electronic transport via the molecule itself.

The possible scenarios emerging from such an experiment are different, involving the relation between several relevant parameters, vide infra, and can be discussed and rationalized on the basis of theories present in the literature [1, 46].

The first published evidence of substrate potential-dependent STM contrast for a metalloprotein [7] reported the variation with substrate potential of the apparent height of azurin measured with respect to a substrate reference level. Imaging a sample region as a function of the substrate potential V_s in the potential range of −300 to +100 mV (versus SCE), surface corrugation changes quite abruptly for values > −100 mV, showing the presence of surface bumps (Fig. 9.4). One can observe that for large enough tunneling current set points (e.g., 2 nA at V_b = +400 mV), azurin apparent height follows a sort of sigmoidal trend with V_s. It is worth noting that the apparent bumps are electronic in nature, since the same experiment performed with an AFM, non-sensitive to tunneling current but providing sample topography via force interaction [41], showed bumps in the whole potential range which did not depend upon V_s.

The picture apparently complicates if one performs a slightly different experiment where the current set-point is lowered (1 nA) and the apparent height variation is now measured against an internal standard, i.e., Zn–azurin. Zn, indeed, cannot undergo a reaction $Zn^{2+} + e^- \rightarrow Zn^{1+}$ but only $Zn^{2+} + 2e^- \rightarrow Zn^0$ at a much lower potential (~ −0.7 V versus SHE). Therefore, as it has been directly demonstrated [8], Zn–azurin does not change its apparent height with V_s in the range of −0.5 to +0.1 mV.

In this case, the apparent height variation of the height-changing molecules with respect to the inert ones shows a resonance-like behavior, as reported in Fig. 9.5.

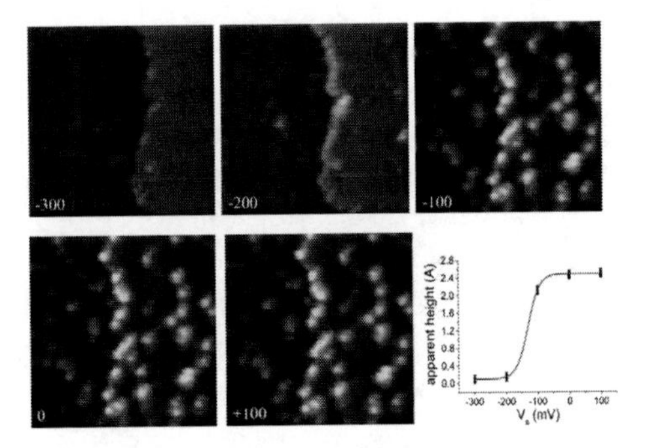

Figure 9.4 Set of ECSTM images of azurin on Au(111) acquired as a function of V_s (indicate on each image in mV versus SCE). Imaging conditions were as follows: I_t = 2 nA, V_{bias} = +400 mV, imaging buffer 50 mM NH_4Ac pH 4.6. Note the sigmoidal shape of the trend of the apparent height as a function of V_s.

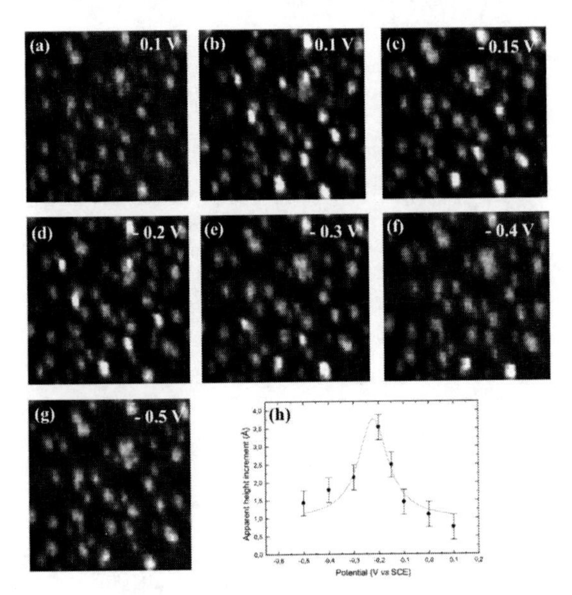

Figure 9.5 ECSTM images of azurin on Au(111) acquired as a function of V_s. Imaging conditions were as follows: I_t = 1 nA, V_{bias} = +400 mV, imaging buffer 50 mM NH_4Ac pH 4.6. Note the resonance-like shape of the apparent height of the molecule marked with a circle in (a) as a function of V_s, measured versus those bumps that do not vary their apparent height. From Alessandrini, A., et al. *Phys. Chem. Chem. Phys.*, 2006, **8**, 4383–4397. Reproduced by permission of the PCCP Owner Societies.

Azurin appears to behave in a different way in the two experiments, featuring a sigmoidal or a resonance-like trend according to the adopted current set-point (2 or 1 nA, respectively).

This apparent discrepancy has been rationalized in the framework of the theories describing the specific case of a redox protein in an ECSTM gap. In any of these theories, one or more electronic levels responsible for the redox behavior of a molecule are involved in electron tunneling between a tip and a metal substrate. Particularly, the two-step coherent electron transfer model by Kutznezov and Ulstrup [47] predicts qualitatively the exact trends observed in the two previous experiments. Their difference is traced back to that between eV_{bias} and λ, i.e., the reorganization energy of the reaction. In case $eV_{\text{bias}} > \lambda$ theory predicts a plateau in the current versus V_s plot, whereas it predicts a resonance-like curve in case $eV_{\text{bias}} < \lambda$, Fig. 9.6.

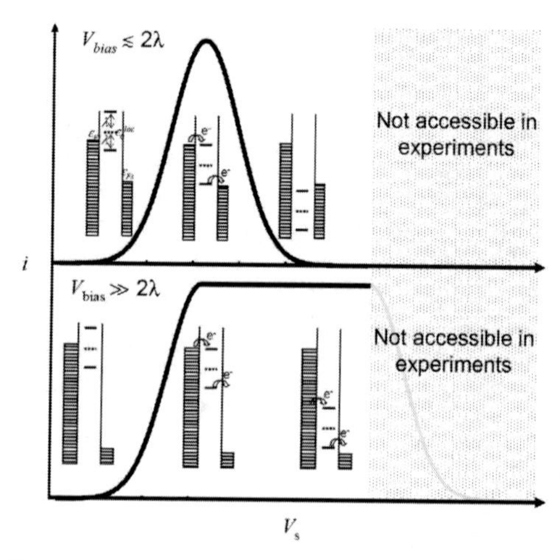

Figure 9.6 The two different trends predicted by two-step electron transfer theory. From Alessandrini, A., et al. *Phys. Chem. Chem. Phys.*, 2006, **8**, 4383–4397, RCS Publishing. Reproduced by permission of the PCCP Owner Societies.

The reason why the two experiments agree with these different trends can be tentatively ascribed to the variation of λ brought about by different current set-points. Indeed, different set-points in STM mean a different tip–sample distance. Such a difference is generally marked for 1 nA variation (although difficult to estimate quantitatively due to the lack of information on the value of β, the tunneling current inverse decay length, at protein's location) due to the exponential dependence of tunneling current on tip–sample separation. One can argue that solvent-related component of λ [48] could be affected in a major fashion by changing tip vertical position (a macroscopic object on the molecular scale). These movements result in a sizable change in the value of λ that, on its turn, can toggle the system between the two different regimes.

It is fair to note, however, that such an explanation, although plausible, operates just at a qualitative level. This is because any theory of ECSTM provides predictions on $I_t = I_t$ (V_s, V_{bias}), and not on the apparent height variation. In order to be more quantitative, one has to get direct access to tunneling current; this, however, requires a different experimental approach, that will be introduced in Section 9.5.1.

It is important to point out that ECSTM experiments allow one to distinguish between redox metalloproteins having the same conformation and differing just in the electronic levels brought about by the metal ion in the redox active site. Figure 9.7 shows the remarkable difference between brighter bumps, ascribable to Cu–azurin, and darker ones, corresponding to non-electroactive, Zn counterpart.

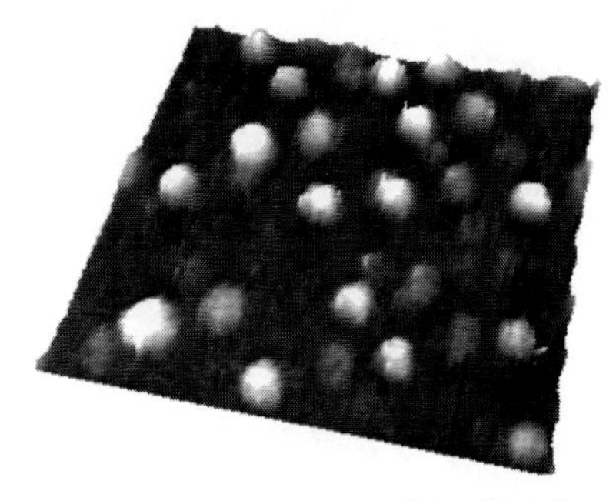

Figure 9.7 ECSTM image of a mixture of Cu (brighter bumps) and Zn–azurin (darker bumps). Imaging conditions were as follows: I_t = 1 nA, V_{bias} = +400 mV, V_s = +200 mV versus SCE, imaging buffer 50 mM NH_4Ac pH 4.6. Image size 89 × 89 nm^2.

9.5.1 ECSTS of Azurin

The need for direct access to tunneling current emerges, thus, as a critical aspect to go further in the experimental analysis and theoretical data interpretation.

In principle, tunneling current can be directly assessed by measuring I_t versus V_{bias} characteristics curves in an off-feedback regime.

Although azurin chemisorbs readily on Au(111) surface by substantially covalent S–Au bonds, repeated scans, needed to image the molecules on the surface, and possible tip–sample interactions during unavoidable, periodic adjustment of tip–sample 3D mutual position required to maintain the correct tip–sample–substrate configuration, can easily damage molecular conformation or even sweep out molecules from the scanned surface. This fact makes $I_t(V_s)$ spectroscopy not easily performable.

In order to circumvent these difficulties, a different experimental arrangement has been suggested. Namely, by exchanging the role of tip and substrate, many of the aforementioned difficulties disappear. The idea is that of letting azurin chemisorb on the surface of an ECSTM tip made of gold. The gold tip has to be almost completely insulated and just its apex is exposed to solution, where metalloproteins can chemisorb. Figure 9.8 reports a TEM image of such a biofunctionalized tip.

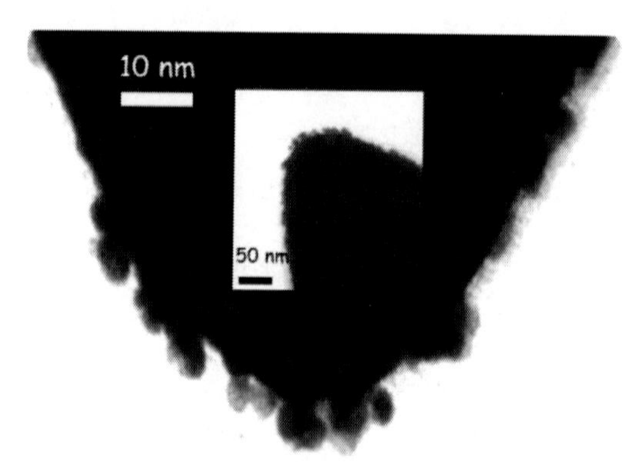

Figure 9.8 TEM image of an azurin-coated, wax-insulated Au tip. Tunneling current is likely to be picked up by the most protruding molecule due to the exponential decay of tunneling current with distance. Adapted with permission from Alessandrini, et al. *Appl. Phys. Lett.*, 2005, 86, 133902. Copyright 2005, American Institute of Physics.

These tips show the typical electrochemical (CV) behavior expected from a nanoelectrode (almost negligible capacitance contribution), and retain redox activity of the chemisorbed molecules [9].

With such probes, it is possible to solve a number of problems affecting the aforementioned spectroscopy-like ECSTM imaging. Indeed, one does not have to search for a molecule on a substrate surface by repeated scans, hence, decreasing the wear effect due to the raster scanning of the tip on the surface. Furthermore, this approach eliminates the need for tracking the molecule, an aspect that is usually very critical in tunneling spectroscopy. The x–y drift will be almost uninfluential on the measured I_t and z-drift will be easily accounted for by waiting for a stabilization period before the beginning of each characteristic curve measurement.

The most relevant advantage of the protein-on-tip approach, however, is that it enables direct access to tunneling current as a function of potential. Indeed, here V_s and V_t flip their role and one will scan V_t and consider I_t as a function of it, rather than of V_s.

Experimentally, one has to engage a certain tunneling current set-point, and, after tunneling junction stabilization (e.g., thermal drift and piezo creep settling down), disable the feedback. Afterwards, one starts sweeping tip voltage while measuring tunneling current, i.e., $I_t(V_t)$ at constant bias.

The result of such measurements on an azurin-coated Au tip is the appearance of a marked resonance in a potential region near to the redox potential of the molecule. That feature is absent in both bare tips and Zn–azurin-coated tips, used as negative control measurements. It is worth of note that the same molecule sustains repeated potential scans without showing a significant degradation in performances. Furthermore, the non-Faradaic nature of the resonance is confirmed in some measurements, which show the co-existence of a resonance and of a genuine Faradaic wave (visible as a shoulder of the bigger tunneling current profile), the latter ascribable to those molecules not on the apex of a tip not perfectly insulated [9]. Figure 9.9a reports these data.

Measuring the $I_t(V_t)$ (or, analogously, $I_t(V_s)$) characteristics for different V_{bias}, Fig. 9.9b, it is possible to obtain a set of curves that are prone to be interpreted in terms of the existing theories.

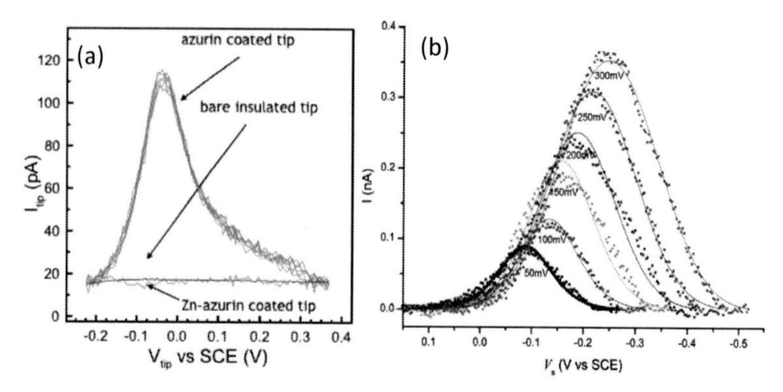

Figure 9.9 (a) $I_t(V_t)$ measured with Au tips. Only Cu–azurin yields current resonances, at variance with bare and Zn-azurin-coated tips; (b) set of $I_t(V_s)$ curves as a function of V_{bias}. (a) Adapted with permission from Alessandrini, et al., *Appl. Phys. Lett.*, 2005, **86**, 133902. Copyright 2005, American Institute of Physics." (b) From Alessandrini, et al. *Phys. Chem. Chem. Phys.*, 2006, **8**, 4383–4397. Reproduced by permission of the PCCP Owner Societies.

The reported data were indeed interpreted by referring to the theory of two-step coherent electron tunneling with partial relaxation [47]. This theoretical approach resulted preferable to other similar ones [1], even if an incoherent (fully relaxed) model cannot be completely figured out since it could be still at play if one hypothesizes a strong electronic coupling between azurin redox electronic levels and those of the leads. Within the mentioned theoretical frame, it is possible to extract values for relevant parameters such as the redox level ε_0 and the reorganization energy λ. Whereas the first is consistent with the value for $E_0^{'}$ estimated from CV, the second appears much lower (0.13 eV) than that measured for the self-exchange reaction (≈0.6 eV) [1]. This discrepancy is still a matter of debate but can be traced back to the ECSTM configuration itself that forces the redox molecule between substrate and tip, two macroscopic objects, in a tiny gap from which most of the solvent is squeezed out. This fact can have a direct impact on the solvent-associated component of the reorganization energy that results decreased.

However, a more recent theoretical review of the reported data, triggered by measurements of Transition Voltage Spectroscopy on azurin chemisorbed on gold in an ECSTM setup [49], has doubted the validity of this interpretation, suggesting a more robust and effective theoretical treatment based on an extended Newns–Anderson model [46]. By that approach, TVS [49] and previous ECSTM/STS [9] results on azurin can be fit and rationalized, overcoming a number of drawbacks of the two step coherent electron tunneling model.

It is now possible to regard the reported results from the standpoint of their relevance for applications. Indeed, the kind of functionality that emerges from the data resembles that of an electronic switch or molecular transistor. As a matter of fact, the behavior of a single azurin molecule in between three metal leads bears a striking analogy to that of a generic single-particle transistor. Generally speaking, the latter takes advantage of a capacitive gate coupled to single-particle levels and able to tune them with respect to the Fermi levels of the two metal leads that play the role of source and drain and that are, as such, biased by an external voltage to drive current through the particle. Figure 9.10 describes that situation. Only when particle levels are properly aligned to the Fermi levels of the leads a sizable current can flow through the particle.

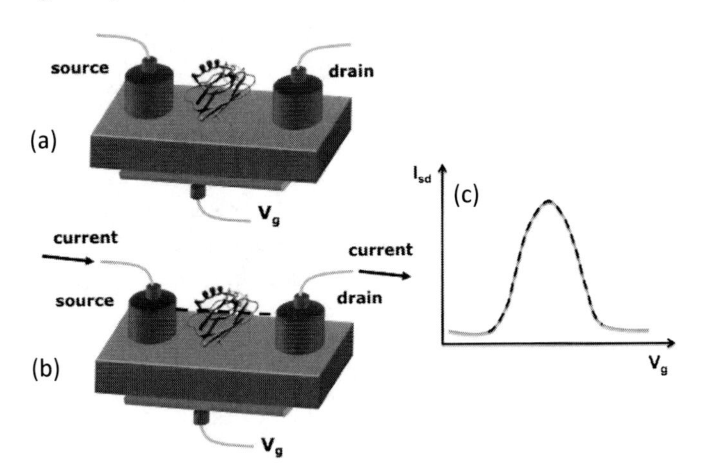

Figure 9.10 Scheme of a single-particle transistor. The role of the single particle in this representation is played by an azurin molecule. (a) Transistor is off; (b) transistor is on; (c) ideal plot of $I_{sd}(V_g)$. Note the analogy between the qualitative trend of the current in this case, with respect to what measured in experiments with azurin-coated tips in ECSTM/STS configuration.

The described situation is very similar to that encountered in case of ECSTS of azurin. Even there, by tuning the electronic redox levels of the molecule to the Fermi levels of the leads results in an enhanced tunneling current flow between electrically biased source and drain (tip and substrate, or vice versa). In this respect, the analogy with a transistor functionality is straightforward. The main difference is the way used for getting the proper tuning of the electronic levels. Indeed, in case of azurin, the gate is electrochemical in nature, rather than capacitive as in the case of a single-particle transistor. This means that it is somehow diffused but very efficient and brought about by the supporting electrolyte that enables potential control at the tip and the substrate by the action of the bi-potentiostat.

Therefore, the behavior of azurin chemisorbed on an ECSTM tip apex is that of a single protein transistor with electrochemical gate.

At this point, one could question the applicability of these kinds of results to the implementation of real devices. We can preliminarily observe that the best on-off ratio one can estimate from the reported data is still quite limited (<500, in the best case), and not comparable with that of more conventional solid-state transistor devices (typical I_{on}/I_{off} > 10,000 for a MOSFET). This is, however, a remarkable result at room temperature, if one considers that similar figures are still far reaching even in graphene-based FET-type transistors [50]. However, as far as transistor performances are concerned, the correct comparison is, to our judgment, with standard solid-state devices rather than with single-particle ones (e.g., single electron transistor-SET) since that is the target of any innovative device that aims at asserting itself as a relevant player in tomorrow (nano)electronics.

Other problems arise in terms of durability and performance reliability of protein based devices, not to think of the need of a wet environment to function. In this respect, it is really hard, and frankly not very realistic, to imagine replacing the current solid-state electronic devices and fabrication strategy with any other technology involving bottom-up assembly starting from functional molecules, albeit promising, and even more stable than proteins. This belief is also supported by the tremendous costs of the current fabrication facilities in electronics industry. These make, in fact, unfeasible any radical change in fabrication strategy, given the substantially unaffordable economic involvement it would require.

In order to find a space for biomolecule-based device concepts, we think that a preferable solution could be that of seeking for niche applications. Among the possible scenarios, ultrasensitive biosensors down to single-molecule level, for instance, could be more appealing and realistic implementations. At any rate, the demonstration of an electronic switch or transistor functionality in a single metalloprotein represents a basic conceptual step towards the general goal of devising a technological control over biological electron transfer phenomena [33].

9.6 ECSTM/STS of the Benzoquinone/Hydroquinone Couple

ECSTM used in molecular electronics investigations has turned out very useful also in single-molecule studies of redox co-factors such as quinone-based molecules. These molecules are characterized by a quite complex, pH-dependent, redox chemistry [50], featuring a nine-membered square matrix describing the possible ET pathways joining the reduced and oxidized forms of the molecule as a function of overpotential and solution pH. Quinones are representative of co-factors that play a prominent role in biological ET, but are also very intriguing for their potential application as multi-level molecular switches in molecular electronics implementations. In case of quinone derivatives, ECSTM shows a substrate potential dependent apparent height (contrast) variation that is related to the appearance of two regions of contrast enhancement [12], at variance with azurin where, as we have seen, only one of these regions is evident. This behavior can be qualitatively observed both in case of 2-(6-mercapt oalkyl)hydroquinone and in that of 4-(2′,5′-dihydroxystyryl)benzyl thioacetate. Its cause can be traced back to the presence of two redox levels that can separately contribute to the tunneling current through the molecule. Whereas the analysis of ECSTM images suggests just a qualitative agreement with the theory, as we have seen in case of azurin, a much more informative experiment is that which provides direct access to tunneling current, enabling, thus, quantitative comparison with theories. The approach followed in case of quinone derivatives has been that of assembling dense SAMs on Au(111) and then measuring tunneling current as a function of substrate potential V_S. This type of measurements requires the STM feedback be disabled (or kept at a minimum, just for compensating for possible drifts), after having reached the desired current set-point. Figure 9.11 reports the corresponding current profiles in case of quinones derivatives with the previously mentioned two different linkers. Both of them show two resonances; interestingly, those measured on OPV-derivatized quinone are much more closely spaced in energy (voltage) pointing to a better coupling between the redox moiety and the metal surface [51, 52].

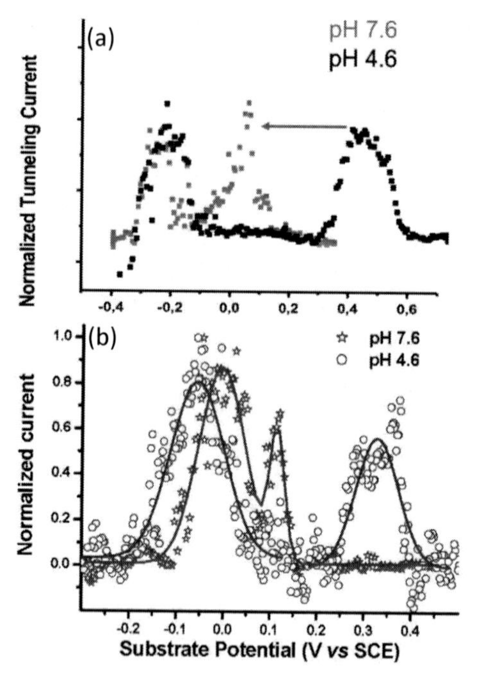

Figure 9.11 Tunneling current as a function of V_s in case of benzoquinone derivatized with (a) alkyl chain, (b) OPV chain. Both measurements at pH 4.6 and 7.6 are reported for the two different linkers. V_{bias} = +100 mV, I_t = 0.5 nA. Adapted with permission from (a) Petrangolini, P., et al. *J. Am. Chem. Soc.*, 2010, **132**, 7445–7453. Copyright 2010, American Chemical Society; and (b) Petrangolini, P., et al. *J. Phys. Chem. C*, 2011, **115**, 19971–19978. Copyright 2011, American Chemical Society.

In both cases, a pH change causes a modulation of the separation between the two regions of tunneling current enhancement. Particularly, by increasing pH from 4.6 to 7.6, the peak distance decreases markedly (from 400 to 100 mV in case of OPV-derivatized quinone). This behavior is readily interpretable in the framework of Laviron theory [50] and supports the idea of a pH control on the functional behavior of these molecules. In particular, in this redox couple, the redox potentials of the reactions that describe the reversible transition from the reduced (hydroquinone) to the oxidized (p-benzoquinone) form vary with proton concentration, since the overall reaction involves the exchange of $2e^-$ and $2H^+$. Increasing solution pH implies decreasing the availability of H^+, shifting thus the equilibrium of the coupled redox reactions. These evidences point to the stabilization of a redox intermediate that is not observable in case of CV measurements, whereas discloses a separate ET behavior for the two redox reactions to which the nine-membered square matrix quoted above [50] condenses in case of equilibrium protonation. Accordingly, increasing pH can be regarded as another way, alternative to substrate potential change, to modulate the quinone redox state.

Besides, it is worth noting that the closer proximity of the redox levels in case of OPV-derivatized quinone at higher pH, enables us testing theoretical predictions about the behavior of two-level redox molecules studied by ECSTM. Indeed, in a work by Kuznetsov and Ulstrup [47] a different trend of tunneling current as a function of V_s is predicted in case that two redox levels are spaced by less than eV_{bias}. In that case, the double peaks of Fig. 9.11 modify in a double plateau trend as a function of V_s due to the contemporary involvement in the energy windows defined by eV_{bias} [51].

Conclusions

In conclusion, it appears quite evident that EC-STM/STS represents a powerful approach to single-molecule characterization of redox adsorbates as far as their transport properties are concerned. The possibility of studying both metalloproteins and co-factors in a liquid, physiologic-like environment, along with the opportunity of controlling the electronic levels involved in the electron flow through those molecules, make the experimental technique at issue a key player in the molecular electronics scenario. Furthermore, powerful theoretical treatments enable now a fine understanding of the mechanisms involved in electron transport via redox molecules sandwiched in an electrochemical, tunneling gap improving, thus, the quantitative description of the phenomenon with major impacts on both fundamental and applied research.

References

1. Alessandrini, A., Corni, S., Facci, P. (2006). Unraveling single metalloprotein electron transfer by scanning probe techniques, *Phys. Chem. Chem. Phys.,* **8**, 4383–4397.
2. Albrecht, T., Guckian, A., Ulstrup, J., et al. (2005). Transistor-like behavior of transition metal complexes, *Nano Lett.,* **5**, 1451–1455; Lie, D. Y. C., and Wang, K. L. (2001). Si/SiGe processing, in *Semiconductors and Semimetals* 73, eds. Willardson, R., and Weber, E., Chapter 4 (Academic Press, San Diego), pp. 151–197.
3. Albrecht, T., Moth-Poulsen, K., Christensen, J. B., et al. (2006). In situ scanning tunnelling spectroscopy of inorganic transition metal complexes, *Faraday Discuss.,* **131**, 265–279.
4. Albrecht, T., Guckian, A., Ulstrup, J., et al. (2005). Transistor effects and in situ STM of redox molecules at room temperature, *IEEE Trans. Nanotechnol.,* **4**, 430–434.
5. Li, Z., Han, B., Meszaros, G., et al. (2006). Two dimensional assembly and local redox-activity of molecular hybrid structures in an electrochemical environment, *Faraday Discuss.,* **131**, 121–114.
6. Pobelov, I. V., Li, Z., Wandlowski, T. (2008). Electrolyte gating in redox-active tunneling junctions—an electrochemical STM approach, *J. Am. Chem. Soc.,* **130**, 16045–16054.
7. Facci, P., Alliata, D., Cannistraro, S. (2001). Potential-induced resonant tunneling through a redox metalloprotein probed by electrochemical scanning probe microscopy, *Ultramicroscopy,* **89**, 291–298.
8. Alessandrini, A., Gerunda, M., Canters, G. W., Verbeet, M. P., Facci, P. (2003). Electron tunnelling through azurin is mediated by the active site Cu ion, *Chem. Phys. Lett.,* **376**, 625–630.
9. Alessandrini, A., Salerno, M., Frabboni, S., Facci, P. (2005). Single-metalloprotein wet biotransistor, *Appl. Phys. Lett.,* **86**, 133902–133903.

10. Chi, Q., Zhang, J., Arslan, T., et al. (2010). Approach to interfacial and intramolecular electron transfer of the diheme protein cytochrome c4 assembled on Au(111) surfaces, *J. Phys. Chem. **B***, 114, 5617–5624.

11. Della Pia, E. A., Chi, Q., Jones, D. D., Macdonald, J. E., Ulstrup, J., Elliott, M. (2011). Single-molecule mapping of long-range electron transport for a cytochrome b562 variant, *Nano Lett.*, **11**, 176–182.

12. Petrangolini, P., Alessandrini, A., Berti, L., Facci, P. (2010). An electrochemical scanning tunneling microscopy study of 2-(6-merca ptoalkyl)hydroquinone molecules on Au (111), *J. Am. Chem. Soc.*, **132**, 7445–7453.

13. Binnig, G., Rohrer, H., Gerber, C., and Weibel, E. (1982). Surface studies by scanning tunneling microscopy, *Phys. Rev. Lett.*, **49**, 57–61.

14. Sonnenfeld, R., and Hansma, P. K. (1986). Atomic-resolution microscopy in water, *Science*, **232**, 211–213.

15. Lustenberger, P., Rohrer, H., Christoph, R., Siegenthaler, H. (1988). Scanning tunneling microscopy at potential controlled electrode surfaces in electrolytic environment, *J. Electroanal. Chem.*, **243**, 225–235.

16. Alessandrini, A., and Facci, P. (2013). Electrochemical scanning tunneling microscopy and spectroscopy for single-molecule investigation, *Methods Mol. Biol.*, **991**, 261–273.

17. Siegenthaler, H., and Christoph, R. (1990). In situ scanning tunneling microscopy in electrochemistry, in *Scanning Tunnelling Microscopy and Related Methods*, 184, eds. Behm, R. J., Garcia, N., and Rohrer, H. (Kluwer, Dordrecht, NATO ASI Series E) pp. 315 –333.

18. Thundat, T., Nagahara, L. A., Oden, P. I., Lindsay, S. M., George, M. A., Glaunsinger, W. S. (1990). Modification of tantalum surfaces by scanning tunneling microscopy in an electrochemical cell, *J. Vac. Sci. Technol. A*, **8**, 3537–3541.

19. Bach, C. E., Nichols, R. J., Beckmann, W., Meyer, H., Schulte, A., Besenhard, J. O., Jannakoudakis, P. D. (1993). Effective insulation of scanning tunneling microscopy tips for electrochemical studies using an electropainting method, *J. Electrochem. Soc.*, **140**, 1281–1284.

20. Christoph, R., Siegenthaler, H., Rohrer, H., Wiese, H. (1989). In situ scanning tunneling microscopy at potential controlled Ag(100) substrates, *Electrochim. Acta*, **34**, 1011–1022.

21. Kolb, D. M. (1996). Reconstruction phenomena at metal-electrolyte interfaces, *Prog. Surf. Sci.*, **51**, 109–173.

22. Alessandrini, A., and Facci, P. (2007), Electrochemically assisted scanning probe microscopy: A powerful tool in nano(bio)science, in *The New Frontiers of Organic and Composite Nanotechnology*, eds. Erokhin, V., Ram, M. K., Yavuz, O., Chapter 5 (Elsevier, New York) pp. 237–286.

23. Magnussen, O. M., Hotlos, J., Nichols, R. J., Kolb, D. M., Behm, R. J. (1990). Atomic structure of Cu adlayers on Au(100) and Au(111) electrodes observed by in situ scanning tunneling microscopy, *Phys. Rev. Lett.*, **64**, 2929–2932.

24. Cunha, F., Tao, N. J., Wang, X. W., Jin, Q., Duong, B., D'Agnese, J. (1996). Potential-Induced Phase Transitions in 2,2′-Bipyridine and 4,4′-Bipyridine Monolayers on Au(111) Studied by in Situ Scanning Tunneling Microscopy and Atomic Force Microscopy, *Langmuir*, **12**, 6410–6418.

25. Tao, N. J., and Shi, Z. (1994). Real-time STM/AFM study of electron transfer reactions of an organic molecule: Xanthine at the graphite-water interface, *Surf. Sci.*, **321**, L149-L156.

26. Tao, N. J. (1996). Probing Potential-Tuned Resonant Tunneling through Redox Molecules with Scanning Tunneling Microscopy, *Phys. Rev. Lett.*, **76**, 4066–4069.

27. Gray, H. B., and Ellis, W. (1994). Electron transfer in biology, in *Electron Transport Metalloproteins in Bioinorganic Chemistry*, eds. Bertini, I., et al. (University Science Books, Sausalito, CA).

28. Cowan, J. A. (1997). *Inorganic Biochemistry: An Introduction* (Wiley-VCH, New York).

29. Holm, R. H., Kennepohl, P., Solomon, E. I. (1996). Structural and functional aspects of metal sites in biology, *Chem. Rev.*, **96**, 2239–2314.

30. Devault, D. (1984). *Quantum Mechanical Tunnelling in Biological Systems* (Cambridge University Press, Cambridge).

31. Harrison, P. M. (1985). *Metalloproteins Parts I and II* (Academic Press, New York).

32. Facci, P. (2002). Single metalloprotein at work: Towards a single-protein transistor, in *Nano-Physics & Bio-Electronics: A New Odyssey*, Chakraborty, T., Peeters, F., Sivan, U., eds., Chapter 12 (Elsevier, Amsterdam), pp. 323–339.

33. Facci, P. (2014). *Biomolecular Electronics: Bioelectronics and the Electrical Control of Biological Systems and Reactions* (Elsevier, Oxford).

34. Adam, E. T. (1991). Copper protein structures, *Adv. Protein Chem.*, **42**, 145–197.

35. Bendall, D. S. (1996). Interprotein electron transfer, in *Protein Electron Transfer,* Bendall, D. S., ed. (BIOS Scientific Publisher, Oxford).

36. Nar, H., Messerschmidt, A., Van De Kamp, M., Canters, G. W., and Huber, R. (1991). Crystal structure analysis of oxidized Pseudomonas aeruginosa azurin at pH 5.5 and pH 9.0. A pH-induced conformational transition involves a peptide bond flip, *J. Mol. Biol.,* **221**, 765–772.

37. Brill, A. S. (1977). *Transition Metals in Biochemistry* (Springer, Berlin).

38. Chi, Q., et al. (1999). Electrochemistry of self-assembled monolayers of the blue copper protein *Pseudomonas aeruginosa* azurin on Au(111), *Electrochem. Commun.*, **1**, 91–96.

39. Lide, D. R. (ed.) (1993). *CRC Handbook of Chemistry and Physics* (CRC Press, Boca Raton).

40. Schnyder, B., Kötz, R., Alliata, D., and Facci, P. (2002). Comparison of the self-chemisorption of azurin on gold and on functionalizedoxide surfaces, *Surf. Interface Anal.*, **34**, 40–44.

41. Alessandrini, A., and Facci, P. (2005). AFM: A versatile tool in biophysics, *Meas. Sci. Technol.*, **16**, R65–R92.

42. Sauke, D. J., Metzler, D. E., and Metzler, C. M. (2001). *Biochemistry: The Chemical Reactions of Living Cell,* 2nd ed. (Harcourt/Academic Press, San Diego).

43. Jordan, F., and Patel, M. S. (2004). *Thiamine: Catalytic Mechanisms in Normal and Disease States* (Marcel Dekker, New York).

44. Meyer, J. (2008). Iron-sulfur protein folds, iron-sulfur chemistry, and evolution, *J. Biol. Inorg. Chem.*, **13**, 157–170.

45. Nic, M., Jirat, J., and Kosata, B (eds.) (2006). *Quinones*, IUPAC Compendium of Chemical Terminology, online edition.

46. Bâldea, I. (2013). Important insight into electron transfer in single. Molecule junctions based on redox metalloproteins from transition voltage spectroscopy, *J. Phys. Chem. C,* **117**, 25798–25804.

47. Friis, E. P., Kharkats, Y. I., Kuznetsov, A. M., and Ulstrup, J. (1998). In situ scanning tunneling microscopy of a redox molecule as a vibrationally coherent electronic three-level process, *J. Phys. Chem. A*, **102**, 7851–7859.

48. Marcus, R. A. (1956). On the theory of oxidation-reduction reactions involving electron transfer. I, *J. Chem. Phys.*, **24**, 966–978.

49. Artés, J. M., López-Martnez, M., Giraudet, A., Dez-Pérez, I., Sanz, F., and Gorostiza, P. (2012). Current-voltage characteristics and transition voltage spectroscopy of individual redox proteins, *J. Am. Chem. Soc.*, **134**, 20218–20221.

50. Laviron, E. (1984). Electrochemical reactions with protonations at equilibrium. Part X. The kinetics of the p-benzoquinone/hydroquinone couple on a platinum electrode, *J. Electroanal. Chem.*, **164**, 213–227.

51. Petrangolini, P., Alessandrini, A., Navacchia, M. L., Capobianco, M. L., and Facci, P. (2011). Electron transport properties of single-molecule-bearing multiple redox levels studied by EC-STM/STS, *J. Phys. Chem. C.*, **115**, 19971–19978.

52. Petrangolini, P., Alessandrini, A., and Facci, P. (2013). Hydroquinone-benzoquinone redox couple as a versatile element for molecular electronics, *J. Phys. Chem. C.*, **117**, 17451–17461.

Chapter 10

Electron Transport in Atomistic Nanojunctions from Density Functional Theory in Scattering Approaches

Yu-Chang Chen[a,b]

[a]*Department of Electrophysics, National Chiao Tung University, 1001University Road, Hsinchu 30010, Taiwan*
[b]*Physics Division, National Center for Theoretical Sciences, 1001 University Road, Hsinchu 30010, Taiwan*

yuchangchen@mail.nctu.edu.tw

10.1 Introduction

Molecular electronics has generated a tremendous wave of scientific interest in the past decade due to prospects of device-size reduction offered by atomic-level control of certain physical properties [1]. In addition to the fabrication of electronic device by traditional silicon technology, crossing molecules between electrodes provides an alternative method to build nanoscale electronic device. The concern with metal–molecule–metal tunnel junctions has been growing due to their potential applications in nanoscale devices. This idea inspires a new form of electronics at atomic level, which attracts much interest from multi-discipline researchers, nourishing the hope to trigger technology revolution in the

Molecular Electronics: An Experimental and Theoretical Approach
Edited by Ioan Bâldea

ISBN 978-981-4613-90-3 (Hardcover), 978-981-4613-91-0 (eBook)
www.panstanford.com

electronic industry. This goal is important because the consumer electronics can be more compact in size, more efficient in energy saving, and more stable in functioning due to less heat generation. It has spurred great interest in the fundamental understanding of quantum transport [2]. The path to extreme device miniaturization resulted in the rapid development of molecular electronics, where molecules are used as building blocks to form nanodevices. Molecular electronics offers bottom-up method taking advantage of molecular self-assembly in addition to the traditional silicon technology, which reduces the size of electronics from top-down approach. Molecular electronics with merit to the small size has struggled for the forefront of electronic devices.

Charge transfer occurs between the electrodes and the nanostructure within the Fermi wavelength. When the miniaturization of nanojunctions reaches the atomic scale, the size of the scattering region is comparable to the Fermi wavelength of electron. This causes shifting and broadening of molecular levels. As a direct result, the detailed information of electronic structures is of crucial importance for understanding the quantum transport properties of nanojunctions. Thus, the studies of electronic and thermoelectric quantum transport properties in molecular junctions require the development of new theories. To ensure the functionality of such devices under finite biases, understanding the quantum transport theory at the molecular level is crucially important. Scientists have made considerable efforts to understand the properties of transport electrons traversing the nanostructures bridged via source-drain electrodes as the down-scaling of electronic components is advancing. Novel and unknown physics in the atomistic nature of the quantum transport in the nanoscale junctions open a challenging interdisciplinary field for scientists. Atomistic system opens new playground for electronic and thermoelectric nanodevice application due to the quantum transport properties of electrons and phonons.

10.2 Theory of Electron Transport

The field of nanoscale electronics has generated a tremendous wave of scientific interest in the past decade due to prospects of device-size reduction offered by atomic-level control of certain physical properties. When the down-size of devices approaches nanometer scale, the quantum mechanical properties become more and more important. Influence of the wave nature associated with quantum mechanics has spurred great interest in the fundamental understanding of quantum transport at the nanoscale. Many-body theories allied to first-principles calculations in the framework of density functional theory in scattering approaches are applied to understand the fundamental properties of electron and energy transport related to the detailed electronic structure of these building blocks. Such theory is suitable for a wide range of device properties in nanojunctions junctions operated under finite source-drain bias with gate field in a three-terminal geometry.

The approaches start with the separation of the full Hamiltonian of the system into $H = H_0 + V$, where H_0 is the Hamiltonian due to the bare electrodes modeled as electron jellium separated by a distance and V is the scattering potential of the nanostructured object bridging the semi-infinite electrodes with a planar surface, as depicted in Fig. 10.1 [3]. Compared with the molecule or nanostructure bridging the bimetallic junction, the electrodes are large enough such that we can model them as a tunneling junction. The M-V-M is considered as the unperturbed Hamiltonian and is modeled as electron jellium.

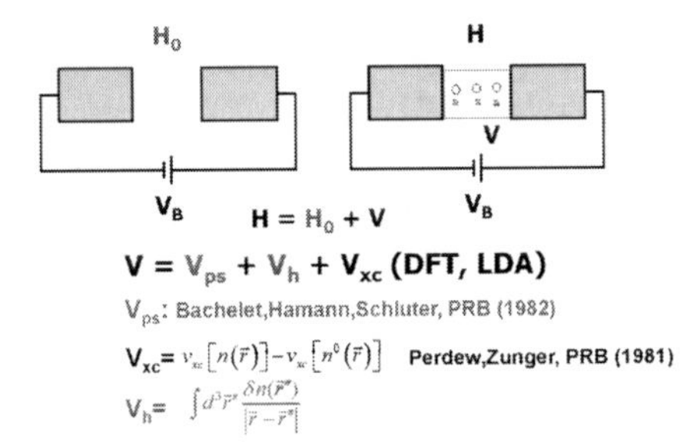

Figure 10.1 Schematics of a typical atomistic junction. The full system ($H = H_0 + V$) can be separated into the bare electrode (H_0) and the scattering region (V). For detailed descriptions of V, see Eq. (10.9).

10.2.1 The Bimetal Junction

First, we consider a metal–vacuum–metal (M–V–M), which consists of two metal surface separated by a vacuum [4, 5]. In the jellium model, conduction electrons in the metal electrode are considered to move in a uniform positive-charge background n^+. The charge density described by the Wigner–Seitz radius r_s, where $\frac{4\pi}{3}r_s^3 = n^+$. Values of r_s for some metals are listed in Table 10.1. We note that electronic charge transfer occurs at the surfaces of metal electrodes, forming dipolar layers on surfaces of electrodes.

Table 10.1 Wigner–Seitz radius of metals

Elements	r_s
Al	2.07
Au	3.01
Na	3.93
K	4.86
Cs	5.62

The dipolar layers generate electric fields, which cause a vacuum potential barrier between two electrodes. The characteristic length scale of charge transfer at the interface is comparable to the Fermi wavelength λ_F. The quantum mechanical wave nature of electrons also causes a Friedel oscillation of charge density in the interface. An external bias, $V_B = (\mu_R - \mu_L)/e$, is also considered in the bimetallic junction, where $\mu_{L(R)}$ is the chemical potential deep in the left (right) electrodes, as shown in Fig. 10.2. The battery causes an additional transfer of charge between two electrodes like a capacitor, generating an electric field in the vacuum region. The above effects are calculated within the framework of density functional theory, where the coupled Poisson and Schrödinger equation are iteratively solved until self-consistency is achieved.

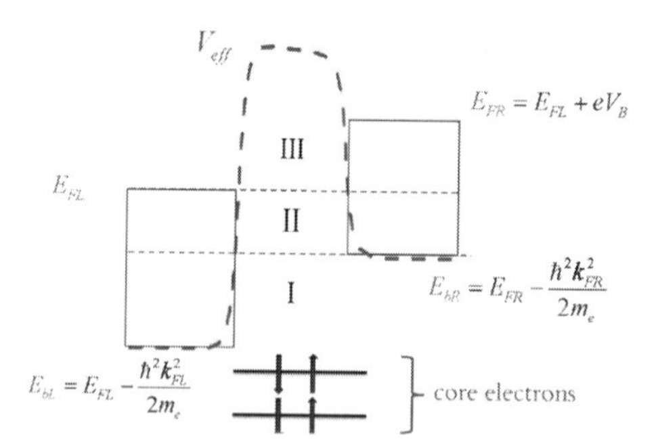

Figure 10.2 Schematics of the band diagram of a bimetal junction with an applied bias V_B. When atoms or molecules are placed in the bimetal junction, partial valence electrons in the scattering region could be localized and for core states.

The wavefunctions of electrons in the bare electrodes are $u_{E\mathbf{K}}^{L(R)}(z)$, which represent electrons with the energy E travel along the z-direction before inclusion of the nanostructured object. The equation $u_{E\mathbf{K}}^{L(R)}(z)$ is calculated by solving the coupled Schrödinger and Poisson equations iteratively until self-consistency is reached, i.e.,

$$\left\{-\frac{e\hbar}{2m_e}\frac{d^2}{dz^2}+v_{\text{eff}}[z,n(z)]-E\right\}u_{E\mathbf{K}}^{\alpha}(z)=0, \tag{10.1}$$

where $n(z)=2\sum_{E}\sum_{\alpha=L,R}|u_{E\mathbf{K}}^{\alpha}(z)|$ and the effective potentials v_{eff} $[z, n(z)]$ comprises the exchange-correlation potential V_{xc} and the electrostatic potential V_{es},

$$v_{\text{eff}}[z, n(z)] = V_{\text{es}}[n(z)] + V_{\text{xc}}[n(z)], \tag{10.2}$$

where the electrostatic potential can be solved by the Poisson equation,

$$\partial_z^2 V_{\text{es}}(z) = 4\pi[n_+ - n(z)], \tag{10.3}$$

where $n(z)$ is the total electron charge density for electrons in the continuum bands of electrodes. We consider the jellium model for the spin electron in the bulk electrodes with $n_+=\frac{1}{3\pi^2}\left(\frac{\sqrt{2mE_{\text{F}}}}{\hbar}\right)^3$. Note that $v_{\text{eff}}(\pm\infty)$ are the bottom of conduction band for electrons deep inside the right and left electrodes. We assume that the conduction band has flat band bottoms for $z\to\pm\infty$, $v_{\text{eff}}(z)$. To solve the coupled Poisson–Schrödinger equation, boundary conditions and several constraints need to be satisfied. For the Schrödinger equation, the right- and left-moving scattering wavefunctions satisfy the scattering boundary conditions inside the electrodes $(z\to\pm\infty)$,

$$u_{E\mathbf{K}}^{L}(z)=\sqrt{\frac{m_e}{2\pi\hbar^2 k_L}}\times\begin{cases} e^{ik_L z}+R_L e^{-ik_L z}, & z<-\infty \\ T_L e^{ik_R z}, & z<\infty \end{cases} \tag{10.4}$$

and

$$u_{E\mathbf{K}}^{R}(z)=\sqrt{\frac{m_e}{2\pi\hbar^2 k_R}}\times\begin{cases} T_R e^{-ik_L z}, & z<-\infty \\ e^{-ik_R z}+R_R e^{ik_R z}, & z>\infty, \end{cases} \tag{10.5}$$

where **K** is the electron momentum in the plane parallel to the electrode surfaces, and z is the coordinate parallel to the direction of the current.

When solving the Poison equation, the electron change density deep inside the electrodes must neutralize the positive background n_+. Moreover, the total charge of the entire system must be neutral. The above constraints of charge neutrality warranty that the $V_{\mathrm{es}}(z)$ calculated by the Poisson equation has a flat band deep inside in the electrodes. With the boundary conditions $\partial_z V_{\mathrm{es}}(z)|_{z=\pm\infty} = 0$, the electrostatic potential $V_{\mathrm{es}}(z)$ due to the distribution of charge can be solve from Eq. (10.3) as

$$V_{\mathrm{es}}(z)=4\pi\int_{-\infty}^{+\infty}|z-z'|[n^{+}(z')-n(z)]dz', \tag{10.6}$$

where the charge density $n(z)$ is calculated using the effective single-particle wavefunctions obtained from solving Eq. (10.1) using boundary conditions in Eqs. (10.4) and (10.5),

$$n(z)=\frac{1}{\pi^2}\sum_{\alpha=L,R}\int_0^{k_{F\alpha}}(k_{F\alpha}^2-k_\alpha^2)u_E^\alpha(z)dk\alpha. \tag{10.7}$$

The numerical calculations start with an initial guess of charge density. The electrostatic potential $V_{\mathrm{es}}(z)$ is calculated according to charge density using Eq. (10.6). The exchange-correlation potential V_{xc} is then added to the electrostatic potential $V_{\mathrm{es}}(z)$ to form an effective potential v_{eff} for effective single-particle electrons. In the framework of DFT, wavefunctions of scattering electrons in the bare junction are solved under the scattering boundary conditions deep inside the electrodes. Those wavefunctions are applied to calculate the new charge density. The charge density calculated by the coupled Poisson–Schrödinger equation may not be conserved. Therefore, the outcome of the calculated charge density is modified through three-step corrections to maintain the charge neutrality for the entire system and the charge neutrality deeply inside the electrodes [6]. In correction 1, charge density in the left- and right-side of the junction is scaled by a constant such that the charge density in deep electrodes neutralizes the positive background, as shown in Fig. 10.3a. In correction 2, a charge Q_{exs} with gaussian shape distribution is added such that the total net charge is 0, as shown in Fig. 10.3b. In correction 3, a suitable gaussian shape charge distribution with zero net charge is applied to maintain an ideal potential difference between the left- and right-bottom of band in metal electrodes, as shown in Figs. 10.4a,b. Figure 10.4c shows the flow chart to calculate the charge density iteratively until the self-consistency is achieved. Figure 10.4d show the converged charge density for bimetal junctions with different r_S. Charge transfer occurs in the interface of the metal materials and vacuum, as shown in Fig. 10.5. It generates dipolar layers with a width around the Fermi wavelength λ_F, generating a potential barrier in the vacuum region. When an external voltage is applied, an additional charge transfer between two metal electrodes occurs, as shown in Fig. 10.6a. The charge separation in parallel bimetal materials generates asymmetric electrostatic potentials due to electric fields similar to a capacitor, as shown in the inset of Fig. 10.6a. Figure 10.6c shows the effective potentials that electrons experience in the framework of DFT with an effective single-particle picture. The electric filed at the center of the junction increases linearly with the bias voltage.

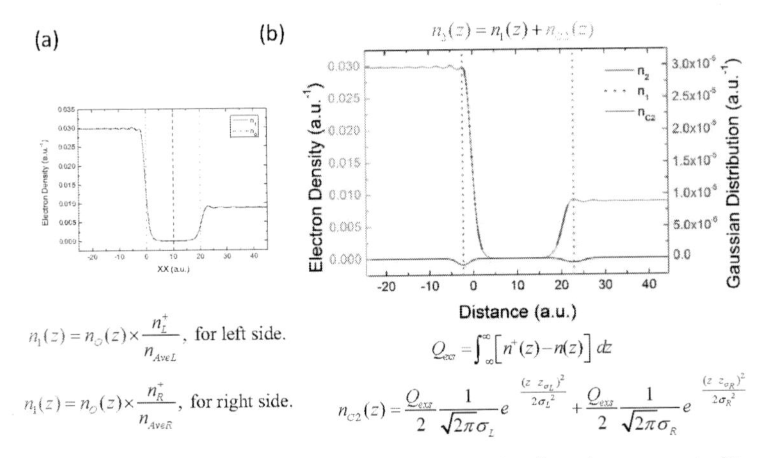

Figure 10.3 Demonstration for how to maintain the charge neutrality: (a) Correction 1: the charge density n_0 is scaled by constants such that $n_1(z = -\infty) \to n_L^+$ and $n_1(z = \infty) \to n_R^+$. (b) Correction 2: a charge Q_{exs} with a gaussian shape distribution is added to $n_1(z)$ such that the total net charge is 0.

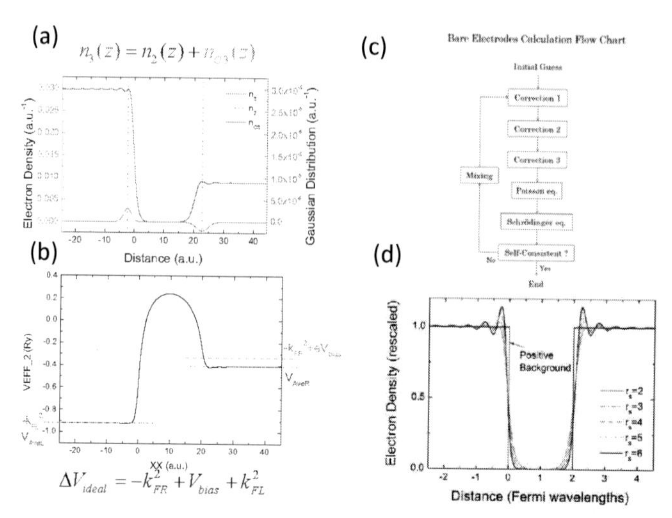

Figure 10.4 Demonstration for how to maintain the correct bandwidth: (a) Correction 3: moving charge from one electrode to another via a suitable gaussian-shape distribution of charge with zero net charge; (b) the band width of metals achieved by correction 3 has correct ideal potential difference $\Delta V_{\text{ideal}} = v_{\text{eff}}(z \to +\infty) - v_{\text{eff}}(z \to -\infty)$. (c) The flow chart to calculate the charge density self-consistently. (d) Charges move from the surface of metal electrodes to vacuum. It generates dipolar layers and creates a potential barrier.

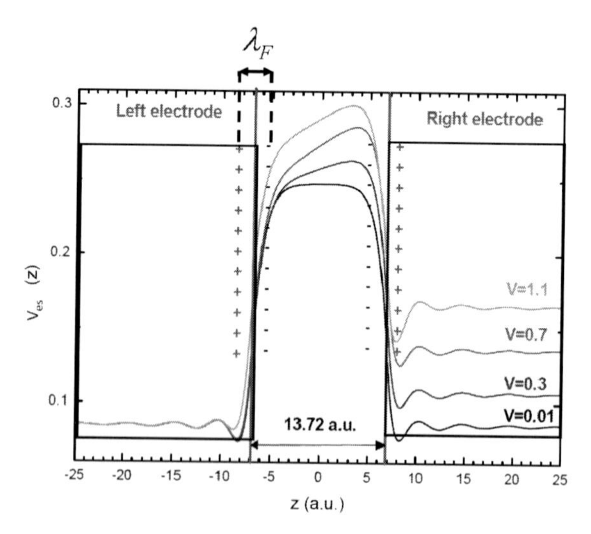

Figure 10.5 Schematics of the electrostatic potentials for various applied voltages.

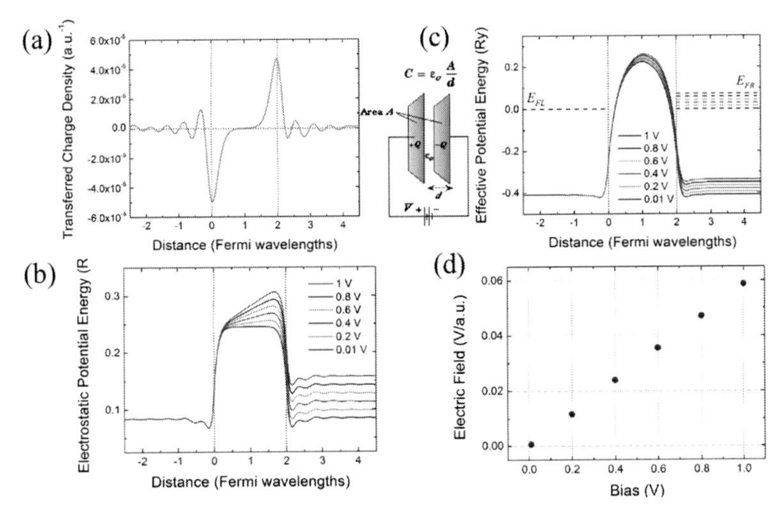

Figure 10.6 Effects of charge transfer on nanojunctions due to external voltages. (a) An additional charge from the right to left surface of electrodes similar to the charge separation in parallel capacitor. (b) The electrostatic potentials due to charge distributions for various biases. (c) The effective potentials for various biases. (d) The induced electric field at the center of the junction as a function of biases.

10.2.2 Inclusion of an Atomistic Nanostructure in the Bimetal Junction

When a molecule is included in the junction, the Lippmann–Schwinger equation is applied to calculate wavefunctions of the entire system (electrodes + molecule) until self-consistency is achieved. The quantum transport theory based on the combination of DFT and Lippmann–Schwinger equation in the scattering approach is equivalent to the theory based on a combination of DFT and non-equilibrium Green's function [8]. For weakly coupled systems such as a molecule weakly adsorbed to electrodes, self-interaction errors may be significant. More elaborate consideration of exchange-correlation energy might be an important ingredient in providing accurate quantitative descriptions in molecular transport calculations [9]. In such case, the electric conductance is also significantly dependent on the contact geometry. The correction to the exchange-correlation kernel due to the Vignale–

Kohn approximation in time-dependent density functional theory (TDCDDFT) is also important in the weakly adsorbed case [10]. The viscous nature of the electron liquid is more important for the molecular junction than for the quantum point contacts [11]. The reason for this is that the correction due to the dynamical corrections depends non-linearly on the gradient of the electron density. For strongly coupled systems such as monatomic chains, the exchange correlation in local-density approximation (LDA), which neglects the dynamical corrections, is remarkably successful when compared to the experiments. In this case, the dependence of conductance on the contact geometry is relatively small due to resonant tunnel of electrons.

The bimetal junction is modeled as an infinitely-large open system. It is considered as an unperturbed Hamiltonian. Inclusion of the nanostructure is considered in the scattering approach. We assume the 3D unperturbed wavefunctions of electrons have the form, $\Psi^{0,L(R)}_{E\mathbf{K}}(\mathbf{r}) = (2\pi)^{-1} e^{i\mathbf{K}\cdot\mathbf{R}} \cdot u^{L(R)}_{E\mathbf{K}}(z)$, where $u^{L(R)}_{E\mathbf{K}}(z)$ is the effective single-particle wavefunction of electron solved self-consistently with the coupled Schrödinger and Poisson equations in the 1D bare electrodes. We note that electrons are free to move in the x–y plane, described by $(2\pi)^{-1}e^{i\mathbf{K}\cdot\mathbf{R}}$. The condition of energy conservation gives $\frac{\hbar^2 k_R^2}{2m_e} = E - \frac{\hbar^2 \mathbf{K}^2}{2m_e} - v_{\text{eff}}(\infty)$ and $\frac{\hbar^2 k_L^2}{2m_e} = E - \frac{\hbar^2 \mathbf{K}^2}{2m_e} - v_{\text{eff}}(-\infty)$ where $v_{\text{eff}}(z)$ is the effective potential comprising the electrostatic potential and exchange correlation energy [7].

Owing to the fact that electrons move at speeds much larger than the speed of nuclei, the electrode-nanostructure-electrode system is considered as a unified coherent quantum system with Born-Oppenheimer approximation (or adiabatic approximation). Only the electronic Schrödinger equation is considered for wavefunctions of electrons, with atoms located at their equilibrium positions. The current-carrying wavefunctions of the total system (jellium + nanostructure) in the continuum are calculated in scattering approaches by solving the Lippmann–Schwinger equation, which is equivalent to solving the electronic Schrödinger equation, iteratively until self-consistency is reached

$$\Psi^{L(R)}_{E\mathbf{K}}(\mathbf{r}) = \Psi^{0,L(R)}_{E\mathbf{K}}(\mathbf{r}) + \int d^3\mathbf{r}_1 \int d^3\mathbf{r}_2 G^0_E(\mathbf{r},\mathbf{r}_1) V(\mathbf{r}_1,\mathbf{r}_2) \Psi^{L(R)}_{E\mathbf{K}}(\mathbf{r}_2), \quad (10.8)$$

where $\Psi_{E\mathbf{K}}^{L(R)}(\mathbf{r})$ stands for the effective single-particle wavefunctions of the entire system, corresponding to propagating electrons incident from the left (right) electrode. The term $V(\mathbf{r}_1, \mathbf{r}_2)$ stands for the scattering potential that the electrons experience when they scatter through the nanojunction,

$$V(\mathbf{r}_1,\mathbf{r}_2) = V_{\text{ps}}(\mathbf{r}_1,\mathbf{r}_2) + \left\{(V_{\text{xc}}[n(\mathbf{r}_1)] - V_{\text{xc}}[n_0(\mathbf{r}_1)]) + \int d\mathbf{r}_3 \frac{\delta n(\mathbf{r}_3)}{|\mathbf{r}_1 - \mathbf{r}_3|}\right\}\delta(\mathbf{r}_1 - \mathbf{r}_2), \qquad (10.9)$$

where $V_{\text{ps}}(\mathbf{r}_1, \mathbf{r}_2)$ is the pseudopotential representing the electron–ion interaction; $n_0(\mathbf{r})$ is the electron density for the pair of biased bare electrodes; $n(\mathbf{r})$ is the combined electron density for the total system, and $\delta n(\mathbf{r}) = n(\mathbf{r}) - n_0(\mathbf{r})$ is their difference. Localized states are solved by direct diagonalization of the Hamiltonian. Both bound states and continuum states are considered: $n(\mathbf{r}) = \mathbf{2}\Sigma_i|\Psi_i|^2 + 2\Sigma_{\alpha=L,R}\int dE \int d^2\mathbf{K}|\Psi_{E\mathbf{K}}^{\alpha}(\mathbf{r})|^2$. The quantity $V_{\text{xc}}[n(\mathbf{r})]$ is the exchange-correlation potential calculated at the level of local-density approximation. The quantity G_E^0 is the outgoing wave Green's function for the bimetallic junction,

$$G_E^0(\mathbf{r}_1,\mathbf{r}_2) = \frac{1}{2\pi^2}\int d^2\mathbf{K}e^{i\mathbf{K}\cdot(\mathbf{R}_1-\mathbf{R}_2)} \times \frac{u_E^L(z_<)u_E^R(z_>)}{W_E}, \qquad (10.10)$$

where $z_<$ $(z_>)$ is the lesser (greater) of z_1 and z_2. The quantity W_E is the Wronskian (which is z independent):

$$W_E = u_E^L(z)\frac{d}{dz}u_E^R(z) - u_E^R(z)\frac{d}{dz}u_E^L(z). \qquad (10.11)$$

We note that the external bias $[V_B = (\mu_R - \mu_L)/e]$ can be considered in the bimetallic junction with different chemical potential deep inside the electrodes. A gate voltage can also be introduced as a capacitor composed of two parallel circular charged disks separated by a certain distance from each other. The axis of the capacitor is perpendicular to the transport direction. One plate is placed close to the molecule, while the other plate, placed far away from the molecule, is set to be the zero reference energy. The simulation box is sufficiently large that the charge density outside the box is unperturbed by the introduction of a group of atoms.

When a molecule is included in the junction, the Lippmann–Schwinger equation is applied to calculate wavefunctions of the entire system (electrodes + a molecule or an atomistic nanostructure) until self-consistency is achieved. Thus, the high-bias case is considered and calculated at the same footing as the zero-bias case using the Lippmann–Schwinger equation.

10.2.3 Solving the Problem in the Plane Wave Basis

In the actual numerical calculations, the plane wave basis is chosen. Here we will assume that the sample is enclosed in a box of size $(2L_x) \times (2L_y) \times (2L_z)$. The total wavefunctions to be solved in plane wave representation are expressed within the box as

$$\psi_{E\mathbf{K}}^{R(L)}(\mathbf{r}) = \sum_{\mathbf{n}} \Psi_{E\mathbf{Kn}}^{L(R)} e^{i\mathbf{k}_n \cdot \mathbf{r}}, \tag{10.12}$$

where $\mathbf{n} = (n_x, n_y, n_z)$ is an integer vector and $\mathbf{k}_n = \left(\frac{\pi}{L_x}n_x, \frac{\pi}{L_y}n_y, \frac{\pi}{L_z}n_z\right)$. Similarly, the wavefunctions of bare electrodes solved by the coupled Schrödinger and Poisson equations are expressed as

$$\Psi_{E\mathbf{K}}^{0,R(L)}(\mathbf{r}) = \sum_{\mathbf{n}} \Psi_{E\mathbf{Kn}}^{0,L(R)} e^{i\mathbf{k}_n \cdot \mathbf{r}}. \tag{10.13}$$

The scattering potential $V(\mathbf{r}_1, \mathbf{r}_2)$ and Green's function $G_E^0(\mathbf{r}_1, \mathbf{r}_2)$ can also be expressed as

$$V(\mathbf{r}_1, \mathbf{r}_2) = \sum_{\mathbf{n}_1, \mathbf{n}_2} e^{i\mathbf{k}_{\mathbf{n}1} \cdot \mathbf{r}_1} V_{\mathbf{n}_1, \mathbf{n}_2} e^{i\mathbf{k}_{\mathbf{n}2} \cdot \mathbf{r}_2}, \tag{10.14}$$

and

$$G_E^0(\mathbf{r}_1, \mathbf{r}_2) = \sum_{\mathbf{n}_1, \mathbf{n}_2} e^{i\mathbf{k}_{\mathbf{n}2} \cdot \mathbf{r}_1} G_{E,\mathbf{n}_1,\mathbf{n}_2}^0 e^{i\mathbf{k}_{\mathbf{n}2} \cdot \mathbf{r}_2}, \tag{10.15}$$

where $\mathbf{n}_1 = (n_{1x}, n_{1y}, n_{1z})$, $\mathbf{n}_2 = (n_{2x}, n_{2y}, n_{2z})$, $\mathbf{k}_{n_1} = \left(\frac{\pi}{L_x}n_{1x}, \frac{\pi}{L_y}n_{1y}, \frac{\pi}{L_z}n_{1z}\right)$, $\mathbf{k}_{n_2} = \left(\frac{\pi}{L_x}n_{2x}, \frac{\pi}{L_y}n_{2y}, \frac{\pi}{L_z}n_{2z}\right)$.

When a molecule or an atomistic nanostructure is included in the junction, the Lippmann–Schwinger equation is applied to calculate wavefunctions of the entire system (electrodes + nanostructure) until self-consistency is achieved. The wavefunctions $\Psi_{E\mathbf{K}}^{R(L)}(\mathbf{r})$ can be solved from a system of linear equations.

$$\sum_{\mathbf{n}'} C_{E\mathbf{nn}'}\Psi_{E\mathbf{Kn}'}^{L(R)} = \Psi_{E\mathbf{Kn}}^{0,L(R)}, \tag{10.16}$$

with

$$C_{E\mathbf{nn}'} = \delta_{\mathbf{nn}'} - 8L_xL_yL_z\sum_{\mathbf{n}''} G_{E,\mathbf{n},\mathbf{n}''}^{0} V_{\mathbf{n}''\mathbf{n}'}^{0}. \tag{10.17}$$

We note that solving the Lippmann–Schwinger equation has been transformed into a problem of linear algebra instead of solving the integral equation [3]. The valence electrons of the atoms sandwiched between electrodes can become localized. The states below the bottom of the conduction band are considered as localized states that can be obtained by a direct diagonalization of the full Hamiltonian in a matrix representation in the plane-wave basis, as shown schematically in Fig. 10.2.

10.3 Quantum Transport Properties Calculated by Current-Carrying Wavefunctions

In the following subsections, we introduce how to calculate a variety of important physical quantities in terms of the scattering wave-functions, which are obtained by iteratively solving the Lippmann–Schwinger equation in the framework of a parameter-free Lippmann–Schwinger equation until self-consistency is obtained.

10.3.1 Moments of Current

The properties of steady-state current fluctuations can provide fundamental insight into the nature of electron transport, including the role of the Pauli exclusion principle and the quantum statistics of charge. Here, we present theory for the moments of the current in atomic-scale junctions. Moments of the current originate from the charge quantization and hence are sensitive to the quantum statistics of discrete charge. The moments of the current provide powerful tools in probing fundamental transport properties relevant to quantum correlation of steady-state current at zero temperature. For instance, the second moment—shot noise—defines the quantum fluctuations of the current at zero temperature due to the quantization of charge. In the classical limit as electrons in a conductor drift in a completely uncorrelated way as described by a Poissonian distribution, Shot noise reaches $2eI$, where e is the electron charge and I is the average current (the first moment), of current events.

To study the quantum statistics related to the moments of the currents due to quantization of charge, we define a field operator describing propagating electrons,

$$\hat{\Psi} = \sum_{\alpha,E,\mathbf{K}} a^{\alpha}_{E\mathbf{K}}(t)\Psi^{\alpha}_{E\mathbf{K}}(\mathbf{r}), \tag{10.18}$$

where $\alpha = L$ and R; effective single-particle wavefunctions $\Psi^{E\mathbf{K}}_{L(R)}(\mathbf{r})$ describes electrons incident from the left and right electrodes; $a^{L(R)}_{E\mathbf{K}}(t) = \exp(-i\omega t)a^{L(R)}_{E\mathbf{K}}$; and $a^{L(R)}_{E\mathbf{K}}$ is the annihilation operators of electrons incident from the left (right) reservoir, satisfying the anti-commutation relations,

$$\{a^{\alpha}_{E_1\mathbf{K}_1}, a^{\beta\dagger}_{E_2\mathbf{K}_2}\} = \delta_{\alpha\beta}\delta(E_1 - E_2)\delta(\mathbf{K}_1 - \mathbf{K}_2), \tag{10.19}$$

where $\beta = L$ or R.

The expectation value of the product of electron creation and annihilation operator at thermal equilibrium is given by

$$\left\langle a^{\alpha\dagger}_{E_1\mathbf{K}_1}\, a^{\beta}_{E_2\mathbf{K}_2} \right\rangle = \delta_{\alpha\beta}\delta(E_1 - E_2)\ \delta(\mathbf{K}_1 - \mathbf{K}_2) f^{\alpha}_{E}, \tag{10.20}$$

where the statistics of electrons coming from the left (right) electrodes are determined by the equilibrium Fermi–Dirac distribution function $f^{L(R)}_{E} = 1/\{1 + \exp[(E - \mu_{L(R)})/(k_{\mathrm{B}}T)]\}$ in the left (right) reservoir. Note that the Lippmann–Schwinger equation allows wavefunctions of the entire system satisfies the same continuum normalization condition,

$$\int d\mathbf{r}[\Psi^{R(L)}_{E_1\mathbf{K}_1}(\mathbf{r})]^*\nabla\Psi^{R(L)}_{E_2\mathbf{K}_2}(\mathbf{r})=\delta(E_1-E_2)\delta(\mathbf{K}_1-\mathbf{K}_2). \quad (10.21)$$

10.3.1.1 First moment-current

We construct the current operator with the field operator,

$$\hat{I}(z,t)=\frac{e\hbar}{mi}\int d\mathbf{R}\int d\mathbf{K}_1\int d\mathbf{K}_2[\hat{\Psi}^\dagger\nabla\hat{\Psi}-\nabla\hat{\Psi}^\dagger\hat{\Psi}], \quad (10.22)$$

where $d\mathbf{R} = dxdy$ and $\hat{\Psi}$ is the field operator described in Eq. (10.18). Equation (10.22) can be expressed alternatively as,

$$\hat{I}(z,t)=\frac{e\hbar}{m_e i}\sum_{E_1E_2}\sum_{\alpha\beta}\int d\mathbf{R}\int d\mathbf{K}_1\int d\mathbf{K}_2$$
$$.e^{i(E_1-E_2)t/\hbar}a^{\alpha\dagger}_{E_1\mathbf{K}_1}a^{\beta}_{E_2\mathbf{K}_2}I^{\alpha\beta}_{E_1\mathbf{K}_1,E_2\mathbf{K}_2}(\mathbf{r}), \quad (10.23)$$

where $I^{\alpha\beta}_{E_1\mathbf{K}_1,E_2\mathbf{K}_2}(\mathbf{r})=(\Psi^{\alpha}_{E_1\mathbf{K}_1})^*\nabla\Psi^{\beta}_{E_2\mathbf{K}_2}-\nabla(\Psi^{\alpha}_{E_1\mathbf{K}_1})^*\Psi^{\beta}_{E_2\mathbf{K}_2}$.

The average of current at zero temperature is given by the quantum statistics average,

$$<\hat{I}(z)>=\frac{e\hbar}{m_e i}\int_{E_{FL}}^{E_{FR}}dE\int d\mathbf{R}\int d\mathbf{K}I^{R,R}_{E\mathbf{K},E\mathbf{K}}(\mathbf{r}), \quad (10.24)$$

where Eq. (10.20) is applied to calculate the expectation of Eq. (10.23). Equation (10.24) is exactly the same as the current described by the current density of probability. Note that $<\hat{I}(z)>$ is independent of z for a steady-state current. The averaged current also lacks the overlap integral of wavefunctions with different transverse momenta **K**. The above expression for current can be casted into a Landauer–Büttiker formalism:

$$I=\frac{2e}{h}\int dE[f^R_E(\mu_R,T_R)-f^L_E(\mu_L,T_L)]\tau(E), \quad (10.25)$$

where $\tau(E) = \tau^R(E) = \tau^L(E)$ is a direct consequence of the time-reversal symmetry, and $\tau^{R(L)}(E)$ is the transmission function of the electron related to the probability current density of electrons with energy E incident from the right (left) electrode,

$$\tau^{R(L)}(E) = \frac{\pi\hbar^2}{m_e i}\int d\mathbf{K}\int d\mathbf{R}\, I^{RR(LL)}_{E\mathbf{K},E\mathbf{K}}(\mathbf{r})$$
$$= \mathbf{tr}_{\mathbf{K}}\left[\frac{\pi\hbar^2}{m_e i}\int d\mathbf{R}\, I^{RR(LL)}_{E\mathbf{K}_1,E\mathbf{K}_2}(\mathbf{r})\right], \tag{10.26}$$

where $\mathbf{tr}_{\mathbf{K}}$ means $\int d\mathbf{K}$ by letting $\mathbf{K}_1 = \mathbf{K}_2 = \mathbf{K}$. The differential conductance ($V_B = 0$ such that $\tau^L(E) = \tau^R(E) = \tau(E)$), typically not sensitively related to temperatures in cases where direct tunneling is the major transport mechanism, may be expressed as

$$\sigma = \frac{2e^2}{h}\int\left(-\frac{\partial f_E}{\partial E}\right)\tau(E)dE. \tag{10.27}$$

10.3.1.2 Second moment-shot noise

The second moment of the current correlation spectral function is defined as,

$$S_2(\omega,T;z_1,z_2) = 2\pi\hbar\int d(t_1-t_2)e^{i\omega(t_1-t_2)} < \Delta\hat{I}(z_1,t_1)\Delta\hat{I}(z_2,t_2)>, \tag{10.28}$$

where $\Delta\hat{I}(t) = \hat{I}(t) - \langle\hat{I}\rangle$. Carrying out the average of autocorrelation of the current operators requires the quantum statistical expectation value of the products of $a^{i\dagger}_{E_1}a^{j}_{E_2}a^{k\dagger}_{E_3}a^{l}_{E_4}$. For a Fermi electron gas at equilibrium, the quantum statistical expectation values of products of four operators are given by

$$\left\langle a^{i\dagger}_{E_1\mathbf{K}_1}a^{j}_{E_2\mathbf{K}_2}a^{k\dagger}_{E_3\mathbf{K}_3}a^{l}_{E_4\mathbf{K}_4}\right\rangle = \left\langle a^{i\dagger}_{E_1\mathbf{K}_1}a^{j}_{E_2\mathbf{K}_2}\right\rangle\left\langle a^{k\dagger}_{E_3\mathbf{K}_3}a^{l}_{E_4\mathbf{K}_4}\right\rangle$$
$$-\left\langle a^{i\dagger}_{E_1\mathbf{K}_1}a^{l}_{E_4\mathbf{K}_4}\right\rangle\left\langle a^{j}_{E_2\mathbf{K}_2}a^{k\dagger}_{E_3\mathbf{K}_3}\right\rangle. \tag{10.29}$$

Applying Eq. (10.29) to Eq. (10.28), the shot noise spectral function is given by

$$S_2(\omega,T;z_1,z_2) = 2\pi\hbar\left(\frac{e\hbar}{m_e i}\right)^2\sum_{\alpha,\beta=L,R}\int dE f^{\alpha}_{E+\hbar\omega}(1-f^{\beta}_E)$$
$$\int d\mathbf{K}_1\int d\mathbf{K}_2\int d\mathbf{R}_1\int d\mathbf{R}_2\tilde{I}^{\alpha\beta}_{E+\hbar\omega\mathbf{K}_1,E\mathbf{K}_2}(\mathbf{r}_1)\tilde{I}^{\beta\alpha}_{E\mathbf{K}_1,E+\hbar\omega\mathbf{K}_2}(\mathbf{r}_2), \tag{10.30}$$

where $I^{\alpha\beta}_{E+\hbar\omega\mathbf{K}_1,E\mathbf{K}_2}(\mathbf{r}) = [I^{\beta\alpha}_{E\mathbf{K}_1,E+\hbar\omega\mathbf{K}_2}(\mathbf{r}_2)]^*$.

Shot noise, the second moment of the steady current, is the quantum fluctuation of current autocorrelation at zero temperature due to the charge quantization. For a DC steady current, the shot noise is given by the noise spectral function at $\omega = 0$ and $T = 0$ K [i.e., $S_2 = S_2(\omega = 0, T = 0)$], which gives

$$S_2(z_1, z_2) = 2\pi\hbar\left(\frac{e\hbar}{m_e i}\right)^2 \int_{E_{FL}}^{E_{FR}} dE \int d\mathbf{R}_1 \int d\mathbf{R}_2 \int d\mathbf{K}_1 \int d\mathbf{K}_2 \cdot \tilde{I}^{RL}_{E\mathbf{K}_1, E\mathbf{K}_2}(\mathbf{r}_1) \tilde{I}^{LR}_{E\mathbf{K}_2, E\mathbf{K}_1}(\mathbf{r}_2). \quad (10.31)$$

Equation (10.31) leads to the relationship $S_2^{RR} = S_2^{LL} = -S_2^{LR} = -S_2^{RL}$, where $S_2^{RR} = S_2(z_1 \to \infty, z_2 \to \infty)$, $S_2^{LL} = S_2(z_1 \to \infty, z_2 \to \infty)$, $S_2^{LR} = S_2(z_1 \to -\infty, z_2 \to \infty)$, and $S_2^{RL} = S_2(z_1 \to \infty, z_2 \to -\infty)$ all of which are consequences of current conservation.

10.3.1.3 Third moment-counting statistics

The third moment of the current, counting statics, is defined in the time-unordered way,[1]

$$S_3(\omega, \omega'; z_1, z_2, z_3) = \int d(t_1 - t_3) \int d(t_2 - t_3) e^{i\omega(t_1 - t_3)} e^{i\omega'(t_2 - t_3)} < \Delta\hat{I}(z_1, t_1) \Delta\hat{I}(z_2, t_2) \Delta\hat{I}(z_3, t_3) >, \quad (10.32)$$

where $\Delta\hat{I}(t) = \hat{I}(t) - \langle\hat{I}\rangle$. The spectral function of the third moment is given by

$$\begin{aligned} S_3(\omega, \omega'; z_1, z_2, z_2) &= (2\pi\hbar)^2 \left(\frac{e\hbar}{mi}\right)^3 \int_{E_{FL}}^{E_{FR}} dE \int d\mathbf{R}_1 \int d\mathbf{R}_2 \int d\mathbf{R}_3 \int d\mathbf{K}_1 \int d\mathbf{K}_2 \int d\mathbf{K}_3 \\ &\cdot \sum_{\alpha,\beta,\gamma = L,R} [f_E^\alpha (1 - f_{E+\hbar\omega}^\beta)(1 - f_{E+\hbar\omega+\hbar\omega'}^\gamma) \tilde{I}^{\alpha\beta}_{E\mathbf{K}_1(E+\hbar\omega)\mathbf{K}_2}(\mathbf{r}_1) \\ &\cdot \tilde{I}^{\beta\gamma}_{(E+\hbar\omega)\mathbf{K}_2(E+\hbar\omega+\hbar\omega')\mathbf{K}_3}(\mathbf{r}_2) \tilde{I}^{\gamma\alpha}_{(E+\hbar\omega+\hbar\omega')\mathbf{K}_3 E\mathbf{K}_1}(\mathbf{r}_3) \\ &- f_E^\alpha f_{E-\hbar\omega}^\beta (1 - f_{E+\hbar\omega}^\gamma) \tilde{I}^{\alpha\gamma}_{E\mathbf{K}_1(E+\hbar\omega)\mathbf{K}_2}(\mathbf{r}_1) \\ &\cdot \tilde{I}^{\beta\alpha}_{(E-\hbar\omega')\mathbf{K}_2 E\mathbf{K}_3}(\mathbf{r}_2) \tilde{I}^{\gamma\beta}_{(E+\hbar\omega)\mathbf{K}_3(E-\hbar\omega')\mathbf{K}_1}(\mathbf{r}_3)], \end{aligned} \quad (10.33)$$

[1]This does not effect the second moment, however, it is important for the third and higher moments.

where the quantum statistic expectation value of $\langle\Delta\hat{I}(z_1,t_1)\Delta\hat{I}(z_2,t_2)\Delta\hat{I}(z_3,t_3)\rangle$ is evaluated using the Wick–Bloch–De Dominicis theorem,

$$< \hat{A}_n \hat{A}_{n-1} \ldots \hat{A}_1 >$$

$$= \begin{cases} 0, \text{ for } n = \text{odd}, \\ \sum_{m=1}^{n-1} \eta^{n-m-1} < \hat{A}_n \hat{A}_m >< \hat{A}_{n-1} \ldots \hat{A}_{m+1} \hat{A}_{m-1} \ldots \hat{A}_1 >, \\ \text{for } n = \text{even}, \end{cases} \quad (10.34)$$

where $\hat{A}_i$ denotes either creation or annihilation operators and $\eta = -1(1)$ for Fermions (Bosons). Similarly, the third moment of the steady-state current at zero temperature and zero-frequency [defined as $S_3 = S_3(\omega = 0, \omega' = 0, T = 0)$] is given by

$$S_3(z_1, z_2, z_2)$$
$$= (2\pi\hbar)^2 \left(\frac{e\hbar}{m_e i}\right)^3 \int_{E_{FL}}^{E_{FR}} dE \int d\mathbf{R}_1 \int d\mathbf{R}_2 \int d\mathbf{R}_3$$
$$[\tilde{I}^{RL}_{E\mathbf{K}_1,E\mathbf{K}_2}(\mathbf{r}_1)\tilde{I}^{LL}_{E\mathbf{K}_2,E\mathbf{K}_3}(\mathbf{r}_2)\tilde{I}^{LR}_{E\mathbf{K}_3,E\mathbf{K}_1}(\mathbf{r}_3)$$
$$- \tilde{I}^{RL}_{E\mathbf{K}_1,E\mathbf{K}_3}(\mathbf{r}_1)\tilde{I}^{RR}_{E\mathbf{K}_2,E\mathbf{K}_1}(\mathbf{r}_2)\tilde{I}^{LR}_{E\mathbf{K}_3,E\mathbf{K}_2}(\mathbf{r}_3)]. \quad (10.35)$$

We note that alternative definitions of the Fourier transform, e.g., the integrals with respect to $(t_1 - t_2)$ and $(t_2 - t_3)$ in Eq. (10.32) will lead to different parametrization of frequencies ω and ω' in the spectral function [Eq. (10.33)]. However, for the steady-state current where $\omega = \omega' = 0$, the third moment at zero temperature and zero frequency is independent of the choices of the Fourier transform. Similar to the relationship $S_2^{RR} = S_2^{LL} = -S_2^{LR} = -S_2^{RL}$, for shot noise, the third moment of the current [Eq. (10.35)] leads to

$$S_3^{R(R,L)R} = S_3^{L(R,L)L} = -S_3^{L(R,L)R} = -S_3^{R(R,L)L}, \quad (10.36)$$

all of which are consequences of current conservation.

The conductance, shot noise, and counting statistics are closely related to the electronic structure of nanojunctions. For example, the conductance, the second-, and the third-moments of the current exhibit odd-even oscillation with the number of carbon atoms due to the full and half filled π^* orbital near the Fermi levels, as reported in Ref. [12].

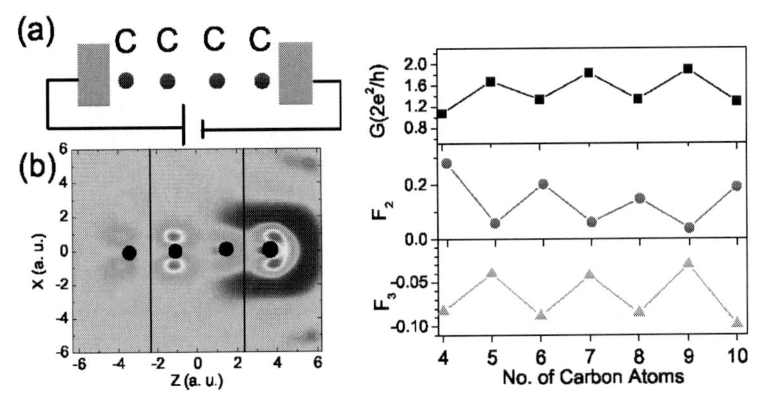

Figure 10.7 Left panel: (a) Schematic of the four-carbon atomic junction. (b) The special distribution of partial charge density for electrons with energies near the Fermi levels at V_B = 0.01 V. Vertical black lines correspond to the edges of the jellium model and circles correspond to atomic position. Right panel: Conductance [(black) square; top panel], the second-order Fano factor [(red) circle; middle panel], and the third-order Fano factor [(green) triangle; bottom panel] of the atomic wires as functions of the number of carbon atoms in the wire at V_B = 0.01 V. Reprinted from Ref. [12].

10.3.2 Current-Induced Forces

Current-induced forces can be decomposed into two components: the direct force and the wind force [2]. The direct force is the electrostatic force acting on the core electron charge due to the electrostatic field induced by the voltage applied across the junction, whereas the wind force is due to the current-carrying electronic states that cause the net momentum transfer from electrons scattered by the ions. The wind force is proportional to the rate of momentum transfer from the electrons to the ions. In microelectronics, this phenomenon is known as electro-migration, which is one of the major mechanisms that frustrate the metallic contacts. The nanojunction may carry much larger current density than the macroscopic counterparts. Consequently, the current-induced forces are of key importance to understand the mechanical instability of nanojunction.

The calculations considering the forces that are applied on the benzene molecule induced by the steady-state current follow the theory developed by Di Ventra, Pantelides, and Lang [13, 14]. It is well known that Hellmann-Feynman theory is the key ingredient of the quantum mechanical treatment of forces acting on nuclei in molecules and solids. The Hellmann-Feynman theory is widely applied for structure optimization in material physics and chemistry. In such calculations, electrons in the electron system are in their instantaneous ground states. However, the conventional Hellmann-Feynman theory fails to apply to molecules and solids in time-dependent external fields and in transport in nanostructures and molecules. So, M. Di Ventra and S. T. Pantelides rigorously derived a general form of the Hellmann-Feynman-like theory considering a fully dynamical description of the many-body problem comprising electrons and ions in the classical limit of ionic motion. According to the above theory, the Pulay terms that appear in Hellmann-Feynman theory vanish in the case of steady-state current [15, 16].

Wavefunctions of the continuum states and localized states are applied to calculate the current-induced forces. With the effective single-particle wavefunctions for transport electrons obtained in the density functional theory in the Lippmann–Schwinger equation, the Pulay-like terms vanish, and the force **F** acting on a given atom at position **R** due to the non-equilibrium current flow at finite biases is given by

$$\mathbf{F} = -\sum_i \left\langle \psi_i \left| \frac{\partial H}{\partial \mathbf{R}} \right| \psi_i \right\rangle - \lim_{\Delta \to 0} \int_\sigma dE \left\langle \psi_\Delta \left| \frac{\partial H}{\partial \mathbf{R}} \right| \psi_\Delta \right\rangle, \tag{10.37}$$

where the first term on the right hand side of Eq. (10.37) is the Hellmann-Feynman contribution to the force due to localized electronic states $|\psi_i\rangle$. The second term is the contribution to the force from the continuum states $|\psi_\Delta\rangle$. The wavefunctions $|\psi_\Delta\rangle$ are the eigendifferentials for each energy interval Δ in the continuum σ:

$$\psi_\Delta = A \int_\Delta dE \psi, \tag{10.38}$$

where A is a normalization constant and ψ's are the effective single-particle wavefunctions in the continuum. In other words, eigendifferentials represent wavefunctions in the continuum space that are square-integrable in the Hilbert space.

Though the current-induced force causes instability in nanojunctions similar to the role of electron migration force in semiconductor device system. However, current-induced force could possibly be engineered to drive rotational motion of an asymmetric single-molecule junction, much like the way a steam of water rotates a waterwheel. It offers a new type of electrical-controllable single-molecule motor driven by the current-induced torque, as reported in Ref. [17].

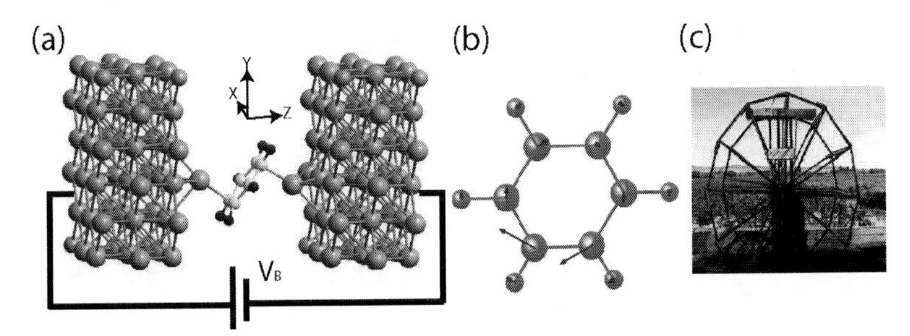

Figure 10.8 (a) Schematic of the Pt/benzene junction as a single-molecule motor; (b) projection of the force vectors on the plane of the benzene molecule; (c) the net torque induced by the curved flow of current streamline tends to rotate the benzene molecule, similar to the way a stream of water rotates a waterwheel. Reprinted from Ref. [17].

10.3.3 Thermoelectric Properties

Research on thermopower is important to understand the renewable energy system that converts waste heat to useful electric power. Thermopower (also called the Seebeck coefficient) describes a thermoelectric phenomenon by which thermal energy is converted into an electric current via a temperature difference. The Seebeck coefficient, defined as dV/dT, where dV is the voltage difference caused by the temperature difference dT, is the most important physical quantity for characterizing thermoelectric properties. In recent years, single-molecule thermoelectric junctions have attracted many theoretical and experimental attentions, thereby providing new opportunities and challenges for exploring the nanoscale renewable energy system. The Seebeck coefficient is related not only to the magnitude but also to the slope of the transmission function in the vicinity of the chemical potentials. Thus, the Seebeck coefficient is superior over the current-voltage characteristic in exploring of the details of electronic structures in single-molecule junctions.

Thermoelectric nanojunctions consist of a nanostructured object sandwiched between source-drain electrodes serving as independent electron reservoirs with distinct chemical potentials $\mu_L(R)$ and independent temperature reservoirs with distinct temperatures $T_L(R)$. The thermoelectric figure of merit ZT depends on the following several physical factors: the Seebeck coefficient (S), the electric conductance (σ), the electronic heat conductance (κ_{el}), and the phononic thermal conductance (κ_{ph}). The thermoelectric efficiency in the nanoscale junctions can thus be described by the dimensionless thermoelectric figure of merit presented as,

$$\text{ZT} = \frac{S^2\sigma}{k_{\text{el}} + k_{\text{ph}}} T, \tag{10.39}$$

where T is the average temperature in the source-drain electrodes. When ZT tends to infinity, the thermoelectric efficiency of nanojunctions will reach the Carnot efficiency. To obtain a large ZT value, the thermoelectric nanojunction would need to have a large value of S, a large value of σ and a small value of the combined heat conductance ($\kappa_{\text{el}} + \kappa_{\text{ph}}$). Thermoelectric devices with a large value of σ are usually accompanied by a large value of κ_{el}, owing to the same proportionality with the transmission function. In case that $\kappa_{\text{el}} = \kappa_{\text{ph}}$, the cancelation between σ and κ_{el} makes the enhancement of the thermoelectric figure of merit ZT quite a challenging task. Although considerable efforts have been exerted to understand the inelastic effects on single-molecular junctions, only a few attempts have been made to study the inelastic effects on the thermoelectric properties of the single-molecule junctions.

10.3.3.1 Seebeck coefficient

The Seebeck coefficients are relevant not only to the magnitude but also to the slope of density of states (DOSs). They can reveal more detailed information about the electronic structures of the materials sandwiched in the nanojunctions beyond what the conductance measurements can provide. Similarly, the Seebeck coefficient for bulk material is also described by a single Fermi level. Nevertheless, molecular tunneling junctions consist of two electrodes as independent electron and heat reservoirs. Thus, it is worthwhile extending the investigation of the Seebeck coefficients to a system with distinct temperatures ($T_{L(R)}$) and chemical potentials ($\mu_{L(R)}$) in the left (right) electrode. We present a theory for two distinct Fermi levels in molecular tunneling junction combining a first-principles approach for the Seebeck coefficient in the two- and three-terminal junctions in non-linear regime [18, 19].

We describe the method for the calculation of the Seebeck coefficient. The left (right) electrode serves as the electron and thermal reservoir with the electron population described by the Fermi–Dirac distribution function. An extra current is induced by an additional infinitesimal temperature (ΔT) and voltage (ΔV) distributed symmetrically across the junctions,

$$\Delta I = I\left(\mu_L, T_L + \frac{\Delta T}{2}; \mu_R, T_R - \frac{\Delta T}{2}\right) + I\left(\mu_L + \frac{e\Delta V}{2}, T_L; \mu_R - \frac{e\Delta V}{2}, T_R\right) - 2I(\mu_L, T_L; \mu_R, T_R), \quad (10.40)$$

After expanding the Fermi–Dirac distribution function to the first order in ΔT and ΔV, we obtain the Seebeck coefficient (defined by $S = \Delta V/\Delta T$) by letting $\Delta I = 0$,

$$S = -\frac{1}{e}\frac{\frac{K_1^L}{T_L} + \frac{K_1^R}{T_R}}{K_0^L + K_0^R}, \quad (10.41)$$

where $K_n^{L(R)} = -\int dE(E - \mu_{L(R)})^n \frac{\partial f_E^{L(R)}}{dE}\tau(E)$, and $\tau(E) = \tau^R(E) = \tau^L(E)$. In this paper, we focus on the zero bias regime with two electrodes at the same temperatures ($\mu_L = \mu_R = \mu$; $T_L = T_R = T$), Eq. (10.41) above can be simplified as,

$$S = \frac{1}{eT}\frac{\int (E-\mu)\frac{\partial f_E}{\partial E}\tau(E)dE}{\int \frac{\partial f_E}{\partial E}\tau(E)dE}. \tag{10.42}$$

In the linear-response and low-temperature regime, the plot of the Seebeck coefficient vs. the temperature is linear. The magnitude of the Seebeck coefficient is proportional to the slope of the transmission function, and inversely proportional to the magnitude of the transmission function at the chemical potential

$$S \approx -\frac{\pi^2 k_B^2}{3e\tau(\mu)}\frac{\partial \tau(\mu)}{\partial E}T. \tag{10.43}$$

Equation (10.42) indicates that the Seebeck coefficient is positive (negative) when the slope of the transmission function is negative (positive). Seebeck coefficient is related to the slope of DOSs, and thus a more sensitive tool to explore the electronic structures of nanojunctions. The electronic structures of nanojunctions are sensitive to the contact geometry between the nanostructure and electrodes.

Recent experiments revealed that surface reconstruction occurs at around 300–400 K in the interface of C_{60} adsorbed on Cu(111) substrate by scanning tunneling microscope (STM) techniques [20]. To understand effects of such reconstruction on thermopower, we investigate the Seebeck coefficients of C_{60} single-molecular junctions without and with surface reconstruction as a function of temperature at different tip-to-molecule heights d from first-principles. The calculations show that a phase transition in the temperature profile of the Seebeck coefficient can happen due to surface reconstruction at high temperature in the C_{60} STM junction [21]. The temperature can be used to manipulate Seebeck coefficients via surface reconstruction by in situ STM experiments. Surface reconstruction may cause a phase transition in the temperature profile of the Seebeck coefficient.

10.3.3.2 Thermal current carried by electrons

There is an analog between the electric current and electron's thermal current. Electric current is calculated according to the probability current. The flow of electrons can also carry energy current. The electron's thermal current, defined as the rate at which thermal energy flows from the right (in to the left) electrode via the probability current, is

$$J_{\text{el}}^{R(L)}(\mu_L,T_L;\mu_R,T_R)$$
$$=\frac{2}{h}\int dE(E-\mu_{R(L)})\tau(E)[f_E(\mu_R,T_R)-f_E(\mu_L,T_L)]. \tag{10.44}$$

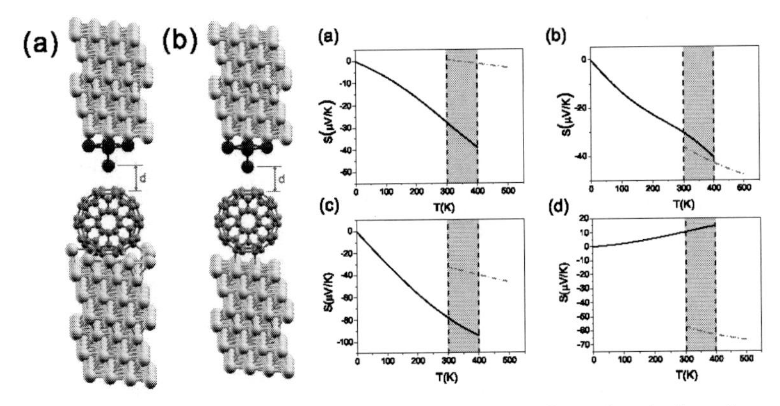

Figure 10.9 Left Panel: The schematics of C_{60} single-molecule junctions in which the C_{60} molecule (bronze) is adsorbed on the Cu(111) surface (yellow) with an STM tip composed of tungsten atoms (blue) (a) with and (b) without surface reconstruction on Cu(111) substrate. d is the tip height. Recent scanning tunneling microscope (STM) experiments reported that interface reconstruction occurs at around 300–400 K for the C_{60} molecule adsorbed on the Cu(111) surface. Surface reconstruction may cause a phase transition in the temperature profile of the Seebeck coefficient. Right Panel: Seebeck coefficients of C_{60} single–molecule junctions without (solid black line) and with (red dashed-dot line) surface reconstruction as a function of the temperature T for various tip heights d = (a) 2.1 Å, (b) 2.3 Å, (c) 2.77 Å, and (d) 3.4 Å. The shaded areas represent the range of temperatures within which surface reconstruction is expected to occur. The left (right) of the shaded areas denote the well-defined unreconstructed (reconstructed) phase). Reprinted from Ref. [21].

The electron's thermal current is the energy current carried by electrons traveling between electrodes driven by dT and dV.

Analogous to the extra current given by Eq. (10.40), the extra electron's thermal current is [22]

$$dJ_{\text{el}} = (dJ_{\text{el}})_T + (dJ_{\text{el}})_V. \tag{10.45}$$

where $(dJ_{\text{el}})_T = J_{\text{el}}(\mu,T;\ \mu,\ T + dT)$ and $(dJ_{\text{el}})_V = J_{\text{el}}(\mu,\ T;\ \mu + edV,\ T)$ are the fractions of electron's thermal current driven by dT and dV, respectively. Note that dV is generated by the Seebeck effect according to the temperature difference dT. Both J_{el} $(\mu,\ T;\ \mu,\ T + dT)$ and $J_{\text{el}}(\mu,\ T;\ \mu + edV,\ T)$ can be calculated using Eq. (10.44).

Given that we define the electron's thermal conductance as $\kappa_{\text{el}} = dJ_{\text{el}}/dT$, the electron's thermal conductance κ_{el} can decomposed into two components:

We obtain the electronic heat conductance (defined by $\kappa_{\text{el}} = \Delta J_{\text{el}}/\Delta T$) after expanding the Fermi–Dirac distribution function to the first order in ΔT and ΔV expressed by:

$$\kappa_{\text{el}} = \frac{2}{h}\left[K_1 eS + \frac{K_2}{T}\right], \tag{10.46}$$

where $K_n = -\int dE(E-\mu)^n \frac{\partial f_E}{\partial E}\tau(E)$. In the low-temperature regime, κ_{el}^T dominates the electron's thermal current. Thus $\kappa_{\text{el}} \approx \kappa_{\text{el}}^T$ is linear in T, that is,

$$\kappa_{\text{el}}(\mu,T) \approx \beta_T T. \tag{10.47}$$

10.3.3.3 Thermal current carried by phonons

It is assumed that the nanojunction is a weak elastic link, with a given stiffness K, that we evaluate from total energy calculations. Two metal electrodes are regarded as the macroscopic bodies under their thermodynamic equilibrium, and are taken as ideal thermal conductors. To leading order in the strength of the weak link, the mechanical link is modeled by a harmonic spring. We then estimate the phononic heat current (J_Q^{ph}) via elastic phonon scattering as [23]

$$J_Q^{\text{ph}} = \frac{2\pi K^2}{\hbar}\int_0^\infty dEEN_L(E)N_R(E)[n_L(E) - n_R(E)], \tag{10.48}$$

where the stiffness $K = AY/l$ is evaluated from the first-principles. The symbol Y is Young's modulus of the junction, and $A(l)$ is the cross section (length) of the nanostructured object sandwiched between electrodes. $N_{L(R)}(E) \simeq CE$, where C is a constant, is the spectral density of phonon states at the left (right) electrode surface. Here, we have assumed that the metal electrodes are thermodynamically in equilibrium as described by the Bose–Einstein distribution function $n_{L(R)} \equiv 1/(e^{E/K_B T_{L(R)}} - 1)$. The rate of thermal energy carried by phonons flowing between two bulk electrodes joined by a nanostructured object is appropriately considered in the weak link model, which is valid for temperatures lower than the Debye temperatures. The range of temperatures in the current study lies within this region (The Debye temperature is 394 K for Al and 170 K for Au, respectively). The phononic heat conductance is defined as

$$\kappa_{\text{ph}} = \left[J_Q^{\text{ph}}\left(T_L + \frac{\Delta T}{2}, T_R - \frac{\Delta T}{2}\right) - J_Q^{\text{ph}}(T_L, T_R) \right] / \Delta T. \tag{10.49}$$

After expanding the Bose–Einstein distribution function in the left (right) electrode to the first order of ΔT in the expression of the phononic heat current, the phononic heat conductance is thus obtained:

$$\kappa_{\text{ph}} = \frac{\pi K^2 C^2}{\hbar} \int_0^{\infty} dE E^3 \sum_{i = L,R} \frac{\partial n_i(E)}{\partial T_i}, \tag{10.50}$$

when $T_R \approx T_L = T$, then Eq. (10.50) can be written as

$$\kappa_{\text{ph}} = \frac{2\pi K^2 C^2}{\hbar} \int_0^{\infty} dE E^3 \frac{\partial n(E)}{\partial T}, \tag{10.51}$$

where $n = 1/[e^{E/(k_B T)} - 1]$. Finally, when $\mu_L \approx \mu_R = \mu$ and $T_R \approx T_L = T$, the thermoelectric figure of merit ZT $= S^2\sigma T/(\kappa_{\text{el}} + \kappa_{\text{ph}})$ can be calculated using Eqs. (10.27), (10.42), (10.46), and (10.51).

10.3.3.4 Thermoelectric figure of merit (ZT)

To gain further insights into the qualitative description of ZT on the dependence of the temperatures and lengths, we expand the Seebeck coefficient S, the electronic heat conductance κ_{el}, and the phononic heat conductance κ_{ph} to their lowest order in temperatures. The above quantities have the following power law expansions:

$$S \approx \alpha T, \tag{10.52}$$

$$\kappa_{\text{el}} \approx \beta T, \tag{10.53}$$

and

$$\kappa_{\text{ph}} \approx \gamma(l) T^3, \tag{10.54}$$

where $\alpha = -\pi^2 k_\text{B}^2 \dfrac{\partial \tau(\mu)}{\partial E} / (3e\tau(\mu))$, $\beta = 2\pi^2 k_\text{B}^2 \tau(\mu)/(3h)$, and $\gamma(l) = 8\pi^5 k_\text{B}^4 C^2 A^2 Y^2 / (15\hbar l^2)$. Here $\tau(E)$ is the transmission function given by Eq. (10.26); μ is the chemical potential; Y is Young's modulus of the junction; $A(l)$ is the cross section (length) of the nanostructured object sandwiched between electrodes; and C is a constant, which is given by the spectral density of phonon states at the left (right) electrode surface: $N_{L(R)}(E) \simeq CE$. Consequently, the thermoelectric figure of merit, $\text{ZT} = S^2\sigma T/(\kappa_{\text{el}} + \kappa_{\text{ph}})$, in the nanojunctions has a simple form,

$$\text{ZT} \approx \frac{\alpha^2 \sigma T^3}{\beta T + \gamma(l) T^3} \rightarrow \begin{cases} (\alpha^2\sigma/\beta)T^2, & \text{for } T \ll T_0, \\ \alpha^2\sigma/\gamma(l), & \text{for } T \gg T_0, \end{cases} \tag{10.55}$$

which is valid in when $T_R \approx T_L = T$ and $V_\text{B} = 0$ V.

We observe a characteristic temperature for ZT in nanojunction: $T_0 = \sqrt{\beta/\gamma(l)}$. When $T << T_0$, the electronic heat current dominates the combined heat current and ZT $\propto T^2$. When $T >> T_0$, the phononic heat current dominates the combined heat current and ZT tends to a saturation value. Moreover, the metallic atomic junctions and the insulating molecular junctions have opposite trend for the dependence of ZT on lengths, that is, ZT increases as the length increases for aluminum atomic junctions, while ZT decreases as the length increases for alkanethiol molecular junctions, as reported in Ref. [19].

10.4 Vibronic Effects

When electrons travel in nanojunctions under a large bias voltage, electrons with large enough energy may hit the atoms in nanostructure, causing the vibrations of nuclei. Such inelastic scattering effects causes small structure at biases $V_B = \hbar\omega_j/e$, at which the electrons have energy large enough to excite the normal mode vibration of molecules. Not all the modes have important contributions to the inelastic profiles. Inelastic scattering occurs when eV_B is larger than the normal mode of smallest energy. The magnitudes of the small structures in the inelastic profiles significantly vary from mode to mode. To understand the rule of mode selection, we investigate how the current density flows through the nanojunction. We observe that the mode selection rule can be connected to the current density and the normal mode vibrations. The important modes in the inelastic profile are characterized by large components of atomic vibrations along the direction of the current density on atoms. We start with the introduction of the Hamiltonian of electron–vibration interactions.

10.4.1 Hamiltonian of Electron–Vibration Interactions

To consider the electron–vibration interactions, we start with the Hamiltonian

$$H = H_{\text{el}} + H_{\text{ion}} + H_{\text{el-ion}}. \tag{10.56}$$

where H_{el} is the electronic part of Hamiltonian,

$$H_{\text{el}} = \sum_i \frac{\mathbf{p}_i^2}{2m_e} + \sum_{i<j} \frac{e^2}{|\mathbf{r}_i - \mathbf{r}_j|}, \tag{10.57}$$

where m_e is the mass of electron, $\mathbf{p}_i$ is the momentum of the i-th electron, and $\mathbf{r}_i$ is the position of the i-th electron.

In Eq. (10.56), H_{ion} is the ionic part of Hamiltonian,

$$H_{\text{ion}} = \sum_i \frac{\mathbf{P}_i^2}{2M_i} + \sum_{i<j} V_{\text{ion}}(\mathbf{R}_i - \mathbf{R}_j), \tag{10.58}$$

where M_i is the mass of the i-th ion, $\mathbf{P}_i$ is the momentum of i-th ion, $\mathbf{R}_i$ is the position of the i-th ion, and $V_{\text{ion}}(\mathbf{R}_i - \mathbf{R}_j)$ is the interaction between the i-th and j–th ions.

$H_{\text{el-ion}}$ describes the interaction between the i-th electron and j-th ions,

$$H_{\text{el-ion}} = \sum_{i,j} V_{ei}(\mathbf{r}_i - \mathbf{R}_j). \tag{10.59}$$

In the previous section, the vibronic coupling is neglected. Only the electronic Schrödinger equation is considered within the Born-Oppenheimer approximation. This problem has been solved when the ions are located in their equilibrium position, i.e.,

$$H^{\text{eq}} = H_{\text{el}} + H_{\text{ion}}^{\text{eq}} + H_{\text{el-ion}}^{\text{eq}}, \tag{10.60}$$

where

$$H_{\text{ion}}^{\text{eq}} = \sum_i \frac{\mathbf{P}_i^2}{2M_i} + \sum_{i<j} V_{\text{ion}}(\mathbf{R}_i^0 - \mathbf{R}_j^0), \tag{10.61}$$

where $\mathbf{R}_i^0$ is the position of the i-th ion at its equilibrium position. $H_{\text{el-ion}}^{\text{eq}}$ describes the interaction between the i-th electron and j–th ions at their equilibrium position,

$$H_{\text{el-ion}}^{\text{eq}} = \sum_{i,j} V_{ei}(\mathbf{r}_i - \mathbf{R}_j^0). \tag{10.62}$$

For atoms at their equilibrium position, the current carrying electron wavefunctions have been solved in the framework of density functional theory in scattering approaches within the Born-Oppenheimer approximation, as described in Section 10.2. The electron–ion interaction $V_{ei}(\mathbf{r}_i - \mathbf{R}_j^0)$ is considered at the level of pseudopotential.

To consider the vibronic coupling, we start with considering small oscillations of ions,

$$\mathbf{R}_i = \mathbf{R}_i^0 + \mathbf{Q}_i, \tag{10.63}$$

where $\mathbf{Q}_i$ is a small deviation of position away from the equilibrium position for the i-th ion represented in the Cartesian coordinate system. Therefore,

$$\begin{aligned} V_{\text{ion}}(\mathbf{R}_i - \mathbf{R}_j) &= V_{\text{ion}}(\mathbf{R}_i^0 - \mathbf{R}_j^0 - (\mathbf{Q}_i - \mathbf{Q}_j)) \\ &\approx V_{\text{ion}}(\mathbf{R}_i^0 - \mathbf{R}_j^0) + H_{\text{osc}}, \end{aligned} \tag{10.64}$$

where the oscillatory part of ions H_{osc} is

$$H_{\text{osc}} = \frac{1}{2}\sum_{i<j}\sum_{\mu,\nu=x,y,z} \mathbf{F}_{\mu\nu}(\mathbf{Q}_i - \mathbf{Q}_j)_\mu(\mathbf{Q}_i - \mathbf{Q}_j)_\nu \mathbf{F}_{\mu\nu}, \tag{10.65}$$

$$= \frac{\partial^2}{\partial \mathbf{R}_\mu \partial \mathbf{R}_\nu} V_{\text{ion}}(\mathbf{R}_i - \mathbf{R}_j)|_{\mathbf{R}_i=\mathbf{R}_i^0;\ \mathbf{R}_j=\mathbf{R}_j^0}.$$

Oscillations of ions can be mapped into a set of independent simple harmonic oscillators via normal coordinates $\{q_j\}$, i.e., $(Q_i)_\mu = \Sigma_j A_{i\mu,j}q_j$. The oscillatory part of ions H_{osc} is diagonalized and have the form of

$$H_{\text{osc}} = \frac{1}{2}\sum_i \dot{q}_i\dot{q}_i + \frac{1}{2}\sum_i \omega_i q_i q_i, \tag{10.66}$$

where ω_i is the frequency of the i-th normal mode. One can introduce a canonical transformation, $\tilde{Q}_i = \sqrt{\frac{\omega_i}{2\hbar}}q_i$ and $\tilde{P}_i = \sqrt{\frac{\omega_i}{2\hbar}}\dot{q}_i$, which transforms H_{osc} to

$$H_{\text{osc}} = \frac{1}{2}\sum_i \hbar\omega_i(\tilde{P}_i^2 + \tilde{Q}_i^2), \tag{10.67}$$

where one has $[\tilde{Q}_j, \tilde{P}_k] = \frac{i}{2}\delta_{jk}$. Using $b_i = \tilde{Q}_i + i\tilde{P}_i$ and $b_i^\dagger = \tilde{Q}_i - i\tilde{P}_i$, the oscillatory part Hamiltonian can be second quantized by phonon creation and annihilation operators and become a set of independent simple harmonic oscillators,

$$H_{\text{osc}} = \frac{1}{2}\sum_i \hbar\omega_j\left(b_j^\dagger b_j + \frac{1}{2}\right). \tag{10.68}$$

Next, we expand the electron–vibration interactions in terms of lowest order in small oscillation, $H_{\text{el-ion}} = H_{\text{el-ion}}^{\text{eq}} + \delta H_{\text{el-ion}}$, where we consider the interaction between electron and ion at the level of non-local pseudopotential,

$$\Delta H_{\text{el-ion}} = \sum_{i,\mu}(\mathbf{Q}_i)_\mu \cdot \frac{\partial}{\partial \mu}V^{ps}(\mathbf{r}, \mathbf{R}_i^0), \tag{10.69}$$

where $\frac{\partial}{\partial \mu} = \frac{\partial}{\partial \mathbf{R}_\mu^0}$ is the derivative with respect to the position of the i-th ion in $\mu = x, y, z$ direction. Using $(Q_i)_\mu = \sum_j A_{i\mu,j}q_j = \sum_j A_{i\mu,j}\sqrt{\frac{\hbar}{2M_i\omega_j}}(b_j + b_j^\dagger), \Delta H_{\text{el-ion}}$ becomes

$$\Delta H_{\text{el-ion}} = \sum_j \sum_i \sum_\mu \sqrt{\frac{\hbar}{2M_i\omega_j}} A_{i\mu,j}(b_j + b_j^\dagger)\frac{\partial}{\partial\mu} V^{ps}(\mathbf{r}, \mathbf{R}_i^0), \tag{10.70}$$

where we have put back the mass of the i-th ions M_i and note that the orthonormal conditions for the canonical transformation between normal coordinates and Cartesian coordinates: $\Sigma_{i,\mu} A_{i\mu,j'}A_{i\mu,j'} = \delta_{j,j'}$. The vibronic coupling can be second quantized by apply the field operator, Eq. (10.18), to Eq. (10.70):

$$H_{\text{el-vib}} = \int \hat{\Psi}^\dagger(\mathbf{r})(\Delta H_{\text{el-ion}})\hat{\Psi}^\dagger(\mathbf{r})d\mathbf{r}. \tag{10.71}$$

Finally, the many-body Hamiltonian of the system under consideration become $H = H_{\text{el}} + H_{\text{vib}} + H_{\text{el-vib}}$, where H_{el} is the electronic part of the Hamiltonian under adiabatic approximations and H_{vib} is the ionic part of the Hamiltonian, which can be casted into a set of independent simple harmonic oscillators via a canonical transformation. $H_{\text{el-vib}}$ is a part of the Hamiltonian for electron–vibration interactions which has the form of,

$$H_{\text{el-vib}} = \sum_{\alpha,\beta,E_1,E_2,j} \left(\sum_{i,\mu} \sqrt{\frac{\hbar}{2M_i\omega_j}} A_{i\mu,j} J^{i\mu,\alpha\beta}_{E_1\mathbf{K}_1,E_2\mathbf{K}_2} \right) a^{\alpha\dagger}_{E_1} a^{\beta}_{E_2}(b_j + b_j^\dagger), \tag{10.72}$$

where α, β = {L, R}; $A_{i\mu,j}$ is a canonical transformation between normal and Cartesian coordinates satisfying $\Sigma_{i,\mu} A_{i\mu,j}A_{i\mu,j'} = \delta_{j,j'}$; b_j is the annihilation operator corresponding to the j-th normal mode ω_j, and $a^{L(R)}$ is the annihilation operator for electrons. The coupling constant $J^{i\mu,\alpha\beta}_{E_1,E_2}$ between electrons and the vibration of the i-th atom in μ $(= x, y, z)$ component can be calculated as,

$$J^{i\mu,\alpha\beta}_{E_1\mathbf{K}_1,E_2\mathbf{K}_2} = \int d\mathbf{r} \int d\mathbf{r}' [\Psi^{\alpha}_{E_1\mathbf{K}_1}(\mathbf{r})]^* [\partial_\mu V^{ps}(\mathbf{r},\mathbf{r}',\mathbf{R}_i)] \Psi^{\beta}_{E_2\mathbf{K}_2}(\mathbf{r}'), \tag{10.73}$$

where $V^{ps}(\mathbf{r}, \mathbf{r}', \mathbf{R}_i)$ is the non-local pseudopotential representing the interaction between electrons and the i-th ion; $\Psi^{\alpha(=L,R)}_{E\mathbf{K}}(\mathbf{r})$ stands for the effective single-particle wavefunction of the entire system corresponding to incident electrons propagated from the left (right) electrode. These wavefunctions are calculated iteratively until convergence and self-consistency are achieved in the framework of DFT combined with the Lippmann–Schwinger equation.

10.4.2 Local Heating

In the nanojunction, the inelastic electron mean free path is large compared to the dimension of the junction. The electron only spends a little time in the junction. Consequently, the electron–vibration interaction only allows a small probability for energy exchange between electrons and molecule vibrations. Although the probability for this is small, local heating can still be substantial due to the large current density and the small size of the junction. There are two major processes that lead to an equilibrium local temperature in a nanojunction. One is due to the electron–vibration interaction that occurs in the atomic region of the junction; the other is due to the dissipation of heat energy to the bulk electrodes via phonon transport. These two processes finally reach thermal equilibrium, such that the equilibrium temperature can be defined in the atomic region.

We show how to evaluate heating in the nanostructure sandwiched between electrodes by the Fermi golden rule due to the eight first-order scattering mechanisms. These scattering processes correspond to electrons incident from the right or left electrode that relax (cool) or excite (heat) the energy level of the normal mode vibrations in the atomic region. The total thermal power generated in the junction can be calculated as the sum over all vibrational modes of the above eight scattering mechanisms [24–27].

The rate of energy absorbed (emitted) by the anchored nanostructures due to incident electrons from the $\beta = \{L, R\}$ electrode and scattered to the $\alpha = \{L, R\}$ electrode via a vibrational mode j is denoted by $W_j^{\alpha\beta,2(1)}$. The total thermal power generated in the junction P is as the sum of contribution from eight first-order scattering processes and from all the vibrational modes of shown in Fig. 10.10,

$$P = \sum_{j\in\text{vib}} \sum_{\alpha=\{L,R\}} \sum_{\beta=\{L,R\}} (W_j^{\alpha\beta,2} - W_j^{\alpha\beta,1}), \tag{10.74}$$

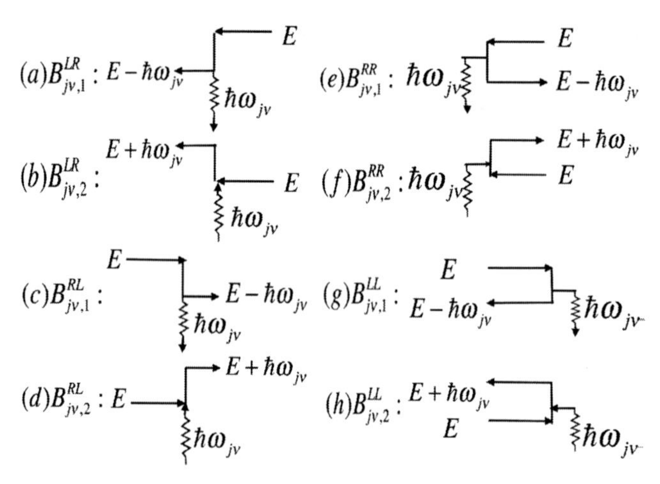

Figure 10.10 Feynman diagrams of the first-order electron–vibration scattering processes considered in this study.

where the power for each process is estimated using the Fermi golden rule for all modes:

$$W_j^{\alpha\beta,1(2)} = \sum_{j=\text{modes}} \sum_{E_i,E_f} \frac{2\pi}{\hbar} \left| \left\langle f \middle| H_{\text{el-vib}} \middle| i \right\rangle \right|^2 (E_f - E_i \pm \hbar\omega_j), \tag{10.75}$$

where the statistic of state $|i\rangle = \left| \{\cdots, f_E^R, \cdots\}; \{\cdots, f_E^L, \cdots\}; \{\cdots, n(\omega_j), \cdots\} \right\rangle$ includes electron states of the left and right electrodes and the local phonon states in the scattering region. Considering the eight scattering processes, one obtains

$$\begin{aligned} & W_j^{\alpha\beta,k}(T_e, T_\omega, V_B) \\ & = \sum_{j=\text{modes}} 2\pi\hbar[\delta_{k,2} + n(\omega_j)] \\ & \cdot \int dE (1 - f^{\alpha}_{E\pm\hbar\omega_j})\, f_E^{\beta} \left| \mathbf{tr}_{\mathbf{K}} \sum_{i,\mu,j} A_{i\mu,j}\, j^{i\mu,\alpha\beta}_{E\pm\hbar\omega_j,\mathbf{K}_1 E \mathbf{K}_2} \right|^2, \end{aligned} \tag{10.76}$$

where k = 2(1) correspond to heating (cooling) processes which excite (relax) the normal mode vibration. Note that choose $E + \hbar\omega_j$ for $k = 1$ and $E - \hbar\omega_j$ for $k = 2$. $\delta_{k,2} = 0(1)$ for $k = 1(2)$.

Most of the heat energy generate in the center scattering region is dissipated to electrodes via the lattice vibration. We estimate the rate of heat dissipated to electrodes via phonon–phonon interactions is calculated using the weak link model, as described in Eq. (10.48). The heating and heat dissipation finally come to equilibrium and reach an effective local temperature T_W. Note that the stiffness $K = AY/l$ of the nanostructure is stimulated by total energy calculations at zero temperature. The magnitude is very large value compared with the bulk materials. It results in very large phononic thermal conductivity compared with the bulk materials. We conjecture that the phononic current could be overestimated by weak-link model due to softening of the weak elastic link in high temperature regime.

10.4.3 Inelastic Current and Thermal Current Carried by Electrons

Similarly, the eight scattering processes contribute to cause small structures in the current-voltage characteristics, known as the inelastic electron tunneling spectroscopy (IETS) [27, 29]. The IETS spectrum is an important tool to explore the electron–vibron interaction in a single molecule junction, which has been studied extensively (for a review, see Ref. [30]). Here, we consider the correction of current due to the eight scattering processes (described in Fig. 10.9) from lowest-order of perturbation theory,

$$\delta I_E(\mu_L, T_L; \mu_R, T_R; T_w)$$
$$= \sum_j \sum_{\alpha,\beta} [\delta_{k,2} + n(\omega_j)] \frac{2e}{h} \int dE[(1 - f^{\alpha}_{E\pm\hbar\omega_j}) f^{\beta}_E \tilde{B}^{\alpha,\beta}_{E\pm\hbar\omega_j, j}], \quad (10.77)$$

where $\alpha = \{L, R\}$; $f_E^{L(R)} = 1/\{\exp[(E - \mu_{L(R)})/(k_B T_{L(R)})] + 1\}$ is the Fermi–Dirac distribution function describing the statistic of electrons deep in the left (right) electrode with temperature $T_L(R)$ and chemical potential $\mu_{L(R)}$; Note that $I^{\alpha\beta}_{E_1\mathbf{K}_1, E_2\mathbf{K}_2} = \int d\mathbf{R}[\Psi^{\alpha^*}_{E_1\mathbf{K}_1}(\mathbf{r})\nabla\Psi^{\beta}_{E_2\mathbf{K}_2}(\mathbf{r}) - \nabla\Psi^{\alpha^*}_{E_1\mathbf{K}_1}(\mathbf{r})\Psi^{\beta}_{E_2\mathbf{K}_2}(\mathbf{r})]$ is calculated from the electronic part of the wavefunctions $\Psi^{L(R)}_{E\mathbf{K}}(\mathbf{r})$. The average number of local phonons is $\langle n(\omega_j)\rangle = 1/\{\exp[\hbar\omega_j/(k_B T_W)] - 1\}$, where T_W is the effective wire temperature. The terms $\tilde{B}^{\alpha,\beta}_{E\pm\hbar\omega_j, j}$ represent the corrections to the elastic current considering the eight first-order scattering processes depicted in Fig. 10.9:

$$\tilde{B}^{L,\beta}_{E\pm\hbar\omega_j,j} = -\mathbf{tr}_{\mathbf{K}}\left[\frac{\pi^2\hbar}{m_e i} I^{RR}_{E\pm\hbar\omega_j\mathbf{K}_1,E\mathbf{K}_2} \cdot \left|\sum_{i,\mu}\pi\sqrt{\frac{\hbar}{2M_i\omega_j}} A_{i\mu,j} J^{i\mu,\alpha R}_{E\pm\hbar\omega_j\mathbf{K}_1,E\mathbf{K}_2}\right|^2\right], \quad (10.78)$$

where $\alpha, \beta = \{L, R\}$. Similarly,

$$\tilde{B}^{R,\beta}_{E\pm\hbar\omega_j,j} = \mathbf{tr}_{\mathbf{K}}\left[\frac{\pi^2\hbar}{m_e i} I^{RR}_{E\pm\hbar\omega_j\mathbf{K}_1,E\mathbf{K}_2} \cdot \left|\sum_{i,\mu}\pi\sqrt{\frac{\hbar}{2M_i\omega_j}} A_{i\mu,j} J^{i\mu,\alpha R}_{E\pm\hbar\omega_j\mathbf{K}_1,E\mathbf{K}_2}\right|^2\right]. \quad (10.79)$$

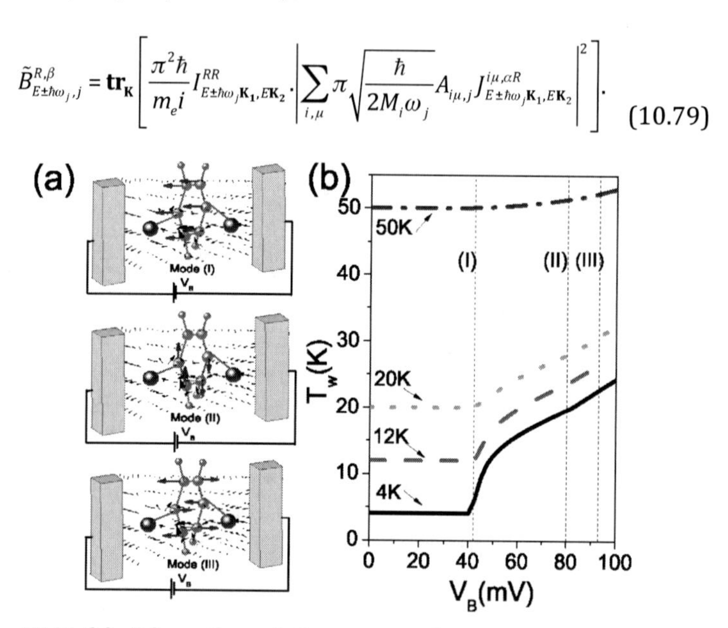

Figure 10.11 (a) Schematics of three most important normal modes [mode (I)–(III) with energy 42, 80, and 93 meV in the Pt-Benzene-Pt single-molecule junction, respectively] in the inelastic profiles. Big (small) red balls represent carbon (hydrogen) atoms, and purple balls represent Pt atoms. Blue arrow lines describe the vibrational pattern of the molecule. Black arrow lines represent the 3D current–density vector plot at $V_B = 0.1$ V. The graph is plotted in log scale to help the visualization of the smaller current density. (b) Local wire temperature T_w vs. V_B with electrode temperatures T_e set as 4, 12, 20, and 50 K. The vertical dotted lines refer to the biases corresponding to modes (I)–(III). Reprinted from Ref. [31].

To simply the calculations, we do the following approximations: $J^{i\mu,\alpha R}_{E\pm\hbar\omega_j\mathbf{K}_1,E\mathbf{K}_2} \approx J^{i\mu,\alpha R}_{E\mathbf{K}_1,E\mathbf{K}_2}$. We also trace out $\mathbf{K}$ for the coupling constant $J^{i\mu,\alpha R}_{E} = \mathbf{tr}_{\mathbf{K}} J^{i\mu,\alpha R}_{E\mathbf{K}_1,E\mathbf{K}_2}$ and for the transmission probability $\tau(E) = \frac{\pi^2\hbar}{m_e i}\mathbf{tr}_{\mathbf{K}}(I^{RR}_{E\mathbf{K}_1,E\mathbf{K}_2})$.

In this approximation, the inelastic current due to electron–vibration interactions is

$$I_{\text{el+vib}}(\mu_L, T_L; \mu_R, T_R; T_w) = \frac{2e}{h}\int dE[(f_E^R - f_E^L) - (\tilde{B}^R - \tilde{B}^L)]\tau(E), \quad (10.80)$$

and

$$J_{\text{el,el+vib}}^{L(R)}(\mu_L, T_L; \mu_R, T_R; T_w) = \frac{2}{h}\int dE[(f_E^R - f_E^L) - (\tilde{B}^R - \tilde{B}^L)]\tau(E)\times(E - \mu_{L(R)}), \quad (10.81)$$

where the transmission function $\tau(E) = \frac{\pi\hbar^2}{mi} \int d\mathbf{R}\int d\mathbf{K}$ $(\Psi_{E\mathbf{K}}^{R*}\nabla\Psi_{E\mathbf{K}}^{R} - \nabla\Psi_{E\mathbf{K}}^{R*}\Psi_{E\mathbf{K}}^{R})$ = $\frac{\pi^2\hbar}{m_e i}\mathbf{tr}_{\mathbf{K}}(I_{E\mathbf{K}_1,E\mathbf{K}_2}^{RR})$ is calculated from the electronic part of the wavefunctions $\Psi_{E\mathbf{K}_{||}}^{R}$. The terms $\tilde{B}^{L(R)}$ represent the corrections to the elastic current considering the eight first-order scattering processes depicted in Fig. 10.9 [31]:

$$\tilde{B}^{\alpha} = \sum_j [\langle|B_{j,k}^{\beta,\alpha}|^2\rangle f_E^{\alpha}(1 - f_{E\pm\hbar\omega_j}^{\beta}) - \langle|B_{j,k}^{\alpha,\alpha}|^2\rangle f_E^{\alpha}(1 - f_{E\pm\hbar\omega_j}^{\alpha})], \quad (10.82)$$

where α, β = {L, R} and $\alpha \neq \beta$. The $B_{j,1(2)}^{RR}$ and $B_{j,1(2)}^{LR}$ denoted in Eq. (10.82) are

$$B_{j,1(2)}^{\alpha,R} = i\pi\sum_{i\mu}\sqrt{\frac{\hbar}{2M_i\omega_j}}A_{i\mu,j}J_{E\pm\hbar\omega_j,E}^{i\mu,\alpha R}D_{E\pm\hbar\omega_j}^{\alpha}\sqrt{\delta + \langle n_j\rangle}, \quad (10.83)$$

where α = {L, R}; and δ = 0 (1) represents the process of phonon emission (absorption). The other two terms in Eq. (10.82) can be obtained by the relations $B_{j,1(2)}^{LL} = -B_{j,1(2)}^{RR}$ and $B_{j,1(2)}^{RL} = -B_{j,1(2)}^{LR}$. Two major processes lead to the equilibrium local temperature in a nanojunction. One is due to the electron vibration interaction that occurs in the atomic region of the junction; the other is due to the dissipation of heat energy to the bulk electrodes via thermal transport. We assume that the energy generated in the atomic region via inelastic electron–vibration scattering and the energy dissipated to the electrodes via thermal current finally reach equilibrium such that a well-defined local temperature can be calculated and measured in the atomic region [28]. The reason for the single temperature for multiple vibrational modes is exactly the same as for that in the bulk system. When the bulk crystal reaches the equilibrium temperature T, the temperature T determines the distributions of the occupations of all phonon branches. In our system, the local temperature in the atomic region determines the number of phonons occupying each phonon mode. The statistical behavior of the multiple vibrational modes and their probabilities in the overall distributions are described by the Bose–Einstein distributions $\langle n_j \rangle = 1/\{\exp[\hbar\omega_j/(k_B T_w)] - 1\}$, where T_w is the effective wire temperature due to local heating and $\langle n_j \rangle$ is the statistical average of the occupation number of the j-th normal mode. This effective local temperature is informative regarding the stability and performance of electronic devices and is thus useful to both theorists and experimentalists. The local temperature T_w is obtained when the power generated in the central region via the electron–vibration interactions balances the rate of thermal energy dissipated to electrodes calculated using the weak-link model [27].

10.4.4 Inelastic Seebeck Effects and ZT

In the bulk system, phonon drag is known to be important in the Seebeck coefficient and ZT. However, effects of electron–phonon interaction on the Seebeck coefficient and ZT are relatively unexplored in atomic/molecular junctions, at least there is still lack of experimental data in literatures. Here we show how to calculate thermoelectric figure of merits ZT in the presence of electron–vibration interactions.

We calculate the inelastic Seebeck coefficient based on the inelastic current described in Eq. (10.80), which is a function of T_L, T_R, T_w, and $V_B = (\mu_R - \mu_L)/e$. We consider an extra current induced by an infinitesimal temperature difference (ΔT) across the junction. This current is counterbalanced by an extra current driven by a voltage (ΔV), which is induced by ΔT via the Seebeck effect, i.e.,

$$I_{\text{el+vib}}(\mu_L, T_L; \mu_R, T_R) = \left[\begin{array}{c} I_{\text{el+vib}}\left(\mu_L, T_L - \dfrac{\Delta T}{2}; \mu_R, T_R + \dfrac{\Delta T}{2}\right) \\ + I_{\text{el+vib}}\left(\mu_L - \dfrac{e\Delta V}{2}, T_L; \mu_R + \dfrac{e\Delta V}{2}, T_R\right) \end{array} \right] \Big/ 2. \quad (10.84)$$

After expanding the above equation to the first order in ΔT and ΔV, we obtain the inelastic Seebeck coefficient (defined as $S_{\text{el+vib}} = \Delta V/\Delta T$),

$$S_{\text{el+vib}} = -\frac{1}{e}\frac{\int dE\left(\frac{\partial \tilde{f}_E^R}{\partial T_R} + \frac{\partial \tilde{f}_E^L}{\partial T_L}\right)\tau(E)}{\int dE\left(\frac{\partial \tilde{f}_E^R}{\partial E} + \frac{\partial \tilde{f}_E^L}{\partial E}\right)\tau(E)}, \tag{10.85}$$

where $\frac{\partial \tilde{f}_E^\alpha}{\partial E} = \frac{\partial f_E^\alpha}{\partial E} - \sum_{j\in \text{vib};k=1,2}(C_{\mu,j,k}^{R\alpha} + C_{\mu,j,k}^{L\alpha})$; $\frac{\partial \tilde{f}_E^\alpha}{\partial T_R} = \frac{\partial f_E^\alpha}{\partial T_R} - \sum_{j\in \text{vib};k=1,2}$ $(C_{T,j,k}^{R\alpha} + C_{T,j,k}^{L\alpha})$; $\alpha = \{L, R\}$ and the terms $C_{\mu,j,1(2)}^{\alpha R}$ and $C_{T,j,1(2)}^{\alpha R}$ due to the electron–vibration interactions,

$$C_{\mu,j,1(2)}^{\alpha R} = \left[f_E^R \frac{\partial f_{E\pm\hbar\omega_{jv}}^\alpha}{\partial E} - (1 - f_{E\pm\hbar\omega_j}^\alpha)\frac{\partial f_E^R}{\partial E}\right]\langle |B_{j,1(2)}^{RR}|^2\rangle; \tag{10.86}$$

$$C_{T,j,1(2)}^{\alpha R} = \left[\frac{E \pm \hbar\omega_j - \mu_\alpha}{T_R} f_E^R \frac{\partial f_{E\pm\hbar\omega_j}^\alpha}{\partial E} - \frac{E - \mu_R}{T_R}(1 - f_{E\pm\hbar\omega_j}^\alpha)\frac{\partial f_E^R}{\partial E}\right] \times \langle |B_{j,1(2)}^{\alpha R}|^2\rangle, \tag{10.87}$$

where $\alpha = \{L, R\}$ and $B_{j,1(2)}^{\alpha\beta}$ are given by Eq. (10.83). The other two terms in Eq. (10.85) can be calculated with the relations $\frac{\partial \tilde{f}_E^L}{\partial T} = \frac{\partial \tilde{f}_E^R}{\partial T}(L \rightleftharpoons R)$ and $\frac{\partial \tilde{f}_E^L}{\partial E} = \frac{\partial \tilde{f}_E^R}{\partial E}(L \rightleftharpoons R)$, where $(L \rightleftharpoons R)$ represents the interchange between R and L. We see that, in the absence of electron–phonon scattering, Eq. (10.85) recovers the elastic Seebeck coefficient described in Ref. [18].

As shown in Eq. (10.81), when a bias is applied to the junction, the inelastic electron thermal currents that flow from the right and into the left electrode are $J_{\text{el,el+vib}}^R$ and $J_{\text{el,el+vib}}^L$, respectively. At finite bias, note that there is no conservation of thermal current between two interfaces since the thermal current can be converted into the electric current. Therefore, we define the inelastic electron's thermal conductance in the junction by taking the average of the inelastic thermal conductance at two interfaces: $\kappa_{\text{el}}^{\text{el+vib}} = (\Delta J_{\text{el.el+vib}}^R/\Delta T + \Delta J_{\text{el.el+vib}}^L/\Delta T)/2$, where

$\Delta J_{\text{el,el+vib}}^R = J_{\text{el}}(\mu_L, T_L; \mu_R, T_R + \Delta T) + J_{\text{el}}(\mu_L, T_L; \mu_R + e\Delta V, T_R) - 2J_{\text{el}}(\mu_L, T_L; \mu_R, T_R)$;

$\Delta J_{\text{el,el+vib}}^L = J_{\text{el}}(\mu_L, T_L + \Delta T; \mu_R, T_R) + J_{\text{el}}(\mu_L + e\Delta V, T_L; \mu_R, T_R) - 2J_{\text{el}}(\mu_L, T_L; \mu_R, T_R)$;

and ΔV is the voltage induced by ΔT via the Seebeck effect. The inelastic electron's thermal conductance after expanding the Fermi–Dirac distribution function in Eq. (10.80) to the first order in ΔT and ΔV is given by

$$\kappa_{\text{el}}^{\text{el+vib}} = \sum_{\alpha,\beta=\{L,R\}} \frac{1}{2h}\left[eS(K_1^{\alpha\beta} - \xi_1) + \frac{K_2^{\alpha\beta}}{T_\alpha} - \xi_2 \right], \tag{10.88}$$

where $K_n^{\alpha\beta} = -\int dE(E-\mu_\alpha) \quad (E-\mu_\beta)^{n-1} \frac{\partial f_E^\beta}{\partial E}\tau(E);\quad \xi_1 = \sum_{\alpha=\{L,R\}} \sum_{j\in\text{vib};k=1,2} \int dE(E-\mu_\alpha)\ (C_{\mu,j,k}^{R\alpha} + C_{\mu,j,k}^{L\alpha})\tau(E);\ \xi_2 = \sum_{\alpha=\{L,R\}} \sum_{j\in\text{vib};k=1,2} \int dE(E-\mu_\alpha)\ (C_{T,j,k}^{R\alpha} + C_{T,j,k}^{L\alpha})\tau(E)$; and $C_{\mu(T),j,k}^{L(R)\alpha}$ are given in Eq. (10.86) and Eq. (10.87). Clearly, in the absence of electron–phonon scattering with $T_L = T_R$, Eq. (10.88) recovers the electron's thermal conductance described in Ref. [19].

We estimate the phonon's thermal conductance using the weak-link model [23]. The phonon's Hamiltonian of the molecular junction is modeled by $H_{ep} = H_L + H_R + \delta H$ where $H_{L(R)} = \sum_n \omega_n^{L(R)} (b_n^{L(R)})^\dagger b_n^{L(R)}$ is the Hamiltonian of the left (right) electrode, where $\omega_n^{L(R)}$ and $b_n^{L(R)}$ are the phonon spectrum and phonon annihilation operator in the left and right electrodes, respectively. The molecule sandwiched between electrodes is modeled by a harmonic spring with stiffness K described by $\delta H = \frac{1}{2}K(u_L^z - u_R^z)^2$, where $u_{L(R)}^z$ is the normal component of the displacement field $\mathbf{u}(\mathbf{r})$ at the surface of the left (right) electrode. The phonon's thermal conductance, approximated by the weak-link model, can be obtained by $\kappa_{\text{ph}} = \frac{\pi K^2 C^2}{h}\int_0^\infty dE E^3 \sum_{i=L,R} \frac{dn_i(E)}{dT_i}$, where the left (right) electrode is modeled as a phonon reservoir described by the Bose–Einstein distribution function $n_{L(R)}$; $N_{L(R)}(E) \approx CE$ where C is a constant, refers to the spectral density of surface phonon states at the left (right) electrode. The stiffness in the Pt/benzene junction has been estimated to be $K = 6.67302 \times 10^{-4}$ eV/a_0^2 by total energy calculations.

The thermoelectric efficiency in the nanoscale junctions is described by the dimensionless thermoelectric figure of merit (ZT). The thermoelectric figure of merit ZT depends on several physical factors: Seebeck coefficient (S), electrical conductance (σ), electronic thermal conductance (κ_{el}), and phononic thermal conductance (κ_{ph}). When the electron–vibration interaction is considered, the ZT becomes $\text{ZT}_{\text{el+vib}} = S^2_{\text{el+vib}}\sigma_{\text{el+vib}}T/(\kappa^{\text{el+vib}}_{\text{el}} + \kappa_{\text{ph}})$ where T is the average temperature in the source-drain electrodes. The inelastic electrical conductance $\sigma_{\text{el+vib}}$ is obtained by differentiating the inelastic current given by Eq. (10.80) with respect to the bias V_{B}; $S_{\text{el+vib}}$ is obtained using Eq. (10.85); $\kappa^{\text{el+vib}}_{\text{el}}$ is obtained using Eq. (10.88); and κ_{ph} is obtained using Eq. (10.51).

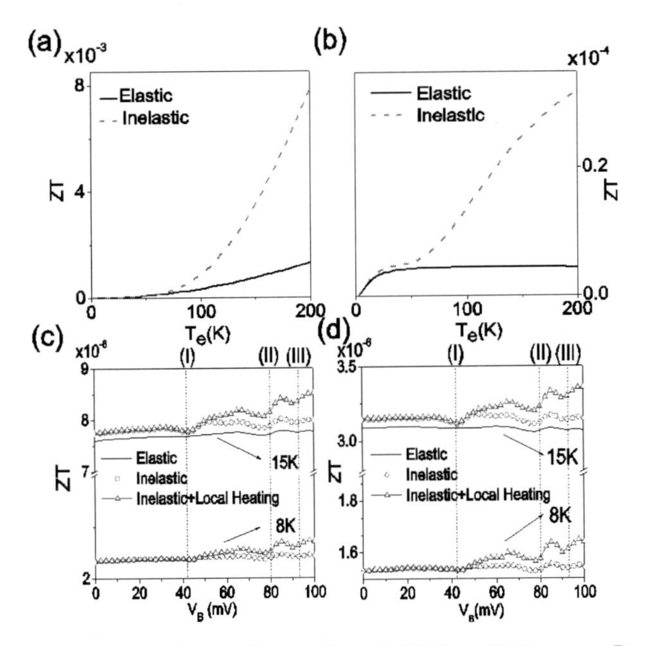

Figure 10.12 Thermoelectric figure of merit ZT for a Pt-Benzene-Pt single-molecule junction (a) without κ_{ph} and (b) with κ_{ph} as a function of the temperature at zero bias for elastic (black solid line) and inelastic (red dashed line) cases. Thermoelectric figure of merit ZT as a function of the source-drain bias V_{B} at T_{e} = 8 and 15 K (c) without κ_{ph} and (d) with κ_{ph}. Reprinted from Ref. [31].

To understand the rule of mode selection for inelastic profiles, the above theory has been applied to investigate the inelastic Seebeck coefficient and inelastic ZT for the Pt-Benzene-Pt single-molecule junction. [31] The relaxed Pt/benzene junction configuration, as shown in Fig. 10.8a, loses mirror symmetry.

This provides an exceptionally excellent system to investigate the rule of mode selection, which may be closely related to how the current density flows through the nanojunction and how normal modes vibrate. The highly tilted benzene molecule causes the streamline flow of the current to curve considerably to one side of the benzene ring, resulting in a non-trivial selection rule highly relevant to the details of the current density. We have identified normal modes that lead to significant structures in the inelastic profile. These modes are characterized by large components of atomic vibrations along the current density direction on top of each individual atom.

10.5 Summary

Molecular electronics attracts much interest from multi-discipline researchers. Scientists have made considerable efforts to understand the properties of nanojunctions as the down-scaling of electronic components is advancing. Single-molecule junctions may have promising use in revolutionizing the design of next-generation electronic device and energy conversion devices at the nanoscale level. The quantum nature due to the size minimization offered by single-molecule junctions may improve the performance of electric and thermoelectric devices beyond the expectation of conventional devices fabricated by traditional silicon technology.

The electronic atomic structure and the wave nature of non-equilibrium electron are essential gradients to understand the quantum transport properties in nanojunctions. Density functional theory combined with the Lippmann–Schwinger equation in scattering approaches provides a powerful tool to explore electronic structure and non-equilibrium electron transport for nanojunctions. Current-carrying wavefunctions obtained self-consistently in the framework of parameter-free DFT + Lippmann–Schwinger equation allow us to investigate many-body physics and quantum interference effects from first-principle approaches. The calculations have been applied to explore quantum transport properties in atomistic nanojunction, e.g., the current-voltage characteristics, current-induced forces, shot noise, counting statistics, Seebeck coefficients, thermoelectric figure of merit, local heating, inelastic vibronic coupling, and novel theories for thermoelectric nanodevices such as thermoelectric nanorefrigerator and thermoelectric self-powered transistor [32, 33].

References

1. Aviram, A., Ratner, M. A. *Chem. Phys. Lett.*, 1974, **29**, 277.
2. Di Ventra, M. *Electrical Transport in Nanoscale Systems*, 1st ed., Cambridge University Press, Cambridge, 2008.
3. Lang, N. D. *Phys. Rev. B*, 1995, **52**, 5335.
4. Lang, N. D. *Solid State Commun.*, 1969, **7**, 1047.
5. Lang, N. D. *Phys. Rev. B*, 1995, **52**, 5335–5342.
6. Ma, C. L., Chen, Y. C., Nghiem, D., Tseng, A., Huang, P. C. *Appl. Phys. A*, 2011, **104**, 325.
7. Di Ventra, M., Lang, N. D. *Phys. Rev. B*, 2001, **65**, 045402.
8. Taylor, J., Guo, H., Wang, J. *Phys. Rev. B*, 2001, **63**, 245407.
9. Thygesen, K. S., Rubio, A. *Phys. Rev. B*, 2008, **77**, 115333.
10. Vignale, G., Di Ventra, M. *Phys. Rev. B*, 2009, **79**, 014201.
11. Sai, N., Zwolak, M., Vignale, G., Di Ventra, M. *Phys. Rev. Lett.*, 2005, **94**, 186810.
12. Liu, Y. S., Chen, Y. C. *Phys. Rev. B*, 2011, **83**, 035401.
13. Di Ventra, M., Pantelides, S. T., Lang, N. D. *Phys. Rev. Lett.*, 2000, **84**, 979–982.
14. Di Ventra, M., Pantelides, S. T., Lang, N. D. *Phys. Rev. Lett.*, 2002, **88**, 046801.
15. Di Ventra, M., Chen, Y. C., Todorov, T. N. *Phys. Rev. B*, 2000, **61**, 16207.
16. Di Ventra, M., Chen, Y. C., Todorov, T. N. *Phys. Rev. Lett.*, 2004, **92**, 176803.
17. Hsu, B. C., Amanatidis, I., Liu, W. L., Tseng, A., Chen, Y. C. *J. Phys. Chem. C*, 2014, **118**, 2245.
18. Liu, Y. S., Chen, Y. C. *Phys. Rev. B*, 2009, **79**, 193101.
19. Liu, Y. S., Chen, Y. R., Chen, Y. C. *ACS Nano*, 2009, **3**, 3497.
20. Pai, W. W., Jeng, H. T., Cheng, C.-M., Lin, C.-H., Xiao, X., Zhao, A., Zhang, X., Xu, Geng, Shi, X. Q., Van Hove, M. A., Hsue, C.-S., Tsuei, K.-D. *Phys. Rev. Lett.*, 2010, **104**, 036103.
21. Hsu, B. C., Lin, C. Y., Hsieh, Y. S., and Chen, Y. C. *Appl. Phys. Lett.*, 2012, **101**, 243103.

22. Liu, Y. S., Hsu, B. C., Chen, Y. C. *J. Phys. Chem. C*, 2011, **115**, 6111.
23. Patton, K. R., Geller, M. R. *Phys. Rev. B*, 2001, **64**, 155320.
24. Chen, Y. C., Zwolak, M., Di Ventra, M. *Nano Lett.*, 2003, **3**, 1691.
25. Chen, Y. C., Zwolak, M., Di Ventra, M. *Nano Lett.*, 2004, **4**, 1079.
26. Chen, Y. C., Zwolak, M., Di Ventra, M. *Nano Lett.* 2005, **5**, 813.
27. Hsu, B. C., Liu, Y. S., Lin, S. H., Chen, Y. C. *Phys. Rev. B*, 2011, **83**, 041404(R).
28. Huang, Z., Xu, B., Chen, Y. C., Di Ventra, M., Tao, N. J. *Nano Lett.*, 2006, **6**, 1240.
29. Chen, Y. C. *Phys. Rev. B*, 2008, **78**, 233310.
30. Galperin, M., Ratner, M. A., Nitzan, A. *J. Phys. Condens. Matter*, 2007, **19**, 103201.
31. Hsu, B. C., Chiang, C. W., Chen, Y. C. *Nanotech.*, 2012, **23**, 275401.
32. Liu, Y. S., Chen, Y. C. *Appl. Phys. Lett.*, 2011, **98**, 213103.
33. Liu, Y. S., Yao, H. T., Chen, Y. C. *J. Phys. Chem. C*, 2011, **115**, 14988.

Chapter 11

Transition Voltage Spectroscopy: An Appealing Tool of Investigation in Molecular Electronics

Ioan Bâldea[a,b]

[a]*Theoretische Chemie, Universität Heidelberg, Im Neuenheimer Feld 229, D-69120 Heidelberg, Germany*
[b]*Institute of Space Sciences, National Institute for Lasers, Plasmas, and Radiation Physics, Bucharest, Romania*

ioan.baldea@pci.uni-heidelberg.de

Transition voltage (V_t) spectroscopy (TVS) represents an appealing tool of molecular electronics. It aims at determining the relative energetic alignment ε_0 of the frontier molecular orbitals by using V_t, the voltage at the minimum of the so-called Fowler–Nordheim plot. A series of theoretical aspects related to TVS will be addressed here. It will be shown that many experimental TVS data can be very accurately described by means of the Newns–Anderson model, which also allows to deduce analytical expressions that can easily be used by experimentalists for processing transport data. The transition voltage can be considered as a molecular signature, because, unlike the ohmic conductance, it is less affected, e.g., by stochastic fluctuations or electron correlations brought about by

Molecular Electronics: An Experimental and Theoretical Approach
Edited by Ioan Bâldea

ISBN 978-981-4613-90-3 (Hardcover), 978-981-4613-91-0 (eBook)
www.panstanford.com

Coulomb interactions at the contacts. Contrary to the initial claim, V_t is not related to a "transition" in a conventional sense; rather, the "transition" is from a linear response regime to a significant nonlinear regime. A critical aspect considered here is the applicability of the so-called Simmons model—the reason why this model is inappropriate for molecular transport will be presented.

11.1 Introduction

In the continuous efforts for miniaturization, using single molecules as active components for future nanoelectronics (Choi et al., 2008; Hybertsen et al., 2008; Lindsay and Ratner, 2007; Reed et al., 1997; Nitzan and Ratner, 2003; Tao, 2006; Song et al., 2009; Song et al., 2011; Venkataraman et al., 2006; Xu and Tao, 2003) appears at present as the only conceivable alternative, which escapes the fundamental limitations of complementary metal-oxide semiconductor (CMOS) technologies.

In nanoelectronic devices, charge transfer between the source and drain electrodes across a nanogap (spatial width d of a few nanometers) can occur via through-bond and through-space processes. An important nanotransport mechanism, to which we will restrict ourselves below, is the coherent electron tunneling.

Currents through vacuum nanojunctions are due to electron tunneling across an energy barrier whose height is basically determined by the metallic work functions. If molecules are inserted into the nanogap and contacted to the electrodes to form molecular junctions, they influence the charge transfer in different ways depending on the molecular properties. If none of the molecular orbitals is sufficiently close to the Fermi energy E_F of the electrodes, one could still consider electrons tunnel through an energy barrier. However, this barrier is different from that in vacuum; the molecules in the nanogap modify the barrier height via polarization effects (dielectric constant $\kappa_r > 1$) and renormalize the electron mass (effective free mass different from free electron mass, $m \neq m_0$). The transport in this case resembles (but is not identical to) the transport through thin metal–semiconductor–metal junctions in traditional electronics. Early attempts made in semiconductor physics (Gundlach, 1966; Simmons, 1963) recognized that the direct determination the barrier height from the measured current–voltage (I–V) characteristics is an important and, at the same time, nontrivial task. The transport mechanism based on the barrier picture is referred to as direct tunneling by some authors, although this term is also employed in a different sense (see below).

In most cases, due to the charge neutrality condition, the electrodes' Fermi level lies between the highest occupied molecular orbital (HOMO) and the lowest unoccupied molecular orbital (LUMO). In such situations, electrons travel from one electrode to another via nonresonant (superexchange (McConnell, 1961; Ratner, 1990; Skourtis and Beratan, 1999)) tunneling, through the tails of the density of states of the molecular orbital (MO), which is closest to the Fermi level. The latter acquires a finite width due to the molecule-electrode couplings. The energetic alignment $\varepsilon_0 \equiv E_{\mathrm{MO}} - E_F$ of the frontier molecular orbitals (HOMO or LUMO) relative to the electrode Fermi energy E_F plays a role similar to the barrier height mentioned above and represents a key parameter of molecular electronic devices.

Transition voltage spectroscopy (TVS) is an appealing method proposed recently (Beebe et al., 2006), which aims at the determination of ε_0. Due to its simplicity, it is becoming increasingly popular among experimentalists working in molecular electronics (Beebe et al., 2008; Chiu and Roth, 2008; Coll et al., 2009; Coll et al., 2011; Fracasso et al., 2013; Guo et al., 2011; Guo et al., 2013a; Guo et al., 2013b; Kim et al., 2011; Lee and Reddy, 2011; Lennartz et al., 2011; Malen et al., 2009; Ricoeur et al., 2012; Song et al., 2009; Song et al., 2011; Smaali et al., 2012; Tan et al., 2010; Trouwborst et al., 2011; Tran et al., 2013; Wang et al., 2011; Yu et al., 2008; Zangmeister et al., 2008.

Below, a series of theoretical aspects related to TVS analyzed in recent work (Araidai and Tsukada, 2010; Bâldea, 2010; Bâldea, 2012a,b; Bâldea and Köppel, 2012a; Bâldea and Köppel, 2012b; Bâldea, 2012f; Bâldea, 2012d,e; Bâldea, 2013a,b,c; Bâldea, 2014a,c, d,e; Chen et al., 2010; Huisman et al., 2009; Markussen et al., 2011; Vilan et al., 2013; Wu et al., 2013a; Wu et al., 2013b) will be considered.

11.2 The Relative Molecular Orbital Alignment Relative to Electrodes' Fermi Energy: A Key Parameter for Charge Transfer

Electron transport through molecular junctions is often dominated by the frontier molecular orbital (HOMO or LUMO, whichever is closest to the electrodes' Fermi energy E_F). As already noted, the corresponding energy offset ε_0 has long been recognized to be a key quantity which controls the charge transfer efficiency (Zahid et al., 2003).

Unfortunately, ε_0 cannot be solely determined from the low bias (ohmic) conductance G, the quantity which is commonly measured experimentally. This can be most easily seen from the expression of G in the case of a single level whose coupling to electrodes (say, substrate s and tip t, to adopt a terminology appropriate for scanning tunneling microscopy STM) is characterized by energy-independent width functions $\Gamma_{s,t}$. This is the Newns–Anderson model (Muscat and Newns, 1978) in the wide-band limit (e.g., (Bâldea, 2010, 2012a; Hall et al., 2000; Metzger et al., 2001; Schmickler, 1986) and references cited therein); see Section 11.5 for further details. The normalized conductance is expressed by

$$\frac{G}{G_0} = N\frac{\Gamma_s\Gamma_t}{\varepsilon_0^2 + (\Gamma_s + \Gamma_t)^2/4} = \overline{N}\frac{\Gamma^2}{\varepsilon_0^2 + \Gamma^2}, \tag{11.1}$$

where $G_0 = 2e^2/h = 77.48$ mS is the conductance quantum and N the (effective) number of molecules mediating the charge transfer. (For the second part of the above formula see also Section 11.5.)

As anticipated, the energy offset ε_0 is not easy to determine. It is a stumbling stone for current (mostly DFT-based) transport theoretical approaches. In principle, it could be easily deduced from I–V data. At biases $V = V_{r\pm}$ where the dominant MO becomes resonant with the electrodes' Fermi levels, $eV_{r\pm} = \pm 2\varepsilon_0$, pronounced current steps or, alternatively, sharp peaks in the differential conductance $\mathcal{G} = \partial I/\partial V$ should appear, as illustrated in Fig. 11.1.

Unfortunately, curves with such sharp steps or peaks appear only in theory. Experimentally, voltages $eV_r \approx 2\varepsilon_0$ are too high and can hardly be sampled. Typical curves measured in experiments look like those presented in Fig. 11.2, not seldom with substantial noise (cf. Fig. 11.2b). The question is how to determine ε_0 from such "innocent," structureless curves.

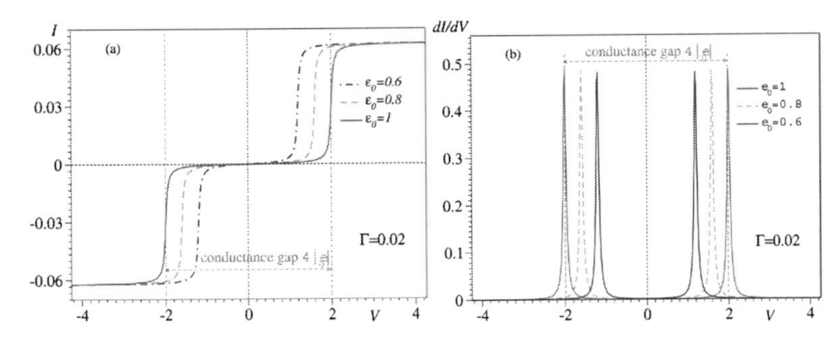

Figure 11.1 Ideal molecular junctions exhibit sharp current (I) steps (panel a) or, alternatively, sharp maxima in the differential conductance (panel b) at biases (V) where the molecular level becomes resonant with the Fermi level of an electrode. For a symmetric potential profile [γ = 0, see Eq. (11.14)], this occurs at $eV_{r\pm} = \pm 2\varepsilon_0$.

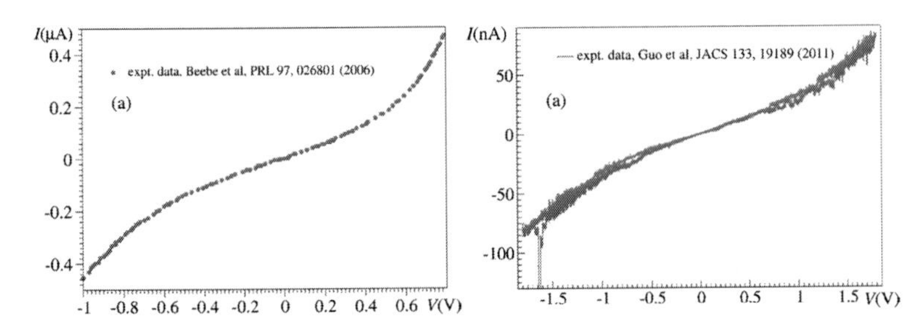

Figure 11.2 I–V measurements in real molecular junctions, as illustrated by the two typical examples presented here, yield structureless curves, which give no straightforward indication on how to estimate the energy offset ε_0. After Beebe et al. (2006) and Guo et al. (2011).

Transition voltage spectroscopy is a method aiming at answering this question. The quantity on which it relies upon, the transition voltage V_t, is defined as the source-drain voltage at the minimum of the so-called Fowler–Nordheim (FN) diagram. The FN diagram, is obtained by straightforwardly recasting the measured I–V curve as a plot of $\log(I/V^2)$ versus $1/V$. The term "transition" is related to the initial claim (Beebe et al., 2006) that the voltage V_t characterizes the crossover between two different tunneling regimes: direct tunneling and field-emission (FN) tunneling. This picture is based on the assumption that the tunneling barrier of the unbiased system ($V = 0$), considered to be rectangular (constant, position-(x-) independent height ε_0 within the nanogap $0 < x < d$, cf. Fig. 11.3a), is tilted by the applied bias $\varepsilon_0 \to \varepsilon_0 - eVx/d$ (cf. Fig. 11.3b,c). The low-bias regime ($eV < \varepsilon_0$), characterized by a trapezoidal barrier (Fig. 11.3b), is also referred to as direct tunneling, but one should be aware that the physical situation it describes is different from the "direct tunneling" mentioned above. In the tunneling barrier picture, the high-bias regime corresponds to a triangular barrier ($eV > \varepsilon_0$, Fig. 11.3c), and so the crossover occurs at the voltage $V = V_t$ given by

$$eV_t = \varepsilon_0. \tag{11.2}$$

In the original TVS paper (Beebe et al., 2006), no calculations have been presented to deduce Eq. (11.2), which has been called the "barrier-shape conjecture" later (Bâldea and Köppel (2012a). As a justification of using Eq. (11.2) to estimate ε_0, Beebe et al. (2006) argued that the triangular barrier mentioned above is similar to a field-emission (Fowler–Nordheim) tunneling (Fowler, 1928; Fowler and Nordheim, 1928; Nordheim, 1928). Employing the curve of $\log(I/V)^2$ versus $1/V$ traces back to the fact that simple Wentzel–Krammers–Brillouin (WKB)-type calculations (Chow, 1963a,b; Duke, 1969; Fowler, 1928; Fowler and Nordheim, 1928; Holm, 1951; Nordheim, 1928; Simmons, 1963; Sommerfeld and Bethe, 1933) yield a linear dependence of $\log(I/V^2)$ on $1/V$ for higher voltages. As illustrated in Fig. 11.3b, a dependence of this kind at higher voltages combined with a ubiquitous ohmic dependence $I \propto V$ at lower voltages is consistent with the occurrence of a minimum of the FN-curve $\log(I/V^2)$ versus $1/V$.

Nevertheless, as shown by independent investigations, a minimum in the FN-curve is necessarily related neither to a change in the barrier shape (Araidai and Tsukada, 2010; Bâldea 2012d;

Huisman et al., 2009) nor to the existence of a field-emission regime (Bâldea, 2012d). Concerning the second aspect, one could mention that the counterpart of the field-emission regime in the case of molecular junctions requires voltages that exceed the width of the metallic conduction band (cf. Fig. 11.4); a situation that can hardly be achieved with nanojunctions fabricated so far.

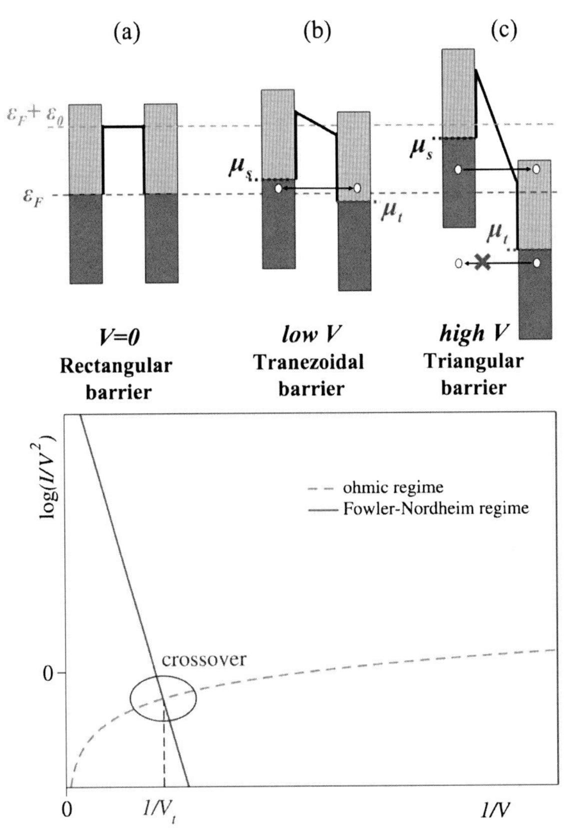

Figure 11.3 Left panel (a–c): schematic representation of the change in the barrier shape under applied bias V. Right panel: the crossover between an ohmic regime ("direct tunneling") $I \propto V$ at low voltages V and a Fowler–Nordheim (FN) regime ("field-emission tunneling") $I \propto V^2 \exp(-\text{const}/V)$ [linear dependence of $\log(I/V^2)$ on $1/V$] at high voltages reflects itself in a minimum of the FN curve $\log(I/V^2)$ versus $1/V$, which represents the transition voltage V_t.

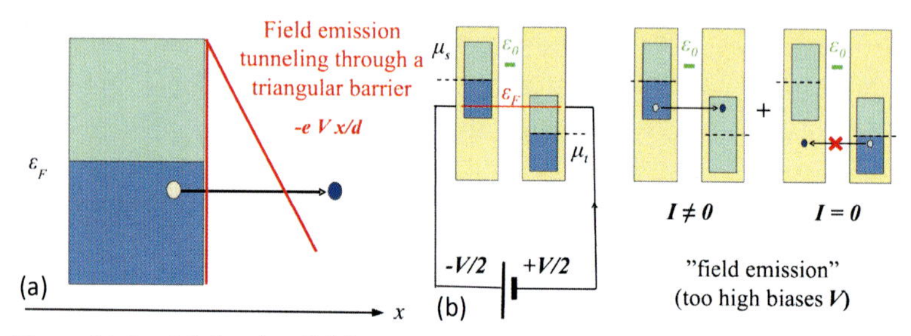

Figure 11.4 (a) In the field-emission regime encountered in traditional vacuum electronics, electrons escape from the metal across a triangular tunneling barrier and do not return in the electrode. (b) In molecular junctions, a similar behavior would be possible only at voltages exceeding the width of the conduction band (Bâldea, 2010), which would completely damage a real junction.

11.3 The Barrier Model Artefacts of the Simmons-Based Approach

Simmons (1963) and others (Chow, 1963a,b; Holm, 1951) found an approximate expression of the electric current due to electron tunneling through a one-dimensional energy barrier $\phi(x)$ expressed by (values $\varepsilon_0 > 0$ and $V > 0$ will be assumed throughout in this section)

$$\phi(x) = \varepsilon_0 - eVx/d + \phi_i(x). \tag{11.3}$$

In addition to the effect of the applied bias V, the tunneling barrier also includes the contribution $\phi_i(x)$ of image charges resulting from the interaction between an electron located at x with the electrodes. In the case of vacuum junctions, ε_0 represents the electrodes' work function.

To give support to the barrier-shape conjecture, Beebe et al. (2006) invoked results obtained by Simmons (Simmons, 1963). In reality, calculations based on the Simmons model do not justify Eq. (11.2). Such calculations have been reported in Huisman et al. (2009) and Trouwborst et al. (2011), which attempted to scrutinize TVS within the Simmons model. Subsequent investigations (Bâldea, 2012f; Bâldea and Köppel, 2012a; Bâldea and Köppel, 2012b; Bâldea, 2012e) revealed serious drawbacks of the "full Simmons" approach of Huisman et al. (2009) and Trouwborst et al. (2011).

Briefly, the TVS analysis of the "full Simmons" barrier model of Refs. (Huisman et al., 2009; Trouwborst et al., 2011) is defective because:

(i) Its starting point, basically the same as that of Simmons (Simmons, 1963), ignores the lateral constriction. The thin films considered by Simmons (Simmons, 1963) are infinite in transverse (y and z) directions. Electrons move freely along the y- and z-axes, and the electric current per unit area J along the x-axis can be expressed as (Bâldea and Köppel, 2012a; Simmons, 1963; Sommerfeld and Bethe, 1933)

$$J = 2\frac{e}{h^3}\int [f(E_p - eV/2) - f(E_p + eV/2)]\frac{\partial E_p}{\partial p_x}\mathcal{T}(p_x;V)dp_x dp_y dp_z$$
$$= \frac{4\pi me}{h^3}\int [f(E - eV/2) - f(E + eV/2)\,\mathcal{T}(E_x;V)dE_x dE_\perp, \qquad (11.4)$$

where m and $-e$ stand for electron mass and charge, respectively, f is the Fermi distribution function, and $\mathcal{T}$ the transmission coefficient. To get the second line in Eq. (11.4), an electrode conduction band with parabolic dispersion in transverse direction $E_\perp = E - E_x = \frac{1}{2m}(p_y^2 + p_z^2)$ has been assumed. The fact that the three-dimensional picture underlying Eq. (11.4) is inadequate for molecular electronics can been readily illustrated by considering the limit $V \to 0$. Equation (11.4) yields an ohmic conductance expressed

$$G \propto \lim_{V\to 0}\frac{dJ(V)}{dV} \propto \int_0^{E_F}\mathcal{T}(E_x, V = 0)dE_x. \qquad (11.5)$$

As visible in Eq. (11.5), the electrons contributing to the ohmic conduction have a kinetic energy E_x of motion across the junction spanning the entire conduction band ($0 \leq E_x \leq E_F$); E_x is not restricted to $E_x \simeq E_F$. This is the direct consequence of the fact that a lateral (y, z) confinement is missing in Eq. (11.4). The states contributing to linear response in Eq. (11.4) do have a total energy close to the Fermi energy, $E_p \equiv E_x + E_y + E_z \simeq E_F$; this is ensured by the difference of the Fermi functions entering there. However, the longitudinal kinetic energy can vary in the range $0 < E_x < E_F$ because the transverse kinetic energies are continuous variables in the range $0 < E_{y,z} < E_F$. This is appropriate for (three-dimensional) insulating/semiconducting films (which may be very thin but have a practically infinite extension in transverse directions, e.g., Simmons (1963), but not for molecular electronics; in particular, not for the atomic-size molecular junctions of Song et al. (2009), where the lateral motion is quantized and only the lowest-energy transverse channel contributes to conduction.

To describe the molecular transport one has to account for lateral confinement and drop $p_{y,z}$-integrations in Eq. (11.4). Then, the counterpart of Eqs. (11.4) is of the well-known Landauer form

$$I = 2\frac{e}{h}\int [f(E_x - eV/2) - f(E_x + eV/2)]\mathcal{T}(E_x;V)dE_x, \qquad (11.6)$$

Equation (11.6) leads to a linear response determined by the states of energy in the tiny interval $E_F - eV/2 < E_x < E_F + eV/2$, and $G \propto \mathcal{T}(E_F; V = 0)$ ("conductance is transmission").

(ii) In Huisman et al. (2009) and Trouwborst et al. (2011), image charge effects are exaggerated, because of using an approximate expression proposed by Simmons (Simmons, 1963), which overestimates the image interaction energy by a factor ~2.

For infinite planar electrodes separated by d, the energy of interaction between a point charge located at x can be expressed in closed analytical form (Sommerfeld and Bethe, 1933)

$$\phi_i(x) = \frac{e^2}{4d\kappa_r}\left[-2\psi(1) + \psi\left(\frac{x}{d}\right) + \psi\left(1 - \frac{x}{d}\right)\right], \qquad (11.7)$$

where ψ is the digamma function and κ_r the dielectric constant.

Instead of $\phi_i(x)$ of the exact Eq. (11.7), Huisman et al. (2009) and Trouwborst et al. (2011) employed the following expression proposed in Simmons (1963):

$$\phi_i^u(x) \approx -1.15\log 2\frac{e^2}{2\kappa_r}\left(\frac{1}{x} + \frac{1}{d-x}\right). \qquad (11.8)$$

In fact, the above expression is erroneous; as remarked by Simmons himself (Simmons, 1964), $\phi_i^u(x)$ should be replaced by

$$\phi_i^c(x) = \frac{1}{2}\phi_i^u(x). \tag{11.9}$$

Unfortunately, it is the wrong formula (11.8) that has been used in the recent works on TVS (Huisman et al., 2009; Trouwborst et al., 2011), and this is another reason why their results are incorrect.

(iii) In the expressions deduced in Simmons (1963) (and others [Chow, 1963a,b; Holm, 1951; Sommerfeld and Bethe, 1933]), which are based on a simplified version of the general WKB approximation, Huisman et al. (2009) and Trouwborst et al. (2011) included in calculations ("full Simmons") terms with significant mathematical contributions only for small barrier widths and heights, i.e., exactly where the WKB method is inapplicable (Bâldea and Köppel, 2012b).

Just because of the inclusion of the terms mentioned above under (iii), which Simmons ignored in his numerical estimations, Huisman et al. (2009) and Trouwborst et al. (2011) "predicted" a maximum of the transition voltage as a function of nanogap size. This is an artefact of an inappropriate mathematical treatment; neither calculations done exactly nor within the full WKB approximation yield such a behavior (Bâldea, 2012f; Bâldea, 2012e; Bâldea and Köppel, 2012b). This is illustrated in Fig. 11.5, which also shows that this artefact is not even related to the lateral constriction.

To avoid misunderstandings (Vilan et al., 2013), it is worth emphasizing that the FN-curve [$\log(I/V^2)$ versus $1/V$] obtained via Simmons-type calculations does have a minimum, which defines a transition voltage V_t. This is not an artefact of Simmons-type calculations; what is an artefact is the incorrect prediction of a maximum of V_t as a function of d.

The popularity of the Simmons approach (Simmons, 1963) is certainly related to the fact that it provided an analytical expression of the current density J

$$J/J_0 = \bar{\varepsilon}_0\{1+\mathcal{O}[(A\bar{\varepsilon}_0^{1/2})^{-1}]\}e^{-A\bar{\varepsilon}_0^{1/2}} - (\bar{\varepsilon}_0 + eV)$$
$$\times\{1+\mathcal{O}[(A(\bar{\varepsilon}_0 + eV)^{1/2})^{-1}]\}e^{-A(\bar{\varepsilon}_0+eV)^{1/2}}$$
$$+\mathcal{O}[e^{-A(\bar{\varepsilon}_0+E_F)^{1/2}}]. \quad (11.10)$$

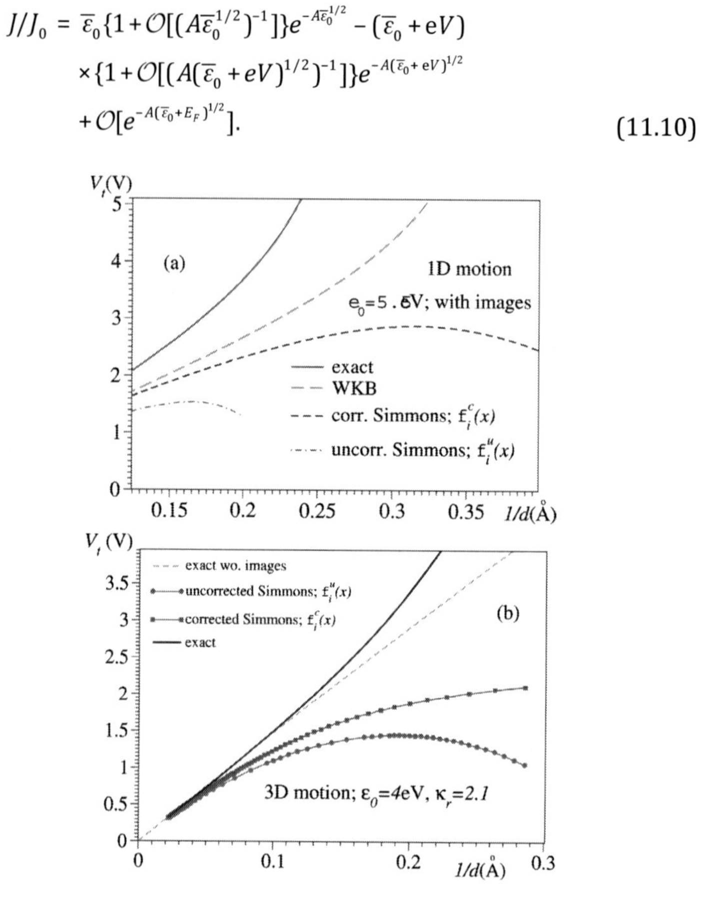

Figure 11.5 Calculations based on the "full Simmons" approach (a simplified version of the WKB method) of Refs. (Huisman et al., 2009; Trouwborst et al., 2011) predict a maximum of the transition voltage V_t as a function of the nanogap size d. This is an artefact of an inappropriate mathematical approximation, as visible from the comparison with the results obtained exactly (without any mathematical approximation) or even within the full WKB approximation. Whether accounting for lateral constriction (one-dimensional electron motion, panels a) or not (three-dimensional electron motion, panel b), this is an artefact of a defective mathematical approach. After Bâldea (2012f) and Bâldea (2012e).

Here $J_0 \approx (e/\hbar)(2\pi\Delta s)^{-2}$ and $A \approx 2\Delta s(2m)^{1/2}/\hbar$, where $\bar{\varepsilon}_0$ and Δs represent renormalizations of ε_0 and d due to image charge effects. Equation (11.10) applies if the Fermi level of the positive electrode lies above the bottom of the negative electrode; otherwise the second term in the RHS ("backward" current) should be omitted.

Equation (11.10) has also been utilized to interpret transport data in molecular junctions. Because this is inappropriate, as discussed above, we will give below the counterpart of Eq. (11.10) obtained by accounting for lateral constriction (Bâldea and Köppel, 2012a):

$$I \simeq G_0 \frac{32}{eE_F}\frac{m}{m_0}[\mathcal{J}(\overline{\varepsilon}_{0-})e^{-a\overline{\varepsilon}_{0-}^{1/2}} - \mathcal{J}(\overline{\varepsilon}_{0+})e^{-a\overline{\varepsilon}_{0+}^{1/2}}], \tag{11.11}$$

Equation (11.11) expresses the current I for voltages not larger than about $eV \sim 1.5\overline{\varepsilon}_0$ (an upper bound safely covering the range investigated experimentally (Song et al., 2009), and values $a\overline{\varepsilon}_0^{1/2} > 8$ (fully compatible with data for octanedithiols (Bâldea and Köppel, 2012a)). Here, $a \equiv 2d\sqrt{2m}/\hbar$, $\overline{\varepsilon}_{0\pm} \equiv \overline{\varepsilon}_0 \pm eV/2$, and $\mathcal{J}(x) \equiv a^{-4}(a^3x^{3/2} + 3a^2x + 6a\sqrt{x} + 6)$.

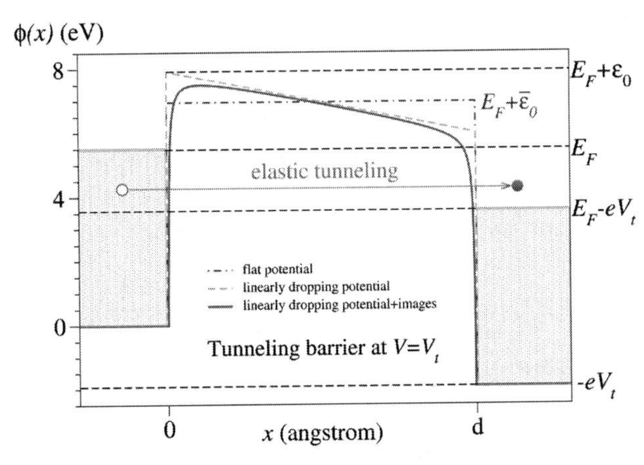

Figure 11.6 The example presented in this figure for $V = V_t$ after Bâldea and Köppel (2012a) illustrates that the tunneling barrier remains trapezoidal at $V = V_t$ and does not change to triangular, contrary to the initial TVS claim (Beebe et al., 2006).

As a word of caution, one should mention that expressions obtained within barrier models like those given above and, e.g., for the conductance attenuation factor $[G \propto \exp(-\beta d)]\beta = 2\sqrt{2m\overline{\varepsilon}_0}/\hbar$ contain an effective mass m, which differs from the free electron mass m_0 in vacuum. Therefore, special care should be paid when estimating the MO energy offset based on such expressions (Bâldea and Köppel, 2012a; Holmlin et al., 2001).

The fact that, contrary to the initial TVS claim (Beebe et al., 2006), the transition voltage calculated exactly within the tunneling barrier model does not necessarily correspond to the change from a trapezoidal to a triangular barrier is illustrated by Fig. 11.6.

11.4 TVS for Vacuum Nanojunctions

The description of molecular transport within the standard tunneling barrier model, Eqs. (11.3) and (11.7), is inherently a crude approximation. Still, one may expect that this picture is appropriate for vacuum nanojunctions. So, it is interesting to compare theoretical results for $V_t = f(d)$ obtained within the barrier model with experimental results recently reported for vacuum break junctions (Trouwborst et al., 2011).

Exact results for V_t can be obtained by numerically solving the Schrödinger equation for an electron moving in the one-dimensional potential $\phi(x)$ of Eq. (11.3) (Bâldea and Köppel, 2012a; Bâldea, 2012f; Bâldea and Köppel, 2012b; Bâldea, 2012e). In the absence of image effects ($\phi_i \equiv 0$), the exact transition voltage approximately exhibits a nearly linear dependence of V_t on $1/d$ (Bâldea, 2012f; Bâldea and Köppel, 2012b). This dependence can be semi-quantitatively described within a WKB-type approximation, which yields (Bâldea, 2012f; Bâldea and Köppel 2012b; Huisman et al., 2009)

$$eV_t \simeq \sqrt{\frac{2\varepsilon_0}{m}}\frac{2\hbar}{d}. \tag{11.12}$$

When the image charge contributions expressed by Eq. (11.7) are exactly included, the transition voltage exhibits an upturn from the straight line $V_t \propto 1/d$ (cf. Fig. 11.5). This contrasts with the experimental curve $V_t = f(1/d)$ (Trouwborst et al., 2011), which exhibits a downturn at small nanogap sizes and evolves into a broad maximum around $d \approx 3$–4Å.

The experimental behavior can be semi-quantitatively described by extending the standard tunneling barrier model to include contributions of electron states (or resonances) at electrodes' surface (Bâldea, 2012f; Bâldea and Köppel, 2012b). The counterpart of Eq. (11.12) is then

$$eV_t \simeq \sqrt{\frac{2\varepsilon_0}{m}}\frac{2\hbar}{d+d_0}. \tag{11.13}$$

Above, d_0 is a new characteristic length, which quantifies effects of surface states and marks the crossover between a linear regime ($V_t \propto 1/d$ for $d >> d_0$) and a plateau ($V_t \approx$ const for $d << d_0$). Estimates based on Eq. (11.13) are in semi-quantitative agreement with experiments (Trouwborst et al. (2011); see Bâldea (2012f) and Bâldea and Köppel (2012b) for details. Recent DFT calculations (Wu et al., 2013a) seem to give microscopic support to this phenomenological picture.

We end this section with the following remark. In the standard monograph on tunneling (Duke, 1969) (p. 61), Duke noted: "The availability of electronic computers has removed both the need and desirability of... approximations [like Simmons']. However, the urge to approximate seems difficult to suppress." As witnessed by recent publications, the Simmons model in particular and tunneling barrier models in general continue to be used in molecular transport. So, it seems indeed unrealistic to assume that this tendency will (ever) stop. The best one can hope is to see such models utilized for molecular junctions in a form as appropriate as possible. Therefore, one should emphasize that, when applied to molecules, Eq. (11.13) does not necessarily imply dependencies $V_t \propto \varepsilon_0^{1/2}$ or, approximately, $V_t \propto 1/d$; the effective mass m in a molecular junction can also have a nontrivial dependence of ε_0 and d. When, along with the lateral constriction, this fact is also taken into account, transport data in molecular junctions can sometimes (Bâldea and Köppel, 2012a) be rationalized even within barrier models.

11.5 Describing TVS within the Newns–Anderson Model

As shown in a series of recent studies (Bâldea, 2010, 2012a,b,d), the single-level model with Lorentzian transmission (sometimes referred to as the Newns–Anderson model (Muscat and Newns,

1978; Schmickler, 1986) turns out to provide a valuable framework to validate TVS as an appealing tool of investigation in molecular electronics. As briefly reviewed below, this model is extremely useful. It can provide excellent fitting curves of the current–voltage measurements in a variety of molecular junctions. Equally important, the Newns–Anderson model allows to deduce simple and accurate analytical formulas, which experimentalists can straightforwardly use to process transport data.

In transport studies based on the Newns–Anderson model (Bâldea, 2010, 2012a,b,d; Medvedev, 2007; Schmickler, 1986) one assumes that the charge transfer through a molecule linked to two electrodes is dominated by a single level. Its energy $\varepsilon_0(V)$ may be shifted by the bias from the value ε_0 of the unbiased system

$$\varepsilon_0(V) = \varepsilon_0 + \gamma eV. \tag{11.14}$$

For positive biases $V > 0$, the level is shifted towards the Fermi level of the negative ($E_F + eV/2$) or of the positive ($E_F - eV/2$) electrode, depending on whether the voltage division factor γ (Bâldea, 2012a; Medvedev, 2007; Zahid et al., 2003) is positive ($0 < \gamma < 1/2$) or negative ($-1/2 < \gamma < 0$), respectively. By describing the electrodes within the ubiquitous wide-band approximation, the width functions $\Gamma_{s,t}$ are energy-independent, and the transmission is Lorentzian. Then the Landauer formula (11.6) allows to express the current in closed analytical form

$$I = \frac{G_0 \bar{N} \varepsilon_0}{\sqrt{\frac{\bar{N}}{g} - 1}} \left\{ \arctan\left[\frac{\varepsilon_0(V) + eV/2}{\varepsilon_0} \sqrt{\frac{\bar{N}}{g} - 1} \right] - \arctan\left[\frac{\varepsilon_0(V) - eV/2}{\varepsilon_0} \sqrt{\frac{\bar{N}}{g} - 1} \right] \right\}. \tag{11.15}$$

In the above expression

$$g = G/G_0 \quad \text{and} \quad \bar{N} = N(1 - 4\delta^2), \tag{11.16}$$

where the quantity

$$\delta \equiv \frac{1}{2} \frac{\Gamma_s - \Gamma_t}{\Gamma_s + \Gamma_t} \tag{11.17}$$

accounts for a possible asymmetry of the molecule–electrode couplings (Bâldea, 2012a; Huisman et al., 2009; Wang et al., 2011). An equivalent form of Eqs. (11.16) and (11.17) is the following:

$$\Gamma_{s,t} = \Gamma(1 \pm 2\delta),\ \Gamma \equiv \frac{\Gamma_s + \Gamma_t}{2} = \frac{|\varepsilon_0|}{\sqrt{\bar{N}/g - 1}}. \tag{11.18}$$

Recasting previously published expressions for I (see, e.g., Bâldea (2012a) and references cited therein) in the form expressed by Eq. (11.15) and the details provided below aim at helping experimentalists to easily fit measured I–V curves.

At low biases, Eq. (11.15) reduces to

$$I = GV = G_0 gV, \tag{11.19}$$

and this allows to straightforwardly extract g. This reduces the number of fitting parameters, and this fact stabilizes the overall fitting procedure, which is important for reliably estimating ε_0; particularly because the experimentally accessible high-V range is limited, and there is a significant ambiguity in determining the best fit for currents I at lower V's.

In cases where the electrode–molecule couplings are sufficiently weak and biases are not too high

$$\Gamma \ll \varepsilon_0(V) \pm eV/2, \tag{11.20}$$

Equation (11.15) acquires a simpler form (Bâldea, 2012a):

$$I \simeq G_0 gV \frac{\varepsilon_0^2}{(\varepsilon_0 + \gamma eV)^2 - (eV/2)^2}. \tag{11.21}$$

Equation (11.21) is not only mathematically simpler than Eq. (11.15); it is the fitting procedure based on it that is easier. In addition to g (which can be directly obtained from the ohmic regime, as noted above) Eq. (1.21) only needs two fitting parameters (ε_0 and γ), revealing that in some cases the energy offset ε_0 and voltage division factor γ can be determined without knowing of coupling asymmetry δ and the effective number of molecules N [both accounted for by in Eq. (11.15)].

Equations (11.1), (11.15), (11.19), and (11.21) are invariant under the transformation $(\varepsilon_0, \gamma) \to (-\varepsilon_0, -\gamma)$. As a practical consequence of this invariance: if the fitting program employed to fit a measured I–V curve yields a mathematical solution where ε_0 is positive (say, $\varepsilon_0^{\rm math} = 0.5$ eV, $\gamma^{\rm math} = 0.07$), the physical solution is a negative energy offset and a voltage division factor with reversed sign ($\varepsilon_0^{\rm phys} = -0.5$ eV, $\gamma^{\rm phys} = -0.07$).

The Newns–Anderson model yields FN curves possessing a minimum (which defines the transition voltage V_t), and this fact clearly demonstrates that TVS is not necessarily related to a change in the shape of the tunneling barrier. Examples of FN curves obtained within this model are depicted in Fig. 11.7. As visible there, the approximate current expressed by Eq. (11.21) is very accurate.

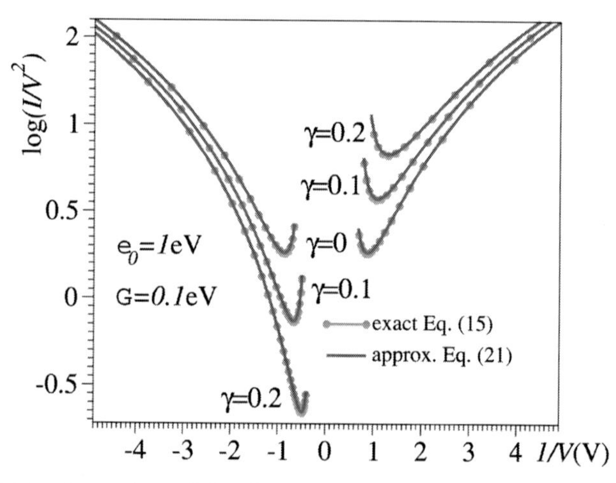

Figure 11.7 Examples of Fowler–Nordheim curves exhibiting minima obtained within the Newns–Anderson model by using the exact Eq. (11.15) and the approximate Eq. (11.21). The parameter values are given in the inset. After Bâldea, 2012a.

Remarkably, as demonstrated in Bâldea (2012a,b,d, 2013c, 2014e), Eq. (11.21) is able, in spite of its simplicity, to excellently reproduce I–V curves measured in a variety of molecular junctions. An illustration is presented in Fig. 11.8.

In the case of an asymmetric potential profile ($\gamma \neq 0$), the FN-minima for positive and negative biases are asymmetrically located: $V_{t+} \neq -V_{t-}$. Equations (11.21) can be used to derive simple analytical expressions for the transition voltages $V_{t\pm}$ (Bâldea, 2012a) in terms of ε_0 and γ. By imposing $\partial[\log(I/V^2)]/\partial(1/V) = 0$, or the mathematically equivalent condition $\partial I/\partial V = 2I/V$ (see also Section 11.9), one gets (Bâldea, 2012a)

$$\chi_{t1} \equiv \varepsilon_0/V_{t1} = -2\gamma + \sqrt{\gamma^2 + 3/4},$$
$$\chi_{t2} \equiv \varepsilon_0/V_{t2} = -2\gamma - \sqrt{\gamma^2 + 3/4}. \qquad (11.22)$$

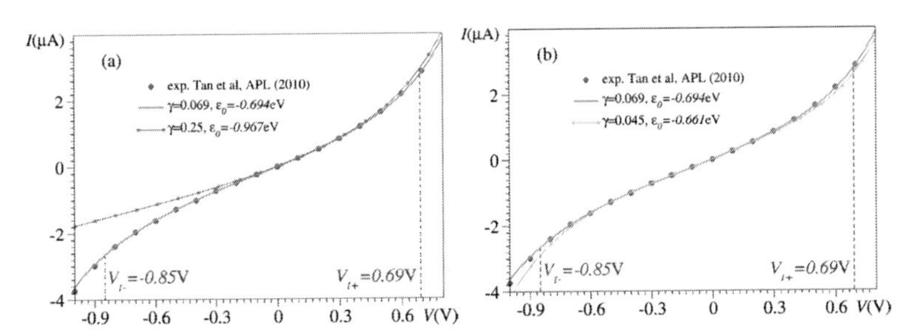

Figure 11.8 The experimental I–V curve measured by Tan et al. (Tan et al., 2010) can be excellently reproduced by the analytical formula of Eq. (11.21). (a) The curve obtained by using the value $\gamma = 0.25$ estimated from DFT calculations (Markussen et al., 2011) completely disagrees with the experimental curve for negative biases (see Bâldea (2012a) for further details). (b) The differences between the curve obtained with the parameters $\varepsilon_0 = -0.694$ eV and $\gamma = 0.069$ obtained by fitting the experimental curve using Eq. (11.21) and that by using the values $\varepsilon_0 = -0.661$ eV and $\gamma = 0.045$ deduced using the experimental values for $V_{t\pm}$ and Eqs. (11.23) and (11.24) are smaller than usual experimental errors. The corresponding values of $\Gamma(\Gamma = 0.124$ eV and $\Gamma = 0.116$ eV, respectively) are indeed smaller but not much smaller than $|\varepsilon_0|$. This fact demonstrates the reliability of the estimates obtained via Eqs. (11.23) and (11.24).

Because $-1/2 \le \gamma \le 1/2$, one can easily convince oneself that $\chi_{t1} > 0$ and $\chi_{t2} < 0$, which shows that the signs of V_{t1} and V_{t2} are opposite. Denoting by $V_{t+}(> 0)$ and $V_{t-}(< 0)$ the transition voltage for positive and negative polarities, $V_{t+} \equiv V_{t1}$ and $V_{t-} \equiv V_{t2}$ for LUMO-mediated transport ($\varepsilon_0 > 0$), while for HOMO-mediated transport ($\varepsilon_0 < 0$) $V_{t+} \equiv V_{t2}$ and $V_{t-} \equiv V_{t1}$. In the HOMO case, $V_{t+} < |V_{t-}|$ for $\gamma > 0$, whereas $V_{t+} > |V_{t-}|$ for $\gamma < 0$.

Equations (11.22) can be inverted

$$|\varepsilon_0| = 2\frac{e|V_{t+}V_{t-}|}{\sqrt{V_{t+}^2 + 10|V_{t+}V_{t-}|/3 + V_{t-}^2}}, \qquad (11.23)$$

$$\gamma = \frac{\text{sign}\,\varepsilon_0}{2} \frac{V_{t+} + V_{t-}}{\sqrt{V_{t+}^2 + 10|V_{t+}V_{t-}|/3 + V_{t-}^2}}. \tag{11.24}$$

The above expressions allow to estimate the MO energy offset ε_0 and the bias asymmetry factor γ in terms of the transition voltages $V_{t\pm}$ for both bias polarities. $V_{t\pm}$ can be straightforwardly extracted from transport measurements.

The dependence of the transition voltages on the voltage division factor γ is shown in Fig. 11.9. Only the range $\gamma > 0$ is presented there because, according to Eq. (11.14), a redefinition of the bias polarity $V \rightarrow -V$ yields a sign change $\gamma \rightarrow -\gamma$; so, e.g., $V_{t\pm}(\gamma) = V_{t-}(-\gamma)$.

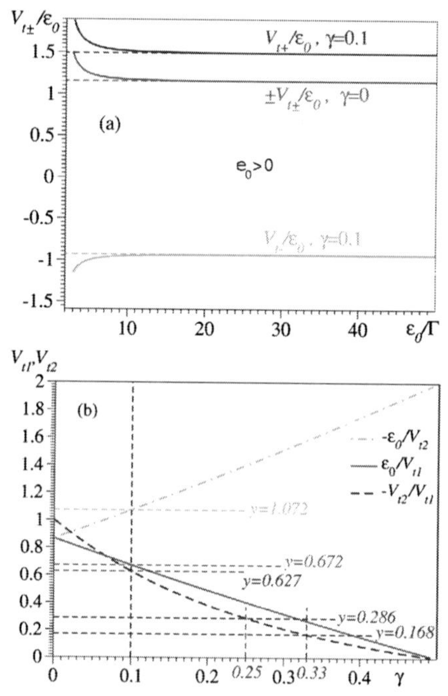

Figure 11.9 Dependence on the bias asymmetry factor γ of the transition voltages V_{t1} and V_{t2} corresponding to biases of opposite polarities: $V_{t1} \equiv V_{t+} > 0$ and $V_{t2} \equiv V_{t-} < 0$ for $\varepsilon_0 > 0$ (LUMO-mediated conduction), and $V_{t2} \equiv V_{t+} > 0$ and $V_{t1} \equiv V_{t-} < 0$ for $\varepsilon_0 < 0$ (HOMO-mediated conduction). Notice the rapid saturation of the curves obtained exactly (thick lines in panel a) to the asymptotic values obtained analytically (thin horizontal lines) via Eqs. (11.22), which are valid $\mathcal{O}(\Gamma/\varepsilon_0)^2$. After Bâldea (2012a).

As expressed by Eqs. (11.22) and illustrated by Fig. 11.9, the bias asymmetry factor γ can be determined from the transition voltage asymmetry ratio $V_{t,\min}/V_{t,\max}$, where

$$V_{t,\min} \equiv \min(V_{t+}, |V_{t-}|), V_{t,\max} \equiv \max(V_{t+}, |V_{t-}|), \tag{11.25}$$

As emphasized in Bâldea (2012a), values as large as γ = 0.25 or even γ = 0.33, as claimed based on DFT-calculations (Markussen et al., 2011), would correspond to asymmetries $V_{t,\min}/V_{t,\max}$ = 0.286 and $V_{t,\min}/V_{t,\max}$ = 0.168, which are unrealistic. Experiments do not show such large asymmetries between biases of opposite polarities. Based on various molecular junctions investigated so far (Bâldea, 2012a,b,d; Beebe et al., 2006; Lee and Reddy, 2011; Tan et al., 2010), one can safely conclude that

$$|\gamma| < 0.1 \text{ is a realistic range.} \tag{11.26}$$

Figure 11.9 shows that for this γ-range

$$|\varepsilon_0| < 1.075 eV_{t,\min}. \tag{11.27}$$

On the other side, in the case of a symmetric potential profile, $\gamma = 0$, $V_t = V_{t+} = -V_{t-}$ [cf. Eq. (11.24)], Eq. (11.23) can be expressed in a very simple form (Bâldea, 2010)

$$eV_t = \frac{2}{\sqrt{3}} |\varepsilon_0| \simeq 1.15 |\varepsilon_0|; |\varepsilon_0| \simeq 0.87 eV_t = 0.87 eV_{t,\min}. \tag{11.28}$$

In view of Eqs. (11.27) and (11.28), one can see that, although the barrier-shape argument cannot be substantiated, the estimate based on the "barrier-shape conjecture," Eq. (11.2),

$$|\varepsilon_0| \approx eV_{t,\min} \tag{11.29}$$

is correct within at least ~13%.

Noteworthy, the transition voltage of smallest magnitude ($V_{t,\min}$) should be employed to estimate $|\varepsilon_0|$. Once again, the DFT-based claim (Chen et al., 2010) that the ratio $|\varepsilon_0|/(eV_t)$ can be as large as 2 is completely unrealistic; if one of the V_t's (say, V_{t2}) is equal to $|\varepsilon_0|/(2e)$, the other is infinite ($|V_{t1}| \to \infty$); see the point $\gamma = 1/2$ in Fig. 11.9.

It is also interesting to compare the asymmetry of the two transition voltages V_{t1} and V_{t2} with the asymmetry of the currents at these voltages $I_{ti} = [I(V)]_{V = Vtj}$ (i = 1, 2). In view of the small values of γ of interest, Eq. (11.26), we restrict ourselves to present the results to the lowest order in γ

$$-\frac{V_{t1}}{V_{t2}} \simeq 1 + \frac{8}{\sqrt{3}}\gamma + \mathcal{O}(\gamma^2), \tag{11.30}$$

$$-\frac{I_{t1}}{I_{t2}} \simeq 1 + \frac{4}{\sqrt{3}}\gamma + \mathcal{O}(\gamma^2). \tag{11.31}$$

The above equations show that the current asymmetry is one half of that in V_t's. One should emphasize that this asymmetry refers to the currents at the transition voltages of opposite polarities, which is different from the current asymmetry $-I(-V)/I(+V)$ at biases of the same magnitude of opposite polarities. The latter can be expressed using Eq. (11.21) as follows:

$$-\frac{I(-V)}{I(+V)} = \frac{(\varepsilon_0 + \gamma V)^2 - (eV/2)^2}{(\varepsilon_0 - \gamma V)^2 - (eV/2)^2}\bigg|_{e|V| \leq |\varepsilon_0|} \simeq 1 + 4\gamma\frac{eV}{\varepsilon_0}. \tag{11.32}$$

As visible in Eq. (11.32), the current asymmetry becomes more pronounced at higher voltages; it is negligible at low voltages.

Equations (11.23), (11.24), and (11.28) apply in cases where $\Gamma << |\varepsilon_0|$. Results for real molecular junctions demonstrated that this condition is often fulfilled (Bâldea, 2012a,b,d). Let us consider the example presented in Fig. 11.8b. Besides the experimental curve (Tan et al., 2010), two theoretical curves are shown there. One curve thereof was computed with the values ε_0 and γ obtained by fitting the experimental I–V curve by using Eq. (11.21). The other theoretical curve was computed using the values of ε_0 and γ deduced from Eqs. (11.23) and (11.24), wherein the values of V_{t+} and V_{t-} have been extracted by recasting the experimental I–V curve as FN curve. Noteworthy, the agreement is very good although the value $\Gamma/|\varepsilon_0| \approx 0.18$ is small, but not very small. To conclude, Eqs. (11.23) and (11.24) or (11.28) can be used to get accurate estimates for real molecular junctions.

The leading order corrections $\mathcal{O}(\Gamma/\varepsilon_0)^2$ can also be expressed in closed analytical form (Bâldea, 2012b,d). Because the general expressions are rather lengthy (they can be found in Bâldea (2012b)), we only give below the result for the case $\gamma = 0$, which reads

$$eV_t = \frac{2}{\sqrt{3}}|\varepsilon_0| + \frac{11}{\sqrt{3}}\frac{\Gamma^2}{3|\varepsilon_0|}. \tag{11.33}$$

11.6 Effect of Stochastic Fluctuations

Whatever the method employed, the nanofabrication of a molecular junction is a statistical event affected by stochastic fluctuations. Statistical data processing is an essential step both for theoretical understanding and for device design. Until recently, statistical analysis was limited to conductance histograms (Jang et al. 2006; Lörtscher et al., 2007; Mishchenko et al., 2010; Xu and Tao, 2003). Recent experimental advances (Guo et al., 2011) allowed to repeatedly form numerous break junctions and measure a large number (~thousands) of full *I–V* curves in a short time, adding thereby a new dimension (namely, the bias *V*) to the ordinary conductance (*G*) measurements at a fixed low bias. Thanks to those achievements, in addition to one-dimensional *G* histograms, not only one-dimensional V_t histograms (Lee and Reddy, 2011; Wang et al., 2011), but also two-dimensional G-V_t histograms (Guo et al., 2011) became available. An example is presented in Fig. 11.10.

Guo et al. (2011) revealed that the quantities G and V_t are affected in a completely different way by stochastic fluctuations. Broad G histograms, characterized by a full width at half maximum comparable to or even larger than the most probable G-value are well known (Jang et al., 2006; Lörtscher et al., 2007; Mishchenko et al., 2010; Xu and Tao, 2003); this clearly indicates a strong impact of fluctuations on G. By contrast, V_t turned out to be much less influenced by fluctuations: typical standard deviations in V_t amount to ~10–20% (Guo et al., 2011; Lee and Reddy, 2011; Wang et al., 2011). According to the initial interpretation based on the Simmons model (Guo et al., 2011), the much stronger impact of fluctuations on G was attributed to the large variability of the electrode–molecule contacts. Soon afterwards that interpretation has been critically analyzed in (Bâldea, 2012d). There, a different

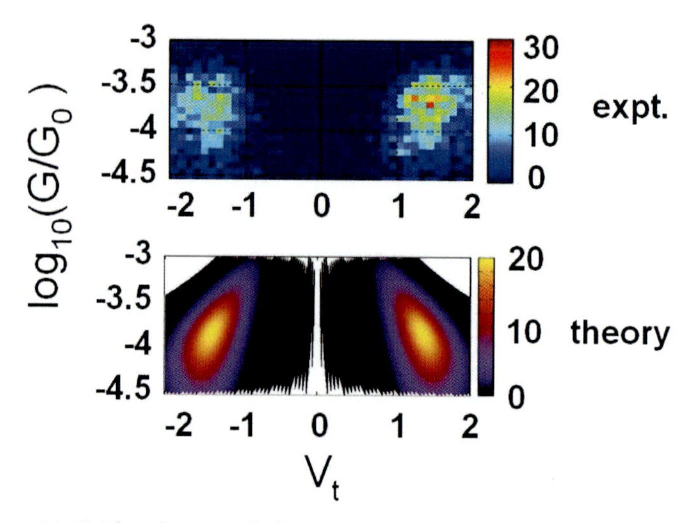

Figure 11.10 The theoretical description based on the Newns–Anderson model (lower panel, after Bâldea (2012d)) allows to quantitatively describe the two-dimensional V_t-G histograms obtained experimentally in molecular junctions based on octanedithiols (upper panel, after Guo et al. (2011); courtesy of the authors of Guo et al. (2011). Notice the decimal logarithm on the y-axis (stochastic fluctuations in G are much larger than those in V_t) and the fact that the x- and y-ranges of both panels coincide.

explanation based on the Newns–Anderson model has been given. As illustrated in Fig. 11.10, the theoretical results are in quantitative agreement with the experimental ones. Without entering into details (which can be found in Bâldea (2012d)), we briefly indicate here the reason for the different behavior. Let us first note that the single-molecule junctions based on alkanedithiols of Guo et al. (2011) satisfy the conditions $\Gamma << \varepsilon_0|$ and $\gamma \simeq 0$ (Bâldea, 2012d). Then Eqs. (11.1) and (11.28) allow to express the relative fluctuations in $G(\delta G/G)$ and $V_t(\delta V_t/V_t)$ as follows:

$$\frac{\delta G}{G} \simeq 2\left|\frac{\delta \Gamma}{\Gamma}\right| + 2\left|\frac{\delta \varepsilon_0}{\varepsilon_0}\right|, \qquad (11.34)$$

$$\frac{\delta V_t}{V_t} \simeq \left|\frac{\delta \varepsilon_0}{\varepsilon_0}\right|. \qquad (11.35)$$

According to the above results, the effect of fluctuations in the MO energy offset on the conductance is two times stronger than on the transition voltage. The analysis of the two-dimensional G–V_t histograms indicated that, for the junctions based on octanedithiols of Guo et al. (2011), $|\delta\Gamma/\Gamma| \simeq 0.25$ and $|\delta\varepsilon_0/\varepsilon_0| \simeq 0.23$; that is, the two relative fluctuations are comparable. So, the larger impact of stochastic fluctuations on G is on one side due to the fact that the effect of ε_0-fluctuations is two times stronger and on the other side that fluctuations in ε_0 and Γ are comparable. Noteworthy, the fact that fluctuations in ε_0 and Γ have a (nearly) equal impact on G is contrary to the initial claim of Guo et al. (2011).

Efforts have also been undertaken to understand the different effect of stochastic fluctuations on G and V_t accounting for electron correlations (Bâldea, 2012b). Model calculations based on realistic parameters indicated a significant impact of electron correlations due to short-range Coulomb interactions at the molecule–electrode contacts (Bâldea, 2012b).

11.7 Analytical Results for *I–V* Spectroscopy

As noted in Section 11.2, to deduce the orbital level alignment one could employ the maxima of the differential conductance $\partial I/\partial V$ versus V (Muralidharan et al., 2006). This procedure, sometimes called current–voltage spectroscopy, can seldom be employed (Lennartz et al., 2011) because of the voltages required are too high.

For completeness, let us mention that the $\partial I/\partial V$-extrema can also be deduced analytically. The imposition of $\partial^2 I/\partial V^2 = 0$ in Eq. (11.15) yields a fifth order algebraic equation, which possesses three real roots, corresponding to two maxima $V_{i\pm}$ and one minimum V_{im} (see Fig. 11.11). To order $\mathcal{O}(\Gamma/\varepsilon_0)^2$ they are expressed by

$$X_{r\pm} \equiv |\varepsilon_0|/V_{r\pm} = \pm 1/2 - \gamma,$$

$$X_m \equiv \frac{|\varepsilon_0|}{V_m} = \frac{1}{2} \frac{(1-4\gamma^2)^{2/3}}{(1+2\gamma)^{1/3} - (1-2\gamma)^{1/3}}. \tag{11.36}$$

As visible in Fig. 11.12, the estimates obtained from Eqs. (11.36) are also very accurate.

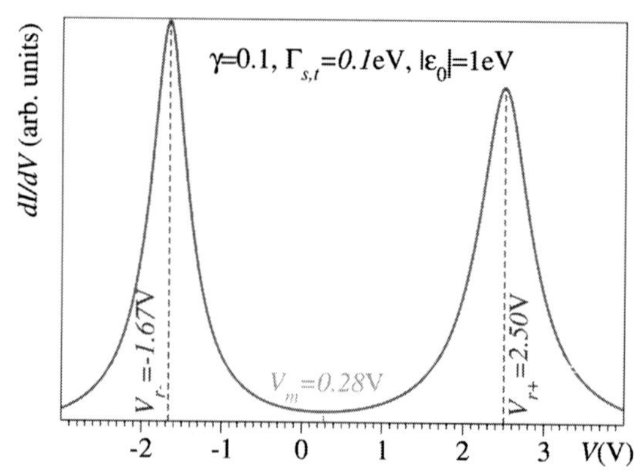

Figure 11.11 A typical curve of the differential conductance for single-level transport and asymmetric bias profile ($\gamma \neq 0$) described within the Newns–Anderson model, which exhibits two maxima at $V_{r\pm}$ related to resonant tunneling and one minimum at V_m. Unlike the case of a symmetric potential profile $\gamma = 0$ presented in Fig. 11.1b, in the present case ($\gamma \neq 0$) the maxima are located asymmetrically ($V_{r+} \neq -V_{r-}$) and the minimum is displaced from the origin ($V_m \neq 0$).

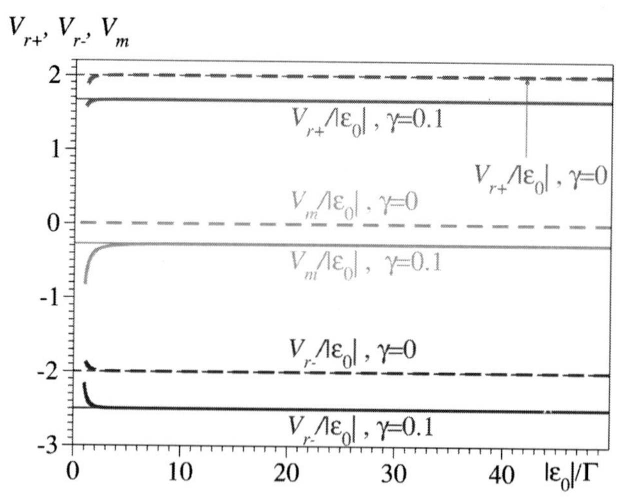

Figure 11.12 Location of the two maxima ($V_{r\pm}$) and the minimum (V_m) of the differential conductance as a function of bias asymmetry γ deduced by solving the equation $\partial^2 I/\partial V^2 = 0$ with the current expressed exactly by Eq. (11.15) (thick lines), and from Eq. (11.36), which is valid $\mathcal{O}(\Gamma/\varepsilon_0)^2$ (thin horizontal lines). Notice the rapid saturation of the exact curves for sufficiently small Γ/ε_0.

11.8 A Remarkably Simple Universality Class

In the symmetric case [$\gamma = 0$, $I(V) = -I(-V)$], using Eqs. (11.28) and (11.19), and the following dimensionless current and voltage variables:

$$\mathcal{I} \equiv \frac{I}{GV_t} \text{ and } \mathcal{V} \equiv \frac{V}{V_t}, \tag{11.37}$$

Eq. (11.21) can be recasted in the form of a universal curve

$$\mathcal{I} = \frac{3\mathcal{V}}{3-\mathcal{V}^2}, \tag{11.38}$$

which is depicted in Fig. 11.13. This is a remarkably simple result, and it can be straightforwardly tested experimentally. A very similar result, formulated as a "law-of-corresponding states" free of any empirical parameters, has been recently reported in a joint theoretical-experimental study (Bâldea et al., 2015).

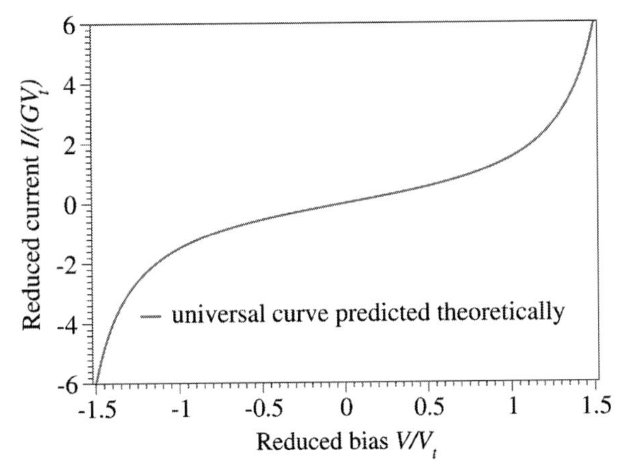

Figure 11.13 According to Eq. (11.38), the current–voltage curves plotted in reduced variables, $\mathcal{V} \equiv V/V_t$ versus $\mathcal{I} = I/(GV_t)$, should collapse on a universal curve, which is depicted in this figure.

11.9 Summary and Outlook

The results of the present analysis of TVS can be summarized as follows:

(i) Contrary to the original claim (Beebe et al., 2006), TVS does imply neither a transition from a trapezoidal tunneling barrier to a triangular one nor a crossover between a direct tunneling regime and a Fowler–Nordheim (field-emission) tunneling regime. In particular, the latter regime would imply voltages, which molecular junctions fabricated so far cannot withstand.

Nevertheless, it is a curiosity that the "barrier-shape conjecture" provides a reasonable estimate of the MO-energy offset $|\varepsilon_0| \approx e|V_{t,\min}|$ for molecular junctions with realistic bias asymmetry [cf. Eq. (11.29)].

(ii) The "transition" (implicitly) meant by TVS occurring at $V = V_t$ is from a linear (ohmic) regime to a regime characterized by a significant nonlinearity of the I–V curve. Mathematically, the following conditions are equivalent:

$$\left.\frac{\partial}{\partial(1/V)}\log\frac{I}{V^2}\right|_{V=V_t} = 0 \Leftrightarrow \left.\frac{\partial}{\partial V}\log\frac{I}{V^2}\right|_{V=V_t} = 0, \tag{11.39}$$

$$\mathcal{G}\,|_{V=V_t} \equiv \left.\frac{\partial I}{\partial V}\right|_{V=V_t} = 2\left.\frac{I}{V}\right|_{V=V_t} \equiv 2G|_{V=V_t}, \tag{11.40}$$

$$\left.\frac{\delta \log I}{\delta \log V}\right|_{V=V_t} = 2; \tag{11.41}$$

each can be taken as a definition of the transition voltage.

Equation (11.40) possesses a clear physical meaning and allows to define the transition voltage as the point where the differential conductance $\mathcal{G}$ is two times larger than the (pseudo-) ohmic conductance G (Bâldea, 2012b; Bâldea and Köppel, 2012a; Bâldea, 2012f).

It might be trivial, but because experimentalists continue to extract V_t by plotting the Fowler–Nordheim quantity $\log(I/V^2)$ as a function of $1/V$, it is still worth mentioning: one can as well use the plot of $\log(I/V^2)$ versus V. Using $1/V$ as abscissa coordinate is only motivated for field emission, because in that case $\log(I/V^2)$ depends linearly on $1/V$ at high voltages. For molecular electronics this is not the case, as illustrated by Fig. 11.7. This is the reason for explicitly writing the second part of Eq. (11.39).

According to Eq. (11.41), the transition voltage corresponds to the point where the slope of the plot of log I versus log V is equal to two. This is the rationale to use log I–log V-curves to extract V_t (Bâldea, 2010; Choi et al., 2008; Wang et al., 2011). Although not discussed here, we note that V_t can also be considered as a scaling voltage (Vilan et al., 2013).

(iii) We emphasized important drawbacks of the Simmons model. We did not aim to scrutinize that model in general; a series of other critical issues were previously addressed in Miskovsky et al. (1982). Rather, our aim was to demonstrate that this approach yields inadequate results for a specific case, namely, TVS in vacuum nanojunctions with sharp electrodes. In the standard monograph on tunneling, Duke noted in 1969: "All of the published applications of the Holm-Chow-Simmons method ... contain algebraic errors ..." (cf. Duke (1969), p. 62). And elsewhere: "Unfortunately, most of the early calculations for both the metal-vacuum contact and other tunnel junctions contain algebraic errors corrected in subsequent works. Thus, it is appropriate to suggest that the more recent references be consulted first" (cf. p. 33 of Duke, (1969)). The first sentence cited is also confirmed by recent publications (Akkerman et al., 2007; Huisman et al., 2009; Trouwborst et al., 2011; Tuan and Lien, 2009). Caution is advised concerning the second quotation: the Simmons model continues to be employed in its uncorrected form. This obviously traces back to the fact that the corrected prefactor in a footnote of Simmons (1964) is in no way linked to the erroneous formula of Simmons (1963). In recent TVS studies (Huisman et al., 2009; Trouwborst et al., 2011), this model was used even without considering the lateral confinement, as appropriate for atomically sharp electrodes. Therefore, we believe that drawing attention on and correcting such errors is important.

(iv) Many aspects related to TVS can be quantitatively understood within the Newns–Anderson model, for which simple but accurate analytical formulas have been derived. They can be straightforwardly used by experimentalists for data processing. In the light of the results obtained within this model, regarding the conductance as a molecular property should be made with care; G is much more affected by fluctuations in the MO-offset than the transition voltage. The fact that V_t is much less affected is important; it makes V_t a molecular signature, and this validates TVS as a tool of investigating molecular transport.

Studies to fully exploit the TVS advantages should include situations where MO energies are modified in a controlled way. The demonstration of the molecular orbital modulation by an external gate voltage (Song et al., 2009) represented an important case where TVS turned out to be very useful. Another important application not fully explored so far would be to investigate solvent-driven shifts of the MO energies, an interesting effect revealed recently (Bâldea, 2012c; Bâldea et al., 2013).

According to the present knowledge in the field, besides TVS two other methods can be employed to get information on the energy offset of the MO that dominates the charge transfer in molecular junctions: concurrently studying of the conductance and the Seebeck coefficient (Paulsson and Datta, 2003), and studying the length dependence of the conductance in a certain molecular series. As compared to TVS, which is entirely based on electric transport measurements on a single molecule, a disadvantage of the former method is that it also requires thermoelectric measurements. In addition, that method cannot provide information on the potential profile (a)symmetry. The disadvantage of the latter method is that studying one molecule is not enough. Measurements for a molecular series are needed, and it is by no means obvious that/why parameters from one member of the series are transferable to another member, and whether/when, e.g., even-odd effects play a significant role. Using the attenuation factor β to differentiate between different molecular series may be subtle and may require some prior knowledge in situations where TVS provides clear distinctions without any prior knowledge, as revealed by recent studies (Fracasso et al., 2013).

To end, as reemphasized recently (Bâldea, 2015)—where an alternative term, namely, peak voltage spectroscopy (PVS) has been proposed in order to avoid confusions created by the misleading term "transition" (cf. pp. 402 and 424)—TVS should not be uncritically used: a simple relationship between the transition voltage and the molecular orbital energy offset does not hold in any situation (cf. Eqs. (11.12) and (11.13)). But in cases where the Newns-Anderson model applies, this relationship is particularly appealing (cf. Eqs. (11.13) and (11.28)). For this reason, TVS is by now a widely accepted tool of investigation in molecular electronics. It is used for quantifying the molecular orbital energy offset relative to electrodes' Fermi energy and applied along with other methods (ultraviolet photoelectron spectroscopy (UPS) (Kim et al., 2011) and thermopower (Baheti et al., 2008; Guo et al., 2013b; Widawsky et al., 2009, 2012) by numerous groups worldwide. Further applications (not included in this review) comprise, e.g., molecular junctions immersed in solvents and reorganization effects (Bâldea et al., 2013; Bâldea (2013a,b,c, 2014e,b)) and vacuum nanojunctions, wherein puzzling findings were recently reported (Bâldea, 2014a; Sotthewes et al., 2014).

Acknowledgments

Financial support for this work provided by the Deutsche Forschungsgemeinschaft (grant BA 1799/2-1) is gratefully acknowledged.

References

Akkerman, H. B., Naber, R. C. G., Jongbloed, B., van Hal, P. A., Blom, P. W. M., de Leeuw, D. M., and de Boer, B. (2007). Electron tunneling through alkanedithiol self-assembled monolayers in large-area molecular junctions, *Proc. Nat. Acad. Sci.*, **104**(27), 11161–11166, doi:10.1073/pnas.0701472104, URL http://www.pnas.org/content/104/27/11161.abstract.

Araidai, M., and Tsukada, M. (2010). Theoretical calculations of electron transport in molecular junctions: Inflection behavior in Fowler–Nordheim plot and its origin, *Phys. Rev. B*, **81**(23), 235114, doi:10.1103/PhysRevB.81.235114.

Baheti, K., Malen, J. A., Doak, P., Reddy, P., Jang, S.-Y., Tilley, T. D., Majumdar, A., and Segalman, R. A. (2008). Probing the chemistry of molecular heterojunctions using thermoelectricity, *Nano Lett.*, **8**(2), 715–719, doi:10.1021/nl072738l, URL http://pubs.acs.org/doi/abs/10.1021/nl072738l, pMID:18269258.

Bâldea, I. (2010). Revealing molecular orbital gating by transition voltage spectroscopy, *Chem. Phys.*, **377**(1–3), 15–20, doi:DOI:10.1016/j.chemphys.2010.08.009, URL http://www.sciencedirect.com/science/article/B6TFM-50RVNS4-2/2/22cf360cd4f9f4b2f1d3992275f49578.

Bâldea, I. (2012a). Ambipolar transition voltage spectroscopy: Analytical results and experimental agreement, *Phys. Rev. B*, **85**, 035442, doi: 10.1103/PhysRevB.85.035442, URL http://link,aps.org/doi/10.1103/PhysRevB.85.035442.

Bâldea, I. (2012b). Effects of stochastic fluctuations at molecule-electrode contacts in transition voltage spectroscopy, *Chem. Phys.*, **400**(0), 65–71, doi:10.1016/j.chemphys.2012.02.011, URL http://www.sciencedirect.com/science/article/pii/S0301010412000870?v=s5.

Bâldea, I. (2012c). Extending the Newns–Anderson model to allow nanotransport studies through molecules with floppy degrees of freedom, *Europhys. Lett.*, **99**(4), 47002, doi:10.1209/02955075/99/47002, URL http://stacks.iop.org/0295-5075/99/i=4/a= 47002.

Bâldea, I. (2012d). Interpretation of stochastic events in single-molecule measurements of conductance and transition voltage spectroscopy, *J. Am. Chem. Soc.*, **134**(18), 7958–7962, doi:10.1021/ja302248h, URL http://pubs.acs.org/doi/abs/10.1021/ja302248h.

Bâldea, I. (2012e). Transition voltage spectroscopy: Artefacts of the Simmons approach, *J. Phys. Chem. Solids*, **73**(9), 1151–1153, doi:10.1016/j.jpcs.2012.05.006, URL http://www. sciencedirect.com/science/article/pii/S0022369712001679.

Bâldea, I. (2012f). Transition voltage spectroscopy in vacuum break junction: Possible role of surface states, *Europhys. Lett.*, **98**(1), 17010, doi:10.1209/0295-5075/98/17010, URL http://stacks. iop.org/0295-5075/98/i=1/a=17010.

Bâldea, I. (2013a). Demonstrating why DFT-calculations for molecular transport in solvents need scissor corrections, *Electrochem. Commun.*, **36**, 19–21, doi:http://dx.doi.org/10.1016/j.elecom.2013.08.027, URL http://www. sciencedirect. com/science/article/pii/S1388248113003421.

Bâldea, I. (2013b). Important insight into electron transfer in single-molecule junctions based on redox metalloproteins from transition voltage spectroscopy, *J. Phys. Chem. C*, **117**(48), 25798–25804, doi:10.1021/jp408873c, URL http://pubs.acs.org/doi/abs/10.1021/jp408873c.

Bâldea, I. (2013c). Transition voltage spectroscopy reveals significant solvent effects on molecular transport and settles an important issue in bipyridine-based junctions, *Nanoscale*, **5**, 9222–9230, doi:10.1039/C3NR51290H, URL http://dx.doi.org/10.1039/C3NR51290H.

Bâldea, I. (2014a). Concurrent conductance and transition voltage spectroscopy study of scanning tunneling microscopy vacuum junctions, does it unravel new physics? *RSC Adv.*, **4**, 33257–33261, doi:10.1039/C4RA04648J, URL http://dx.doi.org/10.1039/C4RA04648J.

Bâldea, I. (2014b). Electrochemical setup–a unique chance to simultaneously control orbital energies and vibrational properties of single-molecule junctions with unprecedented efficiency, *Phys. Chem. Chem. Phys.*, **16**, 25942–25949, doi:10.1039/C4CP04316B, URL http://dx.doi.org/10.1039/C4CP04316B.

Bâldea, I. (2014c). Quantifying the relative molecular orbital alignment for molecular junctions with similar chemical linkage to electrodes, *Nanotechnology*, **25**(45), 455202, doi:doi:10.1088/0957-4484/25/45/455202, URL http://stacks.iop.org/0957-4484/25/i=45/a=455202.

Bâldea, I. (2014d). A quantum chemical study from a molecular transport perspective: Ionization and electron attachment energies for species often used to fabricate single-molecule junctions, *Faraday Discuss. ASAP*, doi:10.1039/C4FD00101J, URL http://dx.doi.org/10.1039/C4FD00101J.

Bâldea, I. (2014e). Single-molecule junctions based on bipyridine: Impact of an unusual reorganization on the charge transport, *J. Phys. Chem. C*, **118**(16), 8676–8684, doi:10.1021/jp412675k, URL http://pubs.acs.org/doi/abs/10.1021/jp412675k.

Bâldea, I. (2015). Important issues facing model-based approaches to tunneling transport in molecular junctions, *Phys. Chem. Chem. Phys.*, **17**, 20217–20230, doi:10.1039/C5CP02595H, URL http://dx.doi.org/10.1039/C5CP02595H.

Bâldea, I., and Köppel, H. (2012a). Evidence on single-molecule transport in electrostatically-gated molecular transistors, *Phys. Lett. A*, **376**(17), 1472–1476, doi:10.1016/j.physleta.2012.03.021, URL http://www.sciencedirect.com/science/article/pii/S0375960112003040.

Bâldea, I., and Köppel, H. (2012b). Transition voltage spectroscopy in vacuum break junction: The standard tunneling barrier model and beyond, *Phys. Stat. Solidi* (b), **249**(9), 1791–1804, doi:10.1002/pssb.201248034, URL http://dx.doi.org/10.1002/pssb.201248034.

Bâldea, I., Köppel, H., and Wenzel, W. (2013). (4,4′)-bipyridine in vacuo and in solvents: A quantum chemical study of a prototypical floppy molecule from a molecular transport perspective, *Phys. Chem. Chem. Phys.*, **15**, 1918–1928, doi:10.1039/C2CP43627B, URL http://dx.doi.org/10.1039/C2CP43627B.

Bâldea, I., Xie, Z., and Frisbie, C. D. (2015). Uncovering a law of corresponding states for electron tunneling in molecular junctions, *Nanoscale*,

7, 10465–14071, doi:10.1039/C5NR02225H, URL http://dx.doi.org/10.1039/C5NR02225H.

Beebe, J. M., Kim, B., Frisbie, C. D., and Kushmerick, J. G. (2008). Measuring relative barrier heights in molecular electronic junctions with transition voltage spectroscopy, *ACS Nano*, **2**(5), 827–832, doi:10.1021/nn700424u, URL http://dx.doi.org/10.1021/nn700424u.

Beebe, J. M., Kim, B., Gadzuk, J. W., Frisbie, C. D., and Kushmerick, J. G. (2006). Transition from direct tunneling to field emission in metal-molecule-metal junctions, *Phys. Rev. Lett.*, **97**(2), 026801, doi:10.1103/PhysRevLett.97.026801.

Chen, J., Markussen, T., and Thygesen, K. S. (2010). Quantifying transition voltage spectroscopy of molecular junctions: Ab initio calculations, *Phys. Rev. B*, **82**(12), 121412, doi:10.1103/PhysRevB.82.121412.

Chiu, P.-W., and Roth, S. (2008). Transition from direct tunneling to field emission in carbon nanotube intramolecular junctions, *Appl. Phys. Lett.*, **92**(4), 042107, doi:10.1063/1.2838353, URL http://link.aip.org/link/?APL/92/042107/1.

Choi, S. H., Kim, B., and Frisbie, C. D. (2008). Electrical resistance of long conjugated molecular wires, *Science*, **320**(5882), 1482–1486, doi:10.1126/science.1156538, URL http://www.sciencemag.org/cgi/content/abstract/320/5882/1482.

Chow, C. K. (1963a). Effect of insulating-film-thickness nonuniformity on tunnel characteristics, *J. Appl. Phys.*, **34**(9), 2599–2602, doi: DOI:10.1063/1.1729776, URL http://dx.doi.org/doi/10.1063/1.1729776.

Chow, C. K. (1963b). On tunneling equations of holm and stratton, *J. Appl. Phys.*, **34**(8), 2490–2492, doi:DOI:10.1063/1.1702773, URL http://dx.doi.org/doi/10.1063/1.1702773.

Coll, M., Gergel-Hackett, N., Richter, C. A., and Hacker, C. A. (2011). Structural and electrical properties of flip chip laminated metal-molecule silicon structures varying molecular backbone and atomic tether, *J. Phys. Chem. C*, **115**(49), 24353–24365, doi:10.1021/jp208275c, URL http://pubs.acs.org/doi/abs/10.1021/jp208275c.

Coll, M., Miller, L. H., Richter, L. J., Hines, D. R., Jurchescu, O. D., Gergel-Hackett, N., Richter, C. A., and Hacker, C. A. (2009). Formation of silicon-based molecular electronic structures using flip-chip lamination, *J. Am. Chem. Soc.*, **131**(34), 12451–12457, doi:10.1021/ja901646j, URL http://dx.doi.org/10.1021/ja901646j.

Duke, C. B. (1969). *Tunneling in Solids* (Academic Press N. Y.), ISBN 0-12-607770-3, Solid State Physics, supplement no. 10, edited by F. Seitz, D. Turnbull, H. Ehrenreich.

Fowler, R. H. (1928). The restored electron theory of metals and thermionic formulae, *Proc. Roy. Soc. London A*, **117**(778), 549–552, URL http://www.jstor.org/stable/94976.

Fowler, R. H., and Nordheim, L. (1928). Electron emission in intense electric fields, *Proc. Roy. Soc. London A*, **119**(781), 173–181, URL http://www.jstor.org/stable/95023.

Fracasso, D., Muglali, M. I., Rohwerder, M., Terfort, A., and Chiechi, R. C. (2013). Influence of an atom in egain/ga2o3 tunneling junctions comprising self-assembled monolayers, *J. Phys. Chem. C*, **117**(21), 11367–11376, doi:10.1021/jp401703p, URL http://pubs.acs.org/doi/abs/10.1021/jp401703p.

Gundlach, K. H. (1966). Zur berechnung des tunnelstroms durch eine trapezfrmige potentialstufe, *Solid-State Electron.*, **9**(10), 949–957, doi:10.1016/0038-1101(66)90071-2, URL http://www.sciencedirect.com/science/article/pii/0038110166900712.

Guo, S., Arts, J. M., and Diez-Pérez, I. (2013a). Electrochemically-gated single-molecule electrical devices, *Electrochim. Acta*, **110**, 741–753, doi:http://dx.doi.org/10.1016/j.electacta.2013.03. 146, URL http://www.sciencedirect.com/science/article/pii/S0013468613005690, selection of papers from the 63rd Annual Meeting of the International Society of Electrochemistry 19–24 August 2012, Prague, Czech Republic.

Guo, S., Hihath, J., Diez-Pérez, I., and Tao, N. (2011). Measurement and statistical analysis of single-molecule current–voltage characteristics, transition voltage spectroscopy, and tunneling barrier height, *J. Am. Chem. Soc.*, **133**(47), 19189–19197, doi:10.1021/ja2076857, URL http://pubs.acs.org/doi/abs/10.1021/ja2076857.

Guo, S., Zhou, G., and Tao, N. (2013b). Single molecule conductance, thermopower, and transition voltage, *Nano Lett.*, **13**(9), 4326–4332, doi:10.1021/nl4021073, URL http://pubs.acs.org/doi/abs/10.1021/nl4021073.

Hall, L. E., Reimers, J. R., Hush, N. S., and Silverbrook, K. (2000). Formalism, analytical model, and a priori green's-function-based calculations of the current–voltage characteristics of molecular wires, *J. Chem. Phys.*, **112**(3), 1510–1521, doi:http://dx.doi.org/10.1063/1.480696, URL http://scitation.aip.org/content/aip/journal/jcp/112/3/10.1063/1.480696.

Holm, R. (1951). The electric tunnel effect across thin insulator films in contacts, *J. Appl. Phys.*, **22**(5), 569–574, doi:10.1063/1.1700008, URL http://link.aip.org/link/?JAP/22/569/1.

Holmlin, R. E., Haag, R., Chabinyc, M. L., Ismagilov, R. F., Cohen, A. E., Terfort, A., Rampi, M. A., and Whitesides, G. M. (2001). Electron transport through thin organic films in metal-insulator-metal junctions based on self-assembled monolayers, *J. Am. Chem. Soc.*, **123**(21), 5075–5085, doi:10.1021/ja004055c, URL http://pubs.acs.org/doi/abs/10.1021/ja004055c.

Huisman, E. H., Guédon, C. M., van Wees, B. J., and van der Molen, S. J. (2009). Interpretation of transition voltage spectroscopy, *Nano Lett.*, **9**(11), 3909–3913, doi:10.1021/nl9021094, URL http://dx.doi.org/10.1021/nl9021094.

Hybertsen, M. S., Venkataraman, L., Klare, J. E., Whalley, A. C., Steigerwald, M. L., and Nuckolls, C. (2008). Amine-linked single-molecule circuits: systematic trends across molecular families, *J. Phys. Condensed Matter*, **20**(37), 374115, doi:10.1088/0953-8984/20/37/374115, URL http://stacks.iop.org/0953-8984/20/i=37/a=374115.

Jang, S.-Y., Reddy, P., Majumdar, A., and Segalman, R. A. (2006). Interpretation of stochastic events in single molecule conductance measurements, *Nano Lett.*, **6**(10), 2362–2367, doi:10.1021/nl0609495, URL http://pubs.acs.org/doi/abs/10.1021/nl0609495.

Kim, B., Choi, S. H., Zhu, X.-Y., and Frisbie, C. D. (2011). Molecular tunnel junctions based on π-conjugated oligoacene thiols and dithiols between Ag, Au, and pt contacts: Effect of surface linking group and metal work function, *J. Am. Chem. Soc.*, **133**(49), 19864–19877, doi:10.1021/ja207751w, URL http://pubs. acs.org/doi/abs/10.1021/ja207751w.

Lee, W., and Reddy, P. (2011). Creation of stable molecular junctions with a custom-designed scanning tunneling microscope, *Nanotechnology*, **22**(48), 485703, doi:10.1088/0957-4484/22/48/485703, URL http://stacks.iop.org/0957-4484/22/i=48/a=485703.

Lennartz, M. C., Atodiresei, N., Caciuc, V., and Karthaeuser, S. (2011). Identifying molecular orbital energies by distance dependent transition voltage spectroscopy, *J. Phys. Chem. C*, **115**(30), 15025–15030, doi:10.1021/jp204240n, URL http://pubs.acs.org/doi/abs/10.1021/jp204240n.

Lindsay, S. M., and Ratner, M. A. (2007). Molecular transport junctions: Clearing mists, *Adv. Mater.*, **19**(1), 23–31, doi:10.1002/adma.200601140, URL http://dx.doi.org/10.1002/adma.200601140.

Lörtscher, E., Weber, H. B., and Riel, H. (2007). Statistical approach to investigating transport through single molecules, *Phys. Rev. Lett.*, **98**, 176807, doi:10.1103/PhysRevLett.98.176807, URL http://link.aps.org/doi/10.1103/PhysRevLett.98.176807.

Malen, J. A., Doak, P., Baheti, K., Tilley, T. D., Segalman, R. A., and Majumdar, A. (2009). Identifying the length dependence of orbital alignment and contact coupling in molecular heterojunctions, *Nano Lett.*, **9**(3), 1164–1169, doi:10.1021/nl803814f, URL http://pubs.acs.org/doi/abs/10.1021/nl803814f.

Markussen, T., Chen, J., and Thygesen, K. S. (2011). Improving transition voltage spectroscopy of molecular junctions, *Phys. Rev. B*, **83**, 155407, doi:10.1103/PhysRevB.83.155407, URL http://link.aps.org/doi/10.1103/PhysRevB.83.155407.

McConnell, H. M. (1961). Intramolecular charge transfer in aromatic free radicals, *J. Chem. Phys.* **35**(2), 508–515, doi:DOI:10.1063/1.1731961, URL http://dx.doi. org/10.1063/1.1731961.

Medvedev, I. G. (2007). Tunnel current through a redox molecule coupled to classical phonon modes in the strong tunneling limit, *Phys. Rev. B*, **76**(12), 125312, doi:10.1103/PhysRevB.76.125312.

Metzger, R. M., Xu, T., and Peterson, I. R. (2001). Electrical rectification by a monolayer of hexadecylquinolinium tricyanoquin-odimethanide measured between macroscopic gold electrodes, *J. Phys. Chem. B*, **105**(30), 7280–7290, doi:10.1021/jp011084g, URL http://dx.doi.org/10.1021/jp0 11084g.

Mishchenko, A., Vonlanthen, D., Meded, V., Bürkle, M., Li, C, Pobelov, I. V., Bagrets, A., Viljas, J. K., Pauly, F., Evers, F., Mayor, M., and Wandlowski, T. (2010). Influence of conformation on conductance of biphenyl-dithiol single-molecule contacts, *Nano Lett.*, **10**(1), 156–163, doi:10.1021/nl903084b, URL http://pubs.acs.org/doi/abs/10.1021/nl903084b, pMID:20025266.

Miskovsky, N. M., Cutler, P. H., Feuchtwang, T. E., and Lucas, A. A. (1982). The multiple-image interactions and the mean-barrier approximation in MOM and MVM tunneling junctions, *Appl. Phys. A: Mater. Sci. Proc.*, **27**, 139–147, doi:10.1007/BF00616664, URL http://dx.doi.org/10.1007/BF 00616664.

Muralidharan, B., Ghosh, A. W., and Datta, S. (2006). Probing electronic excitations in molecular conduction, *Phys. Rev. B*, **73**(15), 155410, doi:10.1103/PhysRevB.73.155410, URL http://link, aps.org/abstract/PRB/v73/e155410.

Muscat, J., and Newns, D. (1978). Chemisorption on metals, *Progr. Surf. Sci.*,

9(1), 1–43, doi:10.1016/0079-6816(78)90005-9, URL http://www.sciencedirect.com/science/article/pii/0079681678900059.

Nitzan, A., and Ratner, M. A. (2003). Electron transport in molecular wire junctions, *Science*, **300**(5624), 1384–1389, doi:10.1126/science.1081572, URL http://www.sciencemag.org/cgi/content/abstract/300/5624/1384.

Nordheim, L. W. (1928). The effect of the image force on the emission and reflexion of electrons by metals, *Proc. Roy. Soc. London A*, **121**(788), 626–639, URL http://www.jstor.org/stable/95122.

Paulsson, M., and Datta, S. (2003). Thermoelectric effect in molecular electronics,*Phys.Rev.B*,**67**,241403,doi:10.1103/PhysRevB.67.241403, URL http://link.aps.org/doi/10.1103/PhysRevB.67.241403.

Ratner, M. A. (1990). Bridge-assisted electron transfer: Effective electronic coupling,*J.Phys.Chem.*,**94**(12),4877–4883,doi:10.1021/j100375a024, URL http://pubs.acs.org/doi/abs/10.1021/j100375a024.

Reed, M. A., Zhou, C., Muller, C. J., Burgin, T. P., and Tour, J. M. (1997). Conductance of a molecular junction, *Science*, **278**(5336), 252–254, doi:10.1126/science.278.5336.252, URL http://www.sciencemag.org/cgi/content/abstract/278/5336/252.

Ricoeur, G., Lenfant, S., Guérin, D., and Vuillaume, D. (2012). Molecule/electrode interface energetics in molecular junction: A transition voltage spectroscopy study, *J. Phys. Chem. C*, **116**(39), 20722–20730, doi:10.1021/jp305739c, URL http://pubs.acs.org/doi/abs/10.1021/jp305739c.

Schmickler, W. (1986). A theory of adiabatic electron-transfer reactions, *J. Electroanal. Chem.*, **204**(1–2), 31–43, doi:10.1016/0022-0728(86)80505-8, URL http://www.sciencedirect.com/science/article/pii/0022072886805058.

Simmons, J. G. (1963). Generalized formula for the electric tunnel effect between similar electrodes separated by a thin insulating film, *J. Appl. Phys.*, **34**(6), 1793–1803, doi:10.1063/1.1702682, URL http://link.aip.org/link/?JAP/34/1793/1.

Simmons, J. G. (1964). Potential barriers and emission? Limited current flow between closely spaced parallel metal electrodes, *J. Appl. Phys.*, **35**(8), 2472–2481, doi:DOI:10.1063/1.1702884, URL http://dx.doi.org/doi/10.1063/1.1702884 (see footnote 8 in this paper).

Skourtis, S. S., and Beratan, D. N. (1999). Theories of structure-function relationships for bridge-mediated electron transfer reactions, *Adv. Chem. Phys.*, **106**, 377–452, doi:10.1002/9780470141656.ch8, URL http://dx. doi. org/10.1002/9780470141656. ch8.

Smaali, K., Clément, N., Patriarche, G., and Vuillaume, D. (2012). Conductance statistics from a large array of sub-10 nm molecular junctions, *ACS Nano*, **6**(6), 4639–4647, doi:10.1021/nn301850g, URL http://pubs.acs.org/doi/abs/10.1021/nn301850g.

Sommerfeld, A., and Bethe, H. (1933). Elektronentheorie der metalle, in Geiger and Scheel (eds.), *Handbuch der Physik*, vol. **24**(2) (Julius-Springer-Verlag, Berlin), ISBN ISBN-10: 3540038663; ISBN-13:978-3540038665, p. 446.

Song, H., Kim, Y., Jang, Y. H., Jeong, H., Reed, M. A., and Lee, T. (2009). Observation of molecular orbital gating, *Nature*, **462**(7276), 1039–1043, doi:10.1038/nature08639, URL http://dx.doi.org/10.1038/nature08639.

Song, H., Reed, M. A., and Lee, T. (2011). Single molecule electronic devices, *Adv. Mater.*, **23**(14), 1583–1608, doi:10.1002/adma.201004291, URL http://dx.doi.org/10.1002/adma.201004291.

Sotthewes, K., Hellenthal, C., Kumar, A., and Zandvliet, H. (2014). Transition voltage spectroscopy of scanning tunneling microscopy vacuum junctions, *RSC Adv.*, **4**, 32438–32442, doi:10.1039/C4RA04651J, URL http://dx.doi.org/10.1039/C4RA04651J.

Tan, A., Sadat, S., and Reddy, P. (2010). Measurement of thermopower and current–voltage characteristics of molecular junctions to 10.1063/1.3291521ify orbital alignment, *Appl. Phys. Lett.*, **96**(1), 013110, doi:DOI:10.1063/1.3291521, URL http://dx.doi.org/10.1063/1.3291521.

Tao, N. J. (2006). Electron transport in molecular junctions, *Nat. Nano*, **1**(3), 173–181, doi:10.1038/nnano.2006.130, URL http://dx.doi.org/10.1038/nnano.2006.130.

Tran, T. K., Smaali, K., Hardouin, M., Bricaud, Q., Oafrain, M., Blanchard, P., Lenfant, S., Godey, S., Roncali, J., and Vuillaume, D. (2013). A crown-ether loop-derivatized oligothiophene doubly attached on gold surface as cation-binding switchable molecular junction, *Adv. Mater.*, **25**(3), 427–431, doi:10.1002/adma.201201668, URL http://dx.doi.org/10.1002/adma.201201668.

Trouwborst, M. L., Martin, C. A., Smit, R. H. M., Guédon, C. M., Baart, T. A., van der Molen, S. J., and van Ruitenbeek, J. M. (2011). Transition voltage spectroscopy and the nature of vacuum tunneling, *Nano Lett.*, **11**(2), 614–617, doi:10.1021/nl103699t, URL http://pubs.acs.org/doi/abs/10.1021/nl103699t.

Tuan, N. A., and Lien, D. P. (2009). The effect of image force and temperature on electrical characteristics of layer-type MTJS, *J. Phys. Conf. Series*,

187(1), 012089, doi:10.1088/1742-6596/187/1/012089, URL http://stacks.iop.org/1742-6596/187/i=1/a=012089.

Venkataraman, L., Klare, J. E., Nuckolls, C., Hybertsen, M. S., and Steigerwald, M. L. (2006). Dependence of single-molecule junction conductance on molecular conformation, *Nature*, **442**(7105), 904–907, doi:http://dx.doi.org/10.1038/nature05037.

Vilan, A., Cahen, D., and Kraisler, E. (2013). Rethinking transition voltage spectroscopy within a generic Taylor expansion view, *ACS Nano*, **7**(1), 695–706, doi:10.1021/nn3049686, URL http://pubs.acs.org/doi/abs/10.1021/nn3049686.

Wang, G., Kim, Y., Na, S.-I., Kahng, Y. H., Ku, J., Park, S., Jang, Y. H., Kim, D.-Y., and Lee, T. (2011). Investigation of the transition voltage spectra of molecular junctions considering frontier molecular orbitals and the asymmetric coupling effect, *J. Phys. Chem. C*, **115**(36), 17979–17985, doi:10.1021/jp204340w, URL http://pubs.acs.org/doi/abs/10.1021/jp204340w.

Widawsky, J. R., Darancet, P., Neaton, J. B., and Venkataraman, L. (2012). Simultaneous determination of conductance and thermopower of single molecule junctions, *Nano Lett.*, **12**(1), 354–358, doi:10.1021/nl203634m, URL http://pubs.acs.org/doi/abs/10.1021/nl203634m.

Widawsky, J. R., Kamenetska, M., Klare, J., Nuckolls, C., Steigerwald, M. L., Hybertsen, M. S., and Venkataraman, L. (2009). Measurement of voltage-dependent electronic transport across amine-linked single-molecular-wire junctions, *Nanotechnology*, **20**(43), 434009, URL http://stacks.iop.org/0957-4484/20/i=43/a=434009.

Wu, K., Bai, M., Sanvito, S., and Hou, S. (2013a). Origin of the transition voltage in gold-vacuum-gold atomic junctions, *Nanotechnology*, **24**(2), 025203, doi:10.1088/0957-4484/24/2/025203, URL http://stacks.iop.org/0957-4484/24/i=2/a=025203.

Wu, K., Bai, M., Sanvito, S., and Hou, S. (2013b). Quantitative interpretation of the transition voltages in gold-poly (phenylene)thiol-gold molecular junctions, *J. Chem. Phys.*, **139**(19), 194703, doi:http://dx.doi.org/10.1063/1.4830399, URL http://scitation.aip.org/content/aip/journal/j cp/139/19/10.1063/1.4830399.

Xu, B., and Tao, N. J. (2003). Measurement of single-molecule resistance by repeated formation of molecular junctions, *Science*, **301**(5637), 1221–1223, doi:10.1126/science.1087481, URL http://www.sciencemag.org/content/301/5637/1221.abstract.

Yu, L. H., Gergel-Hackett, N., Zangmeister, C. D., Hacker, C. A., Richter, C. A., and Kushmerick, J. G. (2008). Molecule-induced interface states dominate charge transport in sialkylmetal junctions, *J. Phys. Condensed Matter*, **20**(37), 374114, URL http://stacks.iop.org/0953-8984/20/i=37/a=374114.

Zahid, F., Paulsson, M., and Datta, S. (2003). Electrical conduction through molecules, in H. Morkoç (ed.), Advanced Semiconductors and Organic Nano-Techniques, vol. 3, Chap. Electrical Conduction through Molecules (Academic Press), ISBN 0-12-507060-8.

Zangmeister, C. D., Beebe, J. M., Naciri, J., Kushmerick, J. G., and van Zee, R. D. (2008). Controlling charge-carrier type in nanoscale junctions with linker chemistry, *Small*, **4**(8), 1143–1147, doi:10.1002/smll.200800359, URL http://dx.doi.org/10.1002/smll.200800359.

Index